The Topological Classification of Stratified Spaces

Chicago Lectures in Mathematics Series
Robert J. Zimmer, series editor
J. Peter May, Spencer J. Bloch, Norman R. Lebovitz,
William Fulton, and Carlos Kenig, editors

Other *Chicago Lectures in Mathematics* titles available from the University of Chicago Press:

Simplicial Objects in Algebraic Topology, by J. Peter May (1967)
Torsion-Free Modules, by Eben Matlis (1973)
Rings with Involution, by I. N. Herstein (1976)
Theory of Unitary Group Representation, by George V. Mackey (1976)
Infinite-Dimensional Optimization and Convexity, by Ivar Ekeland and Thomas Turnbull (1983)
Commutative Semigroup Rings, by Robert Gilmer (1984)
Navier-Stokes Equations, by Peter Constantin and Ciprian Foias (1988)
Essential Results of Functional Analysis, by Robert J. Zimmer (1990)
Fuchsian Groups, by Svetlana Katok (1992)
Unstable Modules over the Steenrod Algebra and Sullivan's Fixed Point Set Conjecture, by Lionel Schwartz (1994)

Shmuel Weinberger

THE TOPOLOGICAL CLASSIFICATION OF STRATIFIED SPACES

The University of Chicago Press
Chicago and London

Shmuel Weinberger is the Thomas F. Scott Professor of Mathematics at the University of Pennsylvania.

The University of Chicago Press, Chicago 60637
The University of Chicago Press, Ltd., London

Printed in the United States of America
03 02 01 00 99 98 97 96 95 94   1 2 3 4 5
ISBN: 0-226-88566-6 (cloth)
0-226-88567-4 (paper)

Library of Congress Cataloging-in-Publication Data

Weinberger, Shmuel.
The topological classification of stratified spaces / Shmuel Weinberger.
p. cm.—(Chicago lectures in mathematics series)
Includes bibliographical references and index.
1. Topological manifolds. 2. Topological spaces. I. Title.
II. Series: Chicago lectures in mathematics.
QA613.2.W45 1994   94-17071
514′.32—dc20   CIP

♾ The paper used in this publication meets the minimum requirements of the American National Standard for Information Sciences—Permanence of Paper for Printed Library Materials, ANSI Z39.48-1984.

To Devorah Aravah,
who reminds me that not everything is a stratified space

Contents

Preface

These notes stem from a ten-week course I taught in Fall 1989 at the University of Chicago. The goal was to introduce students as rapidly as possible to techniques and problems that are at the forefront of research in the topological theory of stratified spaces.

The group who faithfully attended the course was quite heterogeneous, including second-year graduate students, junior faculty, and experts. Because of this I tried to cover aspects of the material that bear relation to other branches of mathematics, such as algebraic geometry, discrete groups, operator algebras, and algebraic K-theory. However, it created the challenge of being understandable and interesting to too many classes of people.

I assumed a knowledge of a one-year course in algebraic topology (e.g. [Sp]), a certain familiarity with characteristic classes ([MS]), and a feeling for the ideas of PL topology (say the first few chapters of [RS4]). After a while, I found it difficult to avoid the idea of spectra (see e.g., [Ad3]) and the basic constructions one does with them. However, I could not begin to make rigorous proofs with such little background. I quickly found that everyone was happiest when I explained the idea of proofs without giving the details, with the conscientious students looking up other accounts to get the detail they were missing.

This seemed to work well enough, although occasionally it was necessary to take a breather to absorb viscerally the techniques that were only briefly discussed in class. To move this along, I found that giving "exercises", which were only rarely such and which instead developed the theory further and required an understanding of how and why a given theorem works, was very valuable. Most students could not get very many of these without much work, although they helped concentrate the mind when reading the references.

I have decided to follow here the idea of the course, although these notes represent a significant expansion. (Many of the appendices were not covered in the original course, and in Part III, only the bulk of chapters 11 and 12 was covered, due to lack of time, and a sense of exhaustion on my part and on the part of the fearless class.) *This has the implication that what follows is not a work intended to be read cover to cover, but should be read in installments, with significant pauses between chapters, and maybe between sections, to develop familiarity with the material presented.* One good way to do this is to read, when available, the papers and books referred to in the text and notes. (Hopefully, this will be a less Herculean task because of the guidance provided by the text.) Also, the problems interspersed in the text should be grappled with. Many of the ones later in the book involve, in my opinion, publishable results, and the solutions of many of the earlier ones are the main results of published papers written when those methods were newer. (I have roughly indicated where I think the line between the publishable and nonpublishable is by calling some things problems and others exercises.)

I hope that the chapters are independent enough that seasoned mathematicians just interested in small chunks can read those sections and, with a minimal amount of flipping back, and assuming some theorems to be "black boxes", find what they are looking for. By reading the notes sections, which appear at the end of many of the chapters, they should then be able to find more detailed accounts of the parts they are interested in pursuing.

This seems like a good place to warn the reader that these notes sections should not be taken for serious historical scholarship. Through-out the book, I picked references that I thought would be useful, sometimes a late account because the idea is presented most clearly there, or sometimes an early account because I felt that would show what the difficulties involved really are. However, I never picked a reference because I necessarily wanted the reader to believe that was where the idea is from. Indeed, I cannot vouch for the accuracy of my own mathematico-historical ideas, especially regarding events that occurred before by becoming aware (early 1980s), so even the implied historical accounts must be taken with a grain of salt. I apologize to those I have slighted and to those I so mislead.

In the years since these lecture notes were first written there have been a number of important developments, mainly in the direction of foundational papers becoming available, but also some affecting certain of the applications in Part III. In addition, developments in the theory of homology manifolds [BFMW] suggest that a larger context of stratified spaces where the strata are not manifolds would lead to a better theory. I have tried to take all this into account in updating these lecture notes but have decided not to make significant changes in the basic structure to achieve this.

The long delay in these notes' revisions is largely due to my own inadequacies and the many suggestions and valuable criticisms made by friends, coworkers, and colleagues, of whom I mention Jonathan Block, Jeff Cheeger, Gustavo Comezana, Steve Ferry, Mark Goresky, Peter May, Andrew Ranicki, Jonathan Rosenberg, Julius Shaneson, Min Yan, and

Bob Zimmer. Their comments have had a large impact on the revision process.

Mathematically, I have a great many debts. The most pressing are to Sylvain Cappell, Steve Ferry, Frank Quinn, Mel Rothenberg, and Bruce Williams. Capell was my teacher and has been a collaborator and friend since. Our joint work on various problems in group actions (e.g., [CW1], and with Julius Shaneson [CSW]) and on "supernormal spaces" [CW2] was a major motivation for what follows. Ferry has also been a wonderful collaborator and friend. He is one of the creators of the field of "controlled topology", and he has helped me enormously in exploring these notions. A number of Quinn's ideas were basic and profoundly influenced my point of view on the general theory that follows. Both Rothenberg and Williams had many useful conversations with me while I was working out a set of technical details for the theorems that ensue. (Their contribution is then critical but, perhaps, invisible.)

I'd also like to record my thanks to the Courant Institute for many years of hospitality and to those who have attended by courses at Courant and Chicago. Bar Ilan University, through my membership in their Research Institute of the Mathematical Sciences, provided me with an atmosphere wherein I was able to produce at least two earlier drafts. The Sloan Foundation and the National Science Foundation, through grants and the Presidential Young Investigator Award, provided financial support for the research described.

And finally I must thank the people who provided safe haven throughout the storms, and sometimes the storms. My parents were always there for me, encouraging and prodding. The ones who were there, but didn't have to be, are Maidi Katz, Philip Schwartz, Shabsai and Debby Wolfe, and especially, Devorah Segal, who has become my wife. Without her tolerance, understanding, patience, and love I would have given up long ago. I dedicate this work to her.

Introduction

In this introduction, I will, with the help of a few judicious lies, try to motivate the contents of this book. Scattered within are some statements that depict where the theory is not yet in perfect form, at least not to my taste. I will not give references in this introduction; everything will be developed more fully in due course in the body of this work.

0.1. Classification of manifolds

The work of Smale proving the high dimensional Poincaré conjecture that manifolds that are homotopy equivalent to the sphere are homeomorphic to it can be said to begin a long and successful development of a theory of high dimensional manifolds. The first step, the h-cobordism theorem, identifies which manifolds with boundary are cylinders (manifolds of the form $M \times I$); and the second step, surgery theory, endeavors to produce cobordisms between homotopy equivalent manifolds that the h-cobordism theorem can identify as cylinders and, hence, show that the manifolds were, in fact, the same all along.

Both steps of this program are fraught with difficulties, of a different but, after more serious study, somewhat analogous sort. In Smale's h-cobordism theorem, which dealt with simply connected manifolds homotopy equivalent to cylinders, there is no obstruction at all; any high dimensional simply connected manifold homotopy equivalent (as a manifold with boundary) to a cylinder is a cylinder.

However, there is a key moment in his proof where the simplest bit of algebra occurs. A geometric pattern of intersections which has been encoded as a unimodular matrix over the integers is decomposed as a product of elementary matrices. (Elementary matrices correspond to some simple geometric moves that produce the diffeomorphisms of Smale's theorem.)

It is this step that is, in general, impossible. It turns out, as discovered by Barden, Mazur, and Stallings, that the geometry of intersections in a nonsimply connected manifold can be successfully encoded in a matrix over the integral group ring of the fundamental group $\mathbb{Z}\pi$. What happens then is that unimodular matrices over $\mathbb{Z}\pi$ are not necessarily products of elementary matrices; indeed the difference between these is encoded in the algebraic K-group $K_1(\mathbb{Z}\pi)$, and this obstructs (modulo some small technicalities) homotopy cylinders from actually being cylinders.

Actually, the algebraic obstruction to the theorem had already been discovered long before by J. H. C. Whitehead in his theory of simple homotopy types, a beautiful theory that only involves polyhedra; i.e. it is a theory in which manifolds play no special role. This theory divides homotopy types into a finer subclassification that can be related to $K_1(\mathbb{Z}\pi)$. Whitehead's invariant is essentially a measure of how the cells used in describing the cellular chain complex of a space get twisted around during the course of a homotopy equivalence. It is essentially this invariant that had been used in producing the first examples of manifolds that are homotopy equivalent, but not diffeomorphic: the lens spaces.

Indeed, the *h*-cobordism theorem and Whitehead's theory of simple homotopy types are only beginnings in the rich connection between topology and algebraic K-theory, more of which will be discussed below, but for the purposes of this introduction, let us continue in a different direction. All we've discussed thus far allows the possibility (an overly optimistic conjecture of Hurewicz) that simply connected homotopy equivalent manifolds are always diffeomorphic.

This is not at all the case. The simplest source of counterexamples comes from the theory of characteristic classes. Let us consider orthogonal sphere bundles over the 4-sphere. For high enough fiber dimensions one can see that these are classified by the first Pontrjagin class, an integer. On the other hand, several different homotopical arguments can be used to show that there are only finitely many homotopy types among these. (The most conceptual is to try to trivialize the clutching defining the bundle using self-homotopy equivalences of the fiber sphere in the place of orthogonal transformations. The obstruction to doing this lies in the third stable homotopy group of the spheres! This is a finite group, so for the integers divisible by its order, the sphere bundle has the homotopy type of the trivial product bundle.)

There are more subtle examples as well, such as Milnor's exotic differential structures on the sphere, but before going on, I should mention that it is a basic and important theorem of Browder and Novikov that the only rational obstructions to Hurewicz's conjecture, i.e., to the diffeomorphism of homotopy equivalent simply connected manifolds, are the rational Pontrjagin classes. Furthermore, there is a realization theorem: any combination of cohomology classes is "rationally realizable" by a homotopy equivalent manifold if and only if a single relation holds. (This relation is the Hirzebruch signature formula that asserts that the signature of an oriented manifold, which is an invariant defined purely in terms of the cohomology algebra structure and is therefore a homotopy invariant, can be expressed in terms of Pontrjagin classes.)

However, it is impossible to understand this rational result of Browder and Novikov without first understanding the work of Kervaire and Milnor analyzing the finite group of differential structures on the sphere. Their analysis begins with the observation that any smooth manifold homotopy equivalent to the sphere is frameable. That is, its stable normal bundle is trivial. (So far, we're just doing more careful bundle theory than merely characteristic classes.)

Therefore, according to the Pontrjagin-Thom method, every homotopy sphere gives rise to an element of the stable homotopy groups of spheres, for they have described this stable homotopy group as the cobordism classes of manifolds with trivialization of their stable normal bundles. (There is an ambiguity here coming from different framings of the normal bundle, but let's ignore that difficulty here.) The idea for realizing the obstruction is this: if we have a manifold M with trivial normal bundle, then every low dimensional sphere inside the manifold has a trivial normal bundle in M. This means that we can find a neighborhood diffeomorphic to $S^n \times D^c$, where D is a normal disk. One scoops this neighborhood out and glues in a copy of $D^{n+1} \times S^{c-1}$, to produce a manifold with this element of homotopy killed. (This process is called surgery and gives the name to the whole classification theory of manifolds.)

We start at the 0-th homotopy of M, and first make M connected. This surgery essentially just takes the connected sum of the components. Then we inductively represent homotopy groups by embedded spheres (using general position). The Hurewicz theorem and Poincaré duality theorem combine to show that if we succeed in making M highly enough connected, i.e., have vanishing homotopy groups through half the dimension of M, M will be homotopy equivalent to the sphere (and PL homeomorphic to one, by Smale's theorem).

The difficulty in trying to embed spheres and make them disjoint occurs around the middle dimension. (Embedding them is the same as making different sheets in a generic immersion disjoint, so these are the same sorts of problem.) Like in the h-cobordism theorem, this leads us to an intersection type matrix algebra. (Thus, for nonsimply connected manifolds, the group ring $\mathbb{Z}\pi$ enters again.) Here there is a more complicated duality aspect, in that in the middle dimension, surgery doesn't necessarily simplify the manifold, and the condition for doing this involves the transpose of the intersection matrix. Thus, the linear K-theory of the first obstruction is essentially replaced by a (variant of) Hermitian K-theory. But, as in the h-cobordism theorem, if an algebraic obstruction in a functor of the fundamental group vanishes, one can continue the geometric construction – in this case, produce a cobordism to a homotopy equivalence.

Similarly, if this basic invariant vanishes, then one tries to replace the nullcobordism of Pontrjagin-Thom theory by a disk. This is obstructed, but one can also study the indeterminacy of the "coboundary"; I'd better bail out here.

Browder and Novikov extended this analysis to other simply connected manifolds (and each of them, ad hoc, to certain nonsimply connected ones). To classify simply connected manifolds within a homotopy type, one basically has to deal with a fairly complex interaction between bundle theory and quadratic form theory. When we toss in Sullivan's efficient analysis of the bundle theory and Wall's analysis of the algebra in the nonsimply connected case, we get the following exact sequence:

$$\to [\Sigma M : F/Cat] \to L_{n+1}(\mathbb{Z}\pi) \to$$
$$S^{Cat}(M) \to [M : F/Cat] \to L_n(\mathbb{Z}\pi)$$

where $S^{Cat}(M)$ denotes the set of Cat (Cat = differentiable, PL, or Top) manifolds simple homotopy equivalent to M, up to isomorphism. The L-groups are algebraically defined, and F/Cat is a classifying space that encodes the relevant bundle data in the different categories.

However, in its present form, manifold theory certainly does not end. Many questions remain. What does F/O look like? What does F/Top look like? Can you compute $L_n(\mathbb{Z}\pi)$ for different classes of groups?

Another problem concerns the nature of this exact sequence. Most naively, $S^{Cat}(M)$ is just a set, so we have to puzzle a little to decide what exactness means. F/Cat with some work is an infinite loop space, so that $[M : F/Cat]$ gains the structure of an abelian group, but then we discover, horrors, that $[M : F/Cat] \to L_n(\mathbb{Z}\pi)$ is not a group homomorphism. In addition, $[M : F/Cat]$ is naturally contravariantly functorial and $L_n(\mathbb{Z}\pi)$ is covariantly functorial. So $S^{Cat}(M)$ probably should not be functorial at all.

Finally, there remains the problem of relating this classification theory to other direct onslaughts on the problems. What range do basic invariants take when we restrict attention to a homotopy class? (Here we have in mind the Novikov higher signature conjecture, which deals with characteristic classes and eta invariants. But the question is general.) Can one analyze the specific manifolds that arise naturally, like homogenous spaces and symmetric spaces? Also, can one answer purely conceptual questions, such as whether embedding M in W is homotopy invariant in M and W? If one manifold fibers over another, is the same true for anything homotopy equivalent to it? And so on.

Some of these questions were dealt with early in the history of our subject, and these contributions are summarized in Wall's book (and, to some extent, in the body of this one), and some of them remain unsolved to this day. Oddly enough, and the next section of this introduction is devoted to explaining why, we have a much better hold of the answers for topological manifolds than for smooth ones. Indeed, for smooth manifolds, we do not even know if the number of elements in $S^{Diff}(M)$ lying in different orbits of $L_{n+1}(\mathbb{Z}\pi)$ (or, if you prefer, the number of inverse images of different elements of $[M : F/O]$) is independent of the orbit. Another example of our thorough ignorance is that we do not know how many smooth structures there are on the sphere, even if we restrict attention to those that bound parallelizable manifolds (although this can be blamed on the homotopy theorists)!

0.2. Topological manifolds

As I noted in the previous section, the theory of topological manifolds is in much better shape than the theory of smooth manifolds. Before getting to this, I had better clear up a misunderstanding that I'm sure I've just caused. It is not an easier subject, at least prima facie.

The smooth and PL categories in fact were studied earlier with much greater success than the topological category. There are so many tools available there that are hard to discern on examining topological manifolds. Firstly, there are Morse functions, giving rise to handlebody structures and there are tubular neighborhoods of submanifolds (in PL these are triangulations and regular neighborhoods). One continues and discovers deep isotopy theorems for pushing subobjects around. Then one learns the joy of transversality and general position. This combination of basic tools is critical to the developments I sketched above.

And none of these tools seem, at first glance, to exist in the topological category.

However, after a long development, they do exist! This subject is explained in the book of Kirby and Siebenmann, and the proofs actually depend in a large way on the development of the complete theory of PL surgery. The critical fact (due to Hsiang-Shaneson and Wall) is that any PL manifold homotopy equivalent to a torus has a finite cover homeomorphic to the torus. That this is useful for the analysis of topological manifolds is the great insight of Kirby's torus trick, which will play an important behind-the-scenes role in this book. But, I cannot digress to explain this now.

I should mention that nowadays using some amazing constructions of Edwards and Chapman, one can prove this result about tori by pure geometry, without invoking surgery.[1] One cannot advertise this as a simplification, only as an important conceptual alternative.

All of this does not explain the advantages that the topological category has over the others – it only explains that it need not be neglected. (What follows is mainly due to Quinn and Ranicki.) In some sense the source of all of the advantages, as far as manifolds alone are concerned,

[1]However, it is much more difficult to give such a proof for the whole classification of "fake tori".

can be traced to something we've already seen: there is a unique homotopy n-sphere topologically, while smoothly there can be many structures.[2] Indeed, the main result of smoothing theory traces the differences between smooth structures and topological ones to a classifying space whose homotopy groups are isomorphic to the group of differential structures on the sphere (under connected sum).

The uniqueness of the homotopy n-sphere has enormous implications. Using the surgery exact sequence (ignoring low dimensional difficulties that must be dealt with separately), it computes the homotopy groups of F/Top and shows that they are isomorphic to the L-groups of the trivial group.

This ultimately leads to a homotopy equivalence between $\mathbb{Z} \times F/Top$ and a space $\mathbf{L}(e)$, whose homotopy groups are the L-groups of the trivial group e. This space has a natural infinite loop space structure coming straight out of surgery theory, and naturally the map $[M : F/Top] \to L_n(\mathbb{Z}\pi)$ defines a homomorphism using this abelian group structure on $[M : F/Top]$.

More is true. There is a natural orientation for (oriented) topological manifolds, so that $[M : F/Top] \cong H_n(M; \mathbf{L}(e))$. In this way, the surgery exact sequence can be rewritten (aside from a $\mathbb{Z}$ that I must return to)

$$\begin{gathered} \to H_{n+1}(M; \mathbf{L}(e)) \to L_{n+1}(\mathbb{Z}\pi) \to \\ S^{Top}(M) \to H_n(M; \mathbf{L}(e)) \to L_n(\mathbb{Z}\pi). \end{gathered}$$

At this point, one guesses (correctly) that $S^{Top}(M)$ is an abelian group, and that the whole sequence is covariantly functorial. Furthermore, there is a natural fourfold periodicity (due to Siebenmann) relating $S^{Top}(M)$ and $S^{Top}(M \times D^4)$.

All of this leads to a great deal of conceptual clarity and extra computability in the whole topological theory – except for the extra $\mathbb{Z}$ in the equation $\mathbb{Z} \times F/Top \cong \mathbf{L}(e)$. This, for instance, actually mars the "periodicity" I just mentioned. $S^{Top}(M \times D^4)$ might be as much as a $\mathbb{Z}$ larger than $S^{Top}(M)$, although once M has a boundary this does not happen.[3]

Since writing the first (three) versions(s) of this book, the explanation has finally come to light (through work of Bryant, Ferry, Mio, and mine). There seems to be $\mathbb{Z}$'s worth of local models as pretty as Euclidean space itself (at least up to s-cobordism, at this point in time), and surgery theory classifies spaces modeled on these. If we have a boundary, that determines the local model, of course, so nothing changes. However, the periodicity requires the extra $\mathbb{Z}$ because one has unwittingly imposed

[2] While there is a unique PL homotopy sphere in high dimensions, some four-dimensional pathology (Rochlin's theorem) causes the PL theory to lack some of the elegance of the topological case.

[3] This was first observed by Nicas.

an additional extraneous condition on the natural class of spaces being classified.

So we will pretend that the $\mathbb{Z}$ is not there. A pretty way to rewrite the sequence is to use the notation (but it's more than that) $\mathbf{L}(\mathbb{R}^n)$ for n-fold loop space of $\mathbf{L}(e)$. Then the sequence is

$$\to H_0(M; \mathbf{L}(\mathbb{R}^n)) \to L_{n+1}(\mathbb{Z}\pi) \to$$
$$S^{Top}(M) \to H_0(M; \mathbf{L}(\mathbb{R}^n)) \to L_n(\mathbb{Z}\pi).$$

The sequence then takes the form of a local-global principle. $\mathbf{L}(\mathbb{R}^n)$ is the L-theory of the local structure of n-manifold M. $L_n(\mathbb{Z}\pi)$ is a global term, and all of the information regarding manifolds simple homotopy equivalent to M is the result of the way local pieces globally assemble to the homotopy type of M. (The astute reader will realize that this formulation is actually more correct than the lies uttered above for nonorientable manifolds!)

0.3. Stratified spaces

Now, let me turn to the main topic of this book, which is stratified spaces. For the purposes of this introduction, one should just think of a stratified space as being a topological space with singularities that occur in various layers. The difference between layers is a manifold (or, I would prefer, an ANR homology manifold), and the pieces should fit together in a not too irregular fashion. We refer to these layers as pure strata.

To get the gist of things, the reader should be able to see stratified spaces almost everywhere. Any manifold with boundary is a stratified space, with two strata. The bottom stratum is the boundary, and the top stratum is the whole manifold. The pure strata are the boundary and the interior.

Any polyhedron can be stratified in a reasonable, PL-invariant fashion. Another interesting source of stratified spaces comes from embeddings. We artificially create a stratum in the ambient manifold, namely the submanifold. (Thus we can consider embedding theory as a proper subset of the theory of stratified spaces.) Similarly, immersions give rise to at least two interesting stratified spaces: the image of the immersed manifold and the ambient manifold with the image as a union of strata.

Algebraic varieties are naturally stratified spaces.

Many maps give rise to stratified spaces by taking mapping cylinders, or simply by looking at the stratification the range would get viewed as a subpolyhedron (or subspace) of the mapping cylinder. The domain gets a stratification by pulling this stratification back.[4]

When compact groups (or even finite groups) act on a manifold, with manifold fixed sets (this is an assumption for nonsmooth actions), the quotient space gets a stratification, indexed by the orbit types of the group.

A last sourcc of stratified space is compactifications of moduli spaces or other varieties, or even Riemannian manifolds with appropriate geometric assumptions. (This includes, the well-known, but often maligned, one point compactification.)[5]

The lists overlap with each other, and each one suggests its own line of questioning, vocabulary of invariants, and methods of attack. Each theory has its historical successes and outstanding problems. One of the major goals of this book is to view these subjects within one context, so that some of the ideas that naturally would arise in one area or another can be applied to all of them. Indeed, our main theorem could have been guessed on the basis of sufficiently deep analysis of any of these particular stratified spaces and extrapolation to the general case; what happened, however, was that many particular cases were studied, with different special techniques idiosyncratic to the individual problems but analogous in conclusion, and the general framework was ultimately thrust upon us.

One fundamental problem in extending the classification of manifolds to stratified spaces is that we do not have a very good bundle theory for stratified spaces. Since the local structure changes from point to point, it is clear that sheaf theoretic ideas should play a role. Moreover, there is a trivial local-global inseparability built into the theory. By taking the cone on a space, we transfer a global problem on one stratified space to a local one on another. In actuality, though, the process reverses: in an induction on the number of strata global information about spaces with fewer strata feeds in to solve the local problems that occur with more strata. By analogy with the surgery exact sequence for manifolds, one would want a sequence

$$\cdots \to L(X \times I) \to S^{Top}(X) \to H_0(X; \mathbf{L}(loc)) \to L(X)$$

[4]For "most" continuous maps the mapping cylinder is not a stratified space, because of lack of homogeneity, but surprisingly, one can still study the map using ideas from the theory of stratified spaces by thinking of constructions relative to the "bottom stratum", i.e., the target. It is not unreasonable to expect that one can deal with worse singularities if one works relative to those "strata" than one can deal with if such strata are to be grist for the mill themselves, so to speak.

[5]It is a good barometer of the reader's feelings about the generality of the notion of stratified spaces desired to inquire which one point compactifications of which noncompact manifolds will be allowed as stratified spaces.

where $L(X)$ is some sort of inductively defined surgery theory built out of the surgery theory of the strata, and the homology term is a cosheaf homology adapted to the local structure.

The main new result of this book is that this is almost true. There is a deviation that can often be ignored and is always 2-torsion, which is discussed in chapters 6 and 10. But the most important point is that it is quite close to true.

The most important aspect of the theory might be that it gives a universally defined place to put the characteristic classes of all stratified spaces. The fact that these classes together with K-theory (developed by Quinn and Steinberger and West, which I'll explain in chapter 6) and a modified version of surgery theory (adapted from earlier work of Browder and Quinn) suffice for classification problems might well be just gravy. (Although this was and remains my primary motivation.)

0.4. Some examples

Given the abstract nonsense of the previous section, it is perhaps remarkable that anything follows just for free. But some things do.

For instance, consider the embedding theory example. In that case there is one bit of information that one can forget, either the ambient manifold or the submanifold. (From the point of view of characteristic classes, one has simply the characteristic classes of ambient manifold and submanifold, but no relation between them.)

A little calculation (made in chapter 11) shows that if the submanifold has codimension at least three, then this pair of forgetful functors combine to give an equivalence on the L-theory both globally and on the cosheaf level. This implies that $S(W, M) \cong S(W) \times S(M)$. In other words, we have proven a "decomposition theorem" for an "L-cosheaf" just by applying forgetful maps, and then fed it into the theory to get a calculation.

What does this mean? Exactly the following classic beautiful theorem:

THEOREM (BROWDER, CASSON, HAEFLIGER, SULLIVAN, AND WALL). *Topological embeddings of M in W of codimension at least three*[6] *are in a 1-1 correspondence with Poincaré embeddings. In particular, if M embeds in W, any manifold homotopy equivalent to M embeds in any manifold homotopy equivalent to W.*

It is not necessary for the reader to know what a Poincaré embedding is to appreciate the result. Just note that a Poincaré embedding is a purely homotopy theoretic notion. On reflection, the theorem is very

[6] The codimension two situation is much more difficult and complex. For a systematic study of the BCHSW phenomenon in codimension two, in both locally flat and nonlocally flat situations, see [CS1,4,8]. We discuss this, only a little bit, in 11.5.

surprising. While it is true in PL and Top, it is totally false in Diff. We know that the differences of the characteristic classes between M and W are restricted in a more and more severe fashion as the codimension of a putative embedding decreases. Despite the dearth of differentiable embeddings, the topological and PL embeddings are, from the point of view of general theory, just the result of a decomposition theorem, and one which is proven by mere application of forgetful maps.

The original proof went via an analysis of the potential regular neighborhoods, i.e., a "tubular neighborhood theorem," and via a subsequent stability theorem for the homotopy theory of unstable classifying spaces. This second ingredient, when combined with PL Smale-Hirsch immersion theory (due to Haefliger-Poenaru), implies that an immersion of M in W gives rise to one of any manifold homotopy equivalent to M in any manifold homotopy equivalent to W. If we apply the surgery theory, we obtain the conclusion that these new immersions can be taken with identical singularity sets: i.e., the same double points, triple points, etc., mapped in exactly the same way.

There is another remarkable thing about this proof, which is a principle that I would like to highlight in this introduction. The forgetful maps we used in the proof are actually of very different natures. Restriction of attention to the submanifold is perfectly reasonable from the stratified point of view: it is restriction to a stratum. Forgetting the submanifold is highly nontrivial! Ordinarily, one cannot forget a lower piece of stratification and still have a stratified object (recall pure strata are demanded to be manifolds).

The proof of this theorem requires an operation that is entirely natural, indeed trivial, in the context within which one would study the theorem, but that is highly nontrivial from the point of view of general theory. Indeed, in all the applications I know, the denouement is effected by application of ideas that are natural to the subject of application (although, perhaps, as in this case, too trivial to even be worth mentioning) independently of their relevance to the stratified surgery theory, but that are not predicted just on general grounds.

The other lesson to learn is that decomposition theorems are useful. Indeed, many geometric problems can be reduced to suitable decomposition theorems for L-cosheaves and the interactions between different decompositions. An important, if vague, open problem is to determine "reasonable" conditions under which the cosheaves that arise decompose. (I have been quite surprised by some that have!)

Another class of spaces where one can get very far in such an analysis is that of spaces with even codimension singularities. For these, intersection homology defines all of the intersection forms that we are used to from ordinary homology theory for manifolds. This enables one to do all of the quadratic form theory and all of the algebraic surgery theoretic constructions on the singular space, just as if it were not singular at all! In particular, Goresky and MacPherson in their original paper defined L-classes, and Cheeger even set up a signature operator (which is, in the manifold case, away from 2 the basic characteristic class in classification theory).

In our context, this means that for "supernormal spaces", i.e., spaces with the links of all singularities simply connected, one can repeat the argument as in embedding theory and show that $S(X)$ for such a space breaks up as a sum of pieces, one for each component of each stratum, and each of these pieces is given by exactly the surgery exact sequence written in homological form. Thus, if all components of all strata are simply connected, the stratified spaces within a given stratified homotopy type are determined up to finite ambiguity by the L-classes of all of the strata, an exact parallel of the result of Browder and Novikov for simply connected manifolds discussed above. (This result was first proven in joint work with Cappell a few months before the general theory was developed.)

Incidentally, this result shows, computationally, how important the shift to homology is. For manifolds, homology and cohomology are distinguished only in the subtle world of conceptual frameworks, functoriality, and aesthetics. For spaces like these, homology and cohomology are not even isomorphic groups. No subtlety is needed in telling what it is you've actually got!

The last classic cause that I'd like to mention is that of orbifolds, or, even more classically, finite group actions. In that case, again there is a natural a priori notion of the relevant characteristic classes given by the equivariant signature operator (on the ambient manifold and the fixed set of various subgroups). This concrete characteristic class computes the abstract characteristic classes of the general theory, away from the prime 2. For odd order groups, Madsen and Rothenberg developed an equivariant topological surgery theory from a rather different point of view. Their magnificent work also contains (in this case) this characteristic class and equivariant transversality. (In fact, transversality is the centerpiece of their development, on which everything else depends.) Their main application was the result (also proven independently by Hsiang and Pardon) that for odd order groups, any two topologically conjugate linear representations are linearly conjugate.

The topological classification of linear representations was an important source of motivation for all workers in this area. That this is a nontrivial subject is due to the startling examples constructed by Cappell and Shaneson of distinct representations of $\mathbb{Z}_{4n}$, with only a few isotropy types, which are topologically conjugate and therefore have homeomorphic quotients. While these examples predated the results on odd order groups mentioned in the previous paragraph, they also demonstrate that there are restrictions on the continued development of the methods based on transversality: equivariant transversality fails for the even order case (and certainly for positive dimensional groups). Historically, this provided a challenge: to see that these examples arise as part of a systematic theory of stratified spaces, rather than as the result of some complicated explicit constructions. Unfortunately, I cannot report on any significant simplifications of the even order nonlinear similarities from the current perspective (largely because the difficulties are mainly at the prime 2 and quickly reduce to the same hard calculations that arose in the Cappell-Shaneson work). However, the general theory explains quite well the shape of the theory.

Perhaps a better tack is to recognize that transformation groups are a classic subject that has many natural and fascinating problems and to focus on some of these. What are the possible fixed sets of group actions (for some specific group) on some specific manifold? What are all of the actions within a given equivariant homotopy type? There are the replacement questions: If M is the fixed set of an action on W and M' is homotopy equivalent to M, is M' the fixed set of an action on W? On some W'? One also wants to know whether natural actions on Euclidean spaces or spheres and projective spaces (etc., of course) given by linear representations are topologically conjugate when their linear structures are different (the Euclidean case is the nonlinear similarity question, discussed above). There are also questions related to the connection between the group action and the action of subgroups. Which proper set of subgroups determines the actions of the whole group? (For instance, can a compact actions be reconstructed from the actions of the finite subgroups?)

All these questions have received enormous amounts of study, which cannot begin to be surveyed in a short introduction to a book about something else! Some of them will be reviewed and developed in chapter 13.

Other questions in transformation group theory become clear only after more profound analysis. The most important of these, in my opinion, are the following:

The role of p-completion in the geometric theory of group actions is one such problem. That is, the homological study of group actions naturally leads to p-complete homotopy theory, i.e., through mod p homology groups, etc. (Deeper, there is the Sullivan conjecture, proven by H. Miller.) The current geometric technique involves p-local homotopy theory. The gap between the p-local and the p-complete is, roughly speaking, rational homotopy theory, a subject about which much is known. However, it is rare that one can give a complete analysis of the actions because of this gap. As a concrete example, actions of $\mathbb{Z}_p$ on the sphere are pretty well classified because in this case there is no gap between the p-adic and the p-local (at least in codimension other than two and assuming some tameness of the fixed set; see chapter 13). For products of spheres, the problem seems extraordinarily difficult.

Another important conceptual barrier lies in the difference between equivariance and isovariance. The way stratified theory is relevant to transformation groups is via consideration of the quotient space. The notion of morphism that seems relevant is that of *stratified map*, which is a map that preserves the pure strata. (For instance, the notion of equivalence in the case of manifolds with boundary is homotopy equivalence of pairs, not maps of pairs that happen to be homotopy equivalences.) For group actions this means that the maps in addition to being equivariant must be isovariant, i.e., satisfy the condition that $G_{fx} = G_x$. The difference between equivariant and isovariant homotopy theory leads, by pure homotopy theory and structural properties of equivariant surgery theory, to the result that there are many nonhomeomorphic equivariant homotopy equivalent $K(\pi, 1)$'s. (This was called[7] the equivariant Borel conjecture.) The question then arises, to what extent are the notions defined by isovariant techniques actually equivariant? Do they have equivariant definitions or interpretations? Is there equivariant functoriality?

(I should point out that another notion, the gap hypothesis, that arises in many discussions of equivariant surgery, suffices according to a theorem of Browder to bridge the difference between equivariant and isovariant homotopy equivalence. The point is that equivariant homotopy theory is much more tractable than isovariant and is in many ways the more natural. However, the geometrical analysis of the equivariant category only seems possible when the gap hypothesis holds. The isovariant analysis is always possible but is rarely computable without the gap hypothesis, and then it coincides with the equivariant one. Nonetheless, and probably needless to say, I think there are real advantages to the extra generality because of the possibility of ad hoc analysis in special cases.)

Lastly, I would like to point out that despite all of the progress made in the past decades, we still do not know what are the possible fixed sets of PL periodic maps on the disk when the period is composite, say 6 or 15.

0.5. A word about methods

The way we analyze stratified spaces is inductively. To see what is going on consider the situation of singular spaces with two strata.

One method of analyzing such spaces is to work in the PL category. In that case the bottom stratum is a manifold, and the top pure stratum

[7] Unfairly to Borel.

is naturally the interior of a manifold with boundary (the complement of the interior of a regular neighborhood of the singular stratum). Either by definition or as a consequence of the homogeneity or the stratified space, these two strata are glued together by a PL homeomorphism of the boundary of this manifold to a block bundle over the singular stratum. Thus, we'd have to analyze this bundle theory to proceed and integrate all of these data together. For smooth stratified spaces, what arises is genuine fiber bundle theory, and one has to map into unstable classifying spaces. These are incredibly difficult to analyze; indeed the homotopy type of $Diff(M)$ is understood only for some low dimensional manifolds M. In the PL case, where it is block bundles that occur, it turns out that surgery theory is adequate for the analysis (see chapter 3). This enables one to continue the induction, since this method is adequate for block bundles with two strata fibers.

Although we could, we do not explicitly do this induction here. We prove the PL result by a slight variant of it in chapter 8. For the topological category we have to be a bit more indirect because block bundle neighborhoods do not in general exist, nor are they unique when they do exist. (The type of neighborhood that exists is a more complicated type of structure: there is a map of the neighborhood to the bottom stratum so that the inverse image of each open set is homotopy equivalent to what homotopy theory would predict the inverse image to look like; see 10.3.A.)

Our method is instead to redo all of topology with care taken not to distort metrics too much. This is sometimes called controlled topology and has been actively developed by a great number of people (for us the relevant work is mainly due to Chapman, Ferry, Quinn, and Hughes). These methods could easily be adapted to re-prove all the basic foundational results of topological manifolds (section 0.2), but we won't develop this. The topological invariance of rational Pontrjagin classes, for instance, is quite literally a weaker version of the problem of intrinsically computing invariants of the link of the bottom stratum (in a two-stratum singular space) in terms of the manifold given as the top pure stratum.

Unfortunately, there is no comprehensive treatment of the basic ideas of controlled topology yet available besides the original papers. We have tried to partially remedy this in chapter 9, but it is clear that much more is necessary (another book someday, but maybe by somebody else[8]...).

For us, the idea will be that when we remove the bottom stratum, we should put a convenient metric on the complementary open manifold that somehow remembers the missing stratum. If there were a single sin-

[8] Recently, Steve Ferry has made two great strides in this expository direction. He has written a beautiful introductory geometric topology text and an up-to-date exposition of recent research in the CBMS lecture note series.

gular point, the metric would be tubular at infinity and resemble a ray in the large (which naturally compactifies with only one point at infinity). In general, the metric should have the lower strata on the horizon, and the "homotopy link" as "horosphere". Then algebraic K-theory and surgery adapted to this quasi-isometry class of manifolds will be what is relevant to performing the inductive step. (Of course, these techniques are valuable for many other sorts of problems as well—see chapter 9 and its appendices.)

Remarkably, the answer is quite similar to the PL situation. (Indeed, in the original argument I developed, I used a topological analogue of a block structure that exists after crossing with a high dimensional torus, then did the PL analysis, and tried to remove the tori.) Geometrically, what occurs is (after shrinking the metric on the complement back to its original incomplete incarnation) an infinite sequence of surgeries on smaller and smaller spheres within an open subset of the "normal bundle" to the singular set. We like to think of this as microsurgery.

The shrunk-down version of this geometry seems quite close in spirit to microlocal analysis. On the other hand, the blown-up version is close to the type of analyses of elliptic operators with bounded propagation speed on noncompact manifolds conducted by Gromov, Lawson, Connes, Higson, Roe, and others. This connection will be explored a little bit in chapter 9. There are many parallels between the theories, in terms of slightly different proofs of various theorems, such as results on the Novikov conjecture. A hope for the future is that the geometric ideas presented here will merely boil down to a refined analysis of the signature operator but that the framework will be suggestive for other operators, which could lead the way to new sorts of applications. (One might even expect the operators themselves to suggest new interesting stratified spaces on which prolongations exist and which could be used to study the original operator.) We will see . . .

PART I: THE THEORY OF MANIFOLDS

This part summarizes the classification of high dimensional manifolds as it is now understood. The chapters need not be read in order. For the reader who just wants a quick feel for what the theory means, I would recommend thinking only about compact simply connected manifolds. Doing this, one can go straight to chapter 2 on surgery and from there to chapter 4 on applications to read the first four sections. Chapter 3 is a bit technical but is critical for understanding the statements of the general stratified theory: it is largely a reformulation of chapter 2. Chapter 1 is, of course, necessary for the complete picture, and the algebraic K-theoretic invariants that enter there are among the most important in topology. Furthermore, there we study some aspects of the theory of noncompact manifolds that are absolutely critical to our understanding of even the simplest compact stratified spaces.

We remind the student that reading this part is not a substitute for reading the many papers and books dealing with the topic discussed below. This part is a summary, not a detailed development, and for later arguments to even make sense, one has to have a feeling for this material, not a mere ability to quote its contents.

1 Algebraic K-Theory and Topology

This chapter is devoted to some important interactions between algebraic K-theory and topology. We answer algebraically certain very special geometric questions that will later play a surprisingly central role. The first two sections consider some geometric/homotopy theoretic questions in the category of simplicial complexes, and sections 4 and 5 describe variants of these in the setting of manifolds. In the last sections, we give some selected applications to manifolds (generalized Poincaré conjecture), knot theory, and group actions.

1.1. Wall's finiteness obstruction

Much of what we will be interested in involves the translation of homological information into geometric (the reverse being usually much more elementary). We start with finiteness conditions.

When is a CW complex homotopy equivalent to a finite complex?

Historically, this was viewed as a start on the question asked by Borsuk at the 1954 Amsterdam International Congress whether all compact ANR's are homotopically finite complexes (so that all compact topological manifolds, in particular, are). The affirmative answer to this question was a triumph of infinite dimensional topology [We] (see also [Ch1]; and see [KS] for the manifold case).

Another early problem where this comes up is the question of which groups act freely on some sphere. Swan discovered an algebraic construction that produces infinite finite dimensional CW complexes whose universal covers are homotopy equivalent to spheres, and the question arose regarding which of these CW complexes can be taken finite. For more on this, see 4.8 below and the survey [DM].

Moreover, this problem is archetypical of many problems wherein K-theory and topology interact.

PROPOSITION. *If X is a simply connected CW complex with all homology groups finitely generated, X is homotopy equivalent to a CW complex with finitely many cells in each dimension. If the cohomology vanishes above some dimension, then one needs no cells above that dimension. If both hold, then X is homotopically a finite complex.*

Let us build these better homotopy X's one dimension at a time. That is, we will produce spaces X_i of dimension i, with maps $X_i \to Y$ that are an isomorphism on homotopy in dimension $\leq i-1$ and surjective on π_i. By the Hurewicz isomorphism theorem this condition is equivalent to the parallel condition on homology. Furthermore, on taking the mapping cylinder, one realizes that the i-skeleton of X can be taken to be X_i. (Compare the proof [Sp] of the Whitehead theorem that recognizes homotopy equivalences as maps that induce an isomorphism on homotopy groups.)

Let us now build X_{i+1}. Since X is produced by attaching cells to X_i, all that one does is add on the cells necessary to kill the extra H_i and to generate H_{i+1}. The finiteness assumption guarantees that this will occur with a finite number of cells, so the first part is proven. To obtain finite dimensionality, consider the place beyond which H_i vanishes. Note that the kernel of $H_i(X_i) \to H_i(X)$ is a free group. All the lower maps are isomorphisms. Therefore, this kernel is generated by the image of π_i. Attach $i+1$ cells corresponding to generators. A calculation gives a homology isomorphism in all dimensions, the Hurewicz theorem shows that one has an equivalence of homotopy groups, and the Whitehead theorem shows that one has a homotopy equivalence.

The two arguments are compatible, so the last statement follows.

What about the nonsimply connected case?

The first point is that one cannot immediately use homology to detect homotopy information, because of the hypothesis of simple connectivity in the Hurewicz theorem. The standard way to get around this difficulty is to pass to the universal cover of X. (Note that as a consequence of the Whitehead theorem, a map is a homotopy equivalence iff it induces an isomorphism on fundamental groups and is a homotopy equivalence of universal covers.) Then it is of course important to take into account the π-module structure, so that cells that we find upstairs really come from downstairs.

With this in mind the case of finitely many cells in a given dimension is rather straightforward.

The second basic point is then the finite dimensionality. One does not immediately see that the relevant kernel is free. However, Wall observes [Wa3] that if twisted cohomology with all coefficients vanishes, then a little homological algebra shows the projectivity of this module. Now we can get at finite dimensionality. If we had a free module, the above argument would work. At the cost of finiteness we can gain freedom:

Exercise (The Eisenberg Swindle). If P is a projective R-module, then there is a, perhaps infinitely generated, free module F so that $F \oplus P$ is free. (Hint: Let Q be a module for which $P \oplus Q$ is free, and group $P \oplus Q \oplus P \oplus Q \oplus P \oplus \ldots$ in two different ways.)

Now one can wedge onto X_i a bunch of spheres corresponding to the generators of F, which will add F onto the homology kernel. This gives us a free homology kernel, which we can kill by attaching cells.

REMARK. This projectivity is exactly equivalent to a purely geometric condition, that X is dominated by a finite complex $p: K \to X$. This means that there is a finite complex with a homotopy section for p. Such an X is said to be **finitely dominated.**

It is now clear that there should be an obstruction related to the difference between projective and free modules over $\mathbb{Z}\pi$ to obtain a finite complex.

DEFINITION. *If R is a ring, let* $\tilde{\mathbf{K}}_0(\mathbf{R})$ *denote the Grothendieck group of finitely generated projective R-modules, modulo free modules. Notice that this is a covariant functor from rings to abelian groups.*

The algebra of $\tilde{K}_0(R)$ for various R is fascinating and is often related to number theory. See e.g. [Ba, Mi6, Re]. For instance, if R is a Dedekind domain (e.g. the ring of integers in a finite extension of $\mathbb{Q}$), then $\tilde{K}_0(R)$ is naturally isomorphic to the ideal class group of R, i.e., the group of ideals modulo principal ideals.

THEOREM (WALL). *The class of the homology kernel in $\tilde{K}_0(\mathbb{Z}\pi)$ is well defined and vanishes iff X is homotopy equivalent to a finite complex.*

EXERCISE. Construct for each element of $\tilde{K}_0(\mathbb{Z}\pi)$ a complex whose Wall finiteness obstruction is the given element. (Hint: Use the construction of finite dimensional infinite complexes given above and the previous exercise.) For $\pi = \mathbb{Z}_p$, p a prime, the natural map $\mathbb{Z}[\mathbb{Z}_p] \to \mathbb{Z}[\xi]$ (ξ a primitive p-th root of unity) induces an isomorphism on K_0. If the cyclotomic field has nontrivial class group, there are spaces that look homologically like finite complexes but that are not in fact finite (up to homotopy).

One can set up this whole theory in a more algebraic setting of R-chain complexes. The finiteness condition on a chain complex suffices to make it chain homotopy equivalent to a chain complex of finitely generated projective modules. The Euler characteristic of such a complex is well defined in $\tilde{K}_0(R)$, and a little thought shows that it measures the obstruction to making this chain complex chain homotopy equivalent to a complex of finitely generated free modules.

Wall's element for a space is then the corresponding element for the cellular chain complex of its universal cover.

EXERCISE. Using Serre's mod C theory [Sp], set up a parallel theory of $R\pi$ finiteness for R a subring of the rationals. Show that for π finite and $R = \mathbb{Q}$ the finiteness condition is equivalent to the conclusion of the Lefshetz fixed point theorem; i.e., that the Lefshetz number of all nonidentity covering translates on the universal cover is zero. (Hint: A rational representation is a multiple of the regular representation iff the characters of all nonidentity elements vanish.)

EXERCISE/REMARK. If M is a module of finite projective length over R, show that the Euler characteristic of a resolution gives a well-defined element of $K_0(R)$. If M is a finite module (i.e., contains only finitely many elements) of order prime to the order of a finite group π, then it necessarily has finite projective length over $\mathbb{Z}\pi$ (Rim's theorem). A theorem of Swan implies that every element of $\tilde{K}_0(\mathbb{Z}\pi)$ arises this way.

Let $\mathbb{Z}_{(\pi)}$ denote $\mathbb{Z}$ with the primes prime to the order of π inverted. If $f : X \to Y$ is a map which is an isomorphism on $H(\ ;\mathbb{Z}_{(\pi)}\pi)$ (this boils down to an isomorphism on $H(\ ;\mathbb{Z}_{(\pi)}\pi)$ for the regular covers), then the difference of finiteness obstructions is (up to sign) the Euler characteristic of the homology of the map f. This is often a very practical way to compute finiteness obstructions. See [As, Mis, Wei9].

1.2. Simple homotopy theory

Historically, the material of this section was discovered before that of the previous one. We ask the uniqueness question. How many different finite complexes are there homotopy equivalent to a given X?

As things stand this is not a very sensible question, since there are obviously infinitely many different finite complexes. One relatively fancy resolution of this theory is to use infinite dimensional topology and ask how many different Hilbert cube manifolds are there homotopy equivalent to X? This smooths out, at least, dimensional issues, and turns out to be quite sensible (see [Ch1]).

Another more pedestrian way to smooth things out is to embed in Euclidean space and take **regular neighborhoods**. Recall that any polyhedron embedded in a PL manifold has a canonically defined regular neighborhood, i.e., well defined up to isotopy. We can get rid of all choices by stabilizing the Euclidean space. (Any finite polyhedron has a unique embedding up to isotopy in a sufficiently high dimensional Euclidean space by standard general position results.)

Figure 1 shows that rather different looking polyhedra have isomorphic regular neighborhoods.

If one attaches a simplex to a polyhedron along one of its faces, then the regular neighborhood is unchanged (up to isomorphism). (Such a move is called an **elementary expansion**. Its reverse is called an **elementary collapse**.) Furthermore, every regular neighborhood is actually built up out of such a successive union (see Fig. 2).

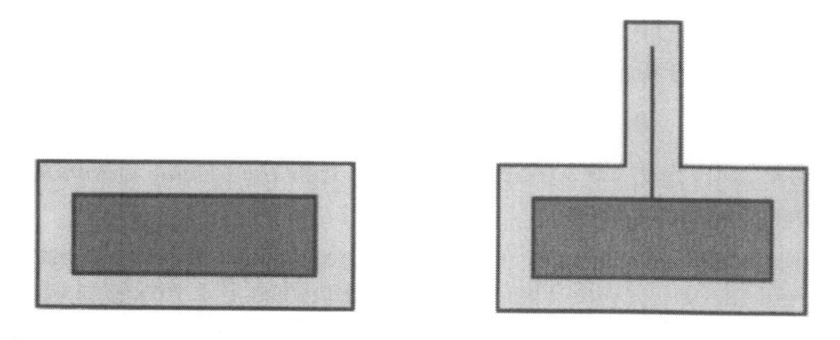

Figure 1. Regular neighborhoods of a space and an expansion of it

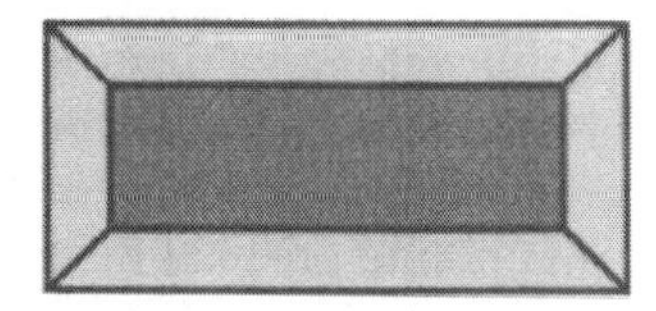

Figure 2. Collapse the two-faces to the polyhedron expanded by the one-faces, and then collapse the one-faces

Therefore, following J.H.C. Whitehead, we make the definition that a homotopy equivalence is **simple** if it is homotopic to the result of a succession of elementary expansions and collapses. We have sketched the following conclusion:

THEOREM (see [RS4]). *A map between finite polyhedra is a simple homotopy equivalence iff, on extending it to a map between regular neighborhoods in a high dimensional Euclidean space, it is homotopic to a PL homeomorphism.*

Now we can be more precise. Which homotopy equivalences are simple?

We can be led to conjecture a complete solution by philosophy alone. Typically, when one has a series of functors, the $(i+1)$st is defined in terms of differences between trivializations of elements of the previous. For instance, by considering the equator of the sphere, one recognizes homotopy groups as being differences between nullhomotopies of lower dimensional spheres.

$\tilde{K}_0(R)$ is made up out of projective modules modulo those which are stably isomorphic to free modules. Therefore $K_1(R)$ should be made out of different stable isomorphisms to free modules, or, equivalently, the automorphisms of free modules. More precisely:

DEFINITION. *$K_1(R)$ is the Grothendieck group of stable automorphisms of finitely generated free modules: i.e., if one has an exact sequence*

$$0 \to (F, \alpha) \to (G, \beta) \to (H, \gamma) \to 0,$$

then $[(F, \alpha)] + [(H, \gamma)] + [(G, \beta)]$. We also view the identity automorphism of a free module as trivial.

Notice that we can view $K_1(R)$ as generated by invertible matrices. The equivalence relation kills elementary matrices (i.e. products of upper triangular matrices) because a matrix of the form

$$\begin{vmatrix} I_n & A \\ 0 & I_m \end{vmatrix}$$

fits into an exact sequence of automorphisms with identity on a kernel R^m and cokernel R^n. This quotient of stable invertible matrices modulo elementary matrices is equivalent (according to "Whitehead's lemma") to the abelianization of the infinite general linear group $GL(R)$.

EXERCISE. Show how to associate an element of $K_1(R)$ to any automorphism of a projective module.

By virtue of the behavior with respect to resolution inherent in forming a Grothendieck group, one can expand, by an artifice similar to Euler characteristic, the definition of $K_1(R)$ to include chain equivalences between based chain complexes or, alternatively, on auto-chain-homotopy equivalences of finite projective chain complexes.

Geometrically, one gets an element of $K_1(\mathbb{Z}\pi)$ from the induced maps on cellular chain complexes on the universal covers of a cellular homotopy equivalence between finite complexes. Unfortunately, this is not quite well defined. The difficulty is that because of picking base points and paths from cells to a fixed base point upstairs, one must mod out by $\pm\pi$. One can (and must, for purposes of well-definedness) check that the map induced by an elementary expansion, or a subdivision, induces the trivial element. (See [Co, Mi3].) This explains the necessity of the following:

THEOREM (WHITEHEAD). *A map $f : X \to Y$ between finite complexes is a simple homotopy equivalence iff an obstruction $\tau(f) \in Wh(\pi) = K_1(\mathbb{Z})/\pm\pi$ defined above vanishes.*

$\tau(f)$ is called the **(Whitehead) torsion** of the homotopy equivalence. The necessity is quite interesting in that it gives examples of homotopy equivalent manifolds that cannot be PL isomorphic.[1] In an appendix we

[1]That one can make sense of the notion of simple homotopy equivalence for smooth manifolds is a consequence of the Cairns-Whitehead theorem triangulating smooth manifolds.

will give concrete examples of how some torsions can be computed. In general, one needs some quite explicit geometry to do this.

The sufficiency is proven by a version of Euler characteristic: one shows that every homotopy equivalence can be replaced by one with only cells in two consecutive dimensions. (This process is called rolling.) The boundary map in the relative cellular chain complex then represents the torsion (up to sign). An explicit geometric construction eliminates the elementary matrices. What's left is well defined.

EXERCISE. Provide an algebraic definition of torsion for $\mathbb{Q}$-homology equivalences. In particular, associate to any finite complex K with $H_*(K, *; \mathbb{Q}) = 0$, by the above process, an element of $K_1(\mathbb{Q})/\{\pm 1\} \cong \mathbb{Q}^*$; namely the rational torsion of the rational homotopy equivalence $K \to *$. Identify this element with the alternating product of the orders of integral homology groups $H_i(K, *; \mathbb{Z})$.

EXERCISE. Show that given a finite complex X and an element of $Wh(\pi)$ there is a finite complex with a map to X realizing the given torsion.

For many groups $Wh(\pi) = 0$. This is true for free groups, free abelian groups, fundamental groups of hyperbolic manifolds (or even torsion-free discrete subgroups of real Lie groups), and others (see [Higman, BHS, Wald3, FJ1]); conceivably, it is true for all torsion-free groups. On the other hand, by comparing to cyclotomic fields, one can show that $Wh(\mathbb{Z}_n) \neq 0$ for $n = 5$ or $n \geq 7$. (See the example at the end of 1.5.)

COROLLARY. *If $Wh(\pi) = 0$ then any two homotopy equivalent compact parallelizable manifolds with fundamental group π are PL homeomorphic after crossing with some disk.*

This is true because their regular neighborhoods are these products, and their regular neighborhoods are isomorphic by Whitehead's theorem.

EXERCISE. Prove the result of Mazur: If $f : M \to N$ pulls back stable tangent bundles, then for some disks $f : M \times D \to N \times D'$ is homotopic (not necessarily rel boundary) to a PL homeomorphism iff f is simple.

We have begun to see an important principle in high dimensional topology: fundamental group and tangential data govern topology. We will see much more of this throughout Part I.

1.2.A. Reidemeister torsion and analytic torsion

One reason that τ is so difficult to compute is that it is an invariant of a map rather than of spaces. The ideal situation is when one has intrinsic invariants of spaces $I(X)$ and then one obtains invariants of maps by the formula

$$I(f) = I(Y) - f_* I(X).$$

Since there are self-homotopy equivalences that induce the identity on fundamental group but have nontrivial torsion, this cannot be possible in general. (For an $n \times n$ matrix one can find such on any space with a bouquet of n i-spheres (i arbitrary) wedged on.) However, there are ways of making torsion more intrinsic in special circumstances. In this appendix, we mention two.

The first intrinsic invariant, Reidemeister torsion, is defined by the algebraic observations mentioned toward the end of the previous section. If one has a based acyclic chain complex, then one can build a natural matrix (well defined up to upper triangular matrices) comparing $\oplus C^{odd} \to \oplus C^{even}$. While no space is ever acyclic in a sense that is immediately useful, because H_0 is never 0, it is possible to arrange this when we have a larger coefficient system.

More concretely, suppose that we have a space X with finite fundamental group π. As $\mathbb{Q}\pi$ is a semisimple $\mathbb{Q}$-algebra it has a Wedderburn decomposition: i.e., $\mathbb{Q}\pi$ breaks up into a sum of pieces corresponding to irreducible rational representations of the group π. Consider the pieces M not occurring in the action of π on the rational homology of the universal cover. (M is a ring.) For instance, if the action is trivial on rational homology, M is the augmentation ideal of $\mathbb{Q}\pi$ (Milnor calls these **special** spaces), while if all representations occur (as in the example yielding all torsions from autohomotopy equivalences), $M = 0$. Then $H_*(X, M) = 0$.

Now, the cellular chain complex of X is based (up to $\pm\pi$ ambiguity) and M is acyclic, so that it is possible to define the **Reidemeister torsion** of X, $R_\tau(X)$, as the torsion of this based acyclic complex; it is an element of $K_1(M)/\pm\pi$.

Now, there are three properties of $K_1(R)$ that help with the calculation. The first is that it has Morita invariance: it does not change on taking matrix rings. (GL(matrix ring) $\cong GL$(of the ring).) The second is that it is additive for products of rings. Finally, for fields F, $K_1(F) \cong F^*$ via determinant; this is an exercise in row operations (=products of elementary matrices) in linear algebra.

Using the Wedderburn decomposition of $\mathbb{Q}\pi$ one gets numbers for each representation not occurring in the rational homology upstairs.

An alternative way to get these numbers is to directly take the representation into $M(\mathbb{Q})$ and apply Morita invariance, which yields something in $K_1(\mathbb{Q})/\{\pm 1\} \cong \mathbb{Q}^{*+}$ (positive rational numbers under multiplication).

It is sometimes more convenient to use complex representations, and then one gets nonzero complex numbers well defined up to certain roots of unity.

The advantage of using representations directly is that one then dispenses with the requirement that π be finite; however, in the infinite group case $R\tau$ does not play the same role in detecting Whitehead torsion. If X is special, then all that is lost by using Reidemeister torsion is elements of finite order. For cyclic groups, for instance, there are none. (See [O1] for the current state of the art in calculation.)

We will not do any examples, but here is how an important one works out:

EXAMPLE (LENS SPACES; see [Co, KMdR]). Consider the sphere as the unit sphere in $\mathbb{C}^i$. Let n be an integer and let a_j for $j = 1, 2, \ldots, i$ be integers relatively prime to n. We define the lens space $L(n; a_1, a_2, \ldots, a_i)$ to be the quotient of the sphere by an action of $\mathbb{Z}_n$ whose generator is given by multiplying the j-th coordinate by $e^{2\pi i a_j/n}$.

All of these spaces have the same homotopy groups by covering space theory. Their homotopy type is determined by the product of the $a_j \bmod n$. When are they diffeomorphic? Certainly, one can change the order of the a's by a permutation, or change the sign of any of them (conjugating by complex conjugation). DeRham showed that this is the only possibility.

The argument for this has two steps. From an explicit cell description of these spaces (coming from an equivariant cell decomposition of the sphere) one computes the Reidemeister torsion to be (up to roots of unity) the product of cyclotomic units:

$$\prod (e^{2\pi i a_j/n} - 1)/(e^{2\pi i/n} - 1).$$

Then one appeals to a multiplicative independence of cyclotomic units type fact called the Franz independence lemma to see that these are never the same for different lens spaces.

We shall on a number of later occasions deal with quotients of disks and spheres for representations that do not give rise to manifolds. Much interesting number theory related to cyclotomic fields enters the topology of various types of locally smooth group actions through the above calculation of torsion.

EXERCISE (see [Mi9]). Show that if X is a finite complex with the integral homology of a circle (e.g. the complement of a knot in the sphere), then $H_*(X; \mathbb{Q}[\mathbb{Z}])$ is a sum of $\mathbb{Q}[\mathbb{Z}]$ torsion modules. The annihilators of these modules are called the Alexander polynomials of X (viewing $\mathbb{Q}[\mathbb{Z}]$ as Laurent polynomials with $\mathbb{Q}$ coefficients). Relate the torsion of the homology equivalence to the circle to these annihilation torsions. (Compare to the exercise in the previous section.)

Use this exercise to show that for codimension two embeddings of one lens space in another, there are restrictions on the Alexander polynomials. (You will need the material from 1.6 to do this part; see [Sm].)

Now we turn for just a moment to **analytic torsion** as defined by Ray and Singer. This is only defined for manifolds and requires the additional data of a Riemannian metric. There are two parts to the idea. Firstly, one does not need a genuinely acyclic complex: it suffices to have a complex whose homology is free and based, or if $\mathbb{C}$ is the coefficient ring, to assume a Hermitian inner product on the homology groups. For Riemannian manifolds, one can obtain such a structure by representing cohomology by harmonic forms and then using the Hodge $*$ operator to define an inner product.

The second idea is the deeper one. One must deal with the infinite dimensionality of the DeRham complex. Basically, one defines for appropriate operators, A:

$$\mathrm{tr}(A^{-s}) = 1/\Gamma(s) \int \mathrm{tr}(e^{-tA})t^{s-1}\,dt$$

for large s. This quantity is the zeta function for the operator A and is the obvious sum in terms of A's eigenvalues. (Formally, imagine that A is an infinite diagonal matrix with its eigenvalues along the diagonal. Then

$$\mathrm{tr}(e^{-tA}) = \sum e^{-t\lambda}.$$

The integration then gives $\sum \int t^{s-1}e^{-t\lambda}\,dt = \Gamma(s) \sum \lambda^{-s}$, which certainly deserves to be called a zeta function.)

The product of the eigenvalues, the determinant, should then be given via

$$-\log\det A = d/ds\,\mathrm{tr}(A^{-s}) \qquad \text{for } s = 0.$$

(In the formal calculation $d/ds\,\mathrm{tr}(A^{-s}) = -\sum \log(\lambda)\,\lambda^{-s}$, so that we get the sum of the $\log\lambda$ on setting $s = 0$.) Through a certain amount of analysis, this trace $\big(d/ds\,\mathrm{tr}(A^{-s})\big)$ can be shown to actually exist for the DeRham complex with coefficients in an acyclic representation (as we did for Reidemeister torsion) for large s and can be meromorphically continued. In this way, the relevant derivative can be computed at 0 and one thus obtains analytically a number which is independent of metric ([RaS]) and is equal to the Reidemeister torsion ([Che3, and Mu]). Needless to say, the real interest in this is because it enables one to generalize the notion of torsion to other elliptic complexes which do not have any direct topological interpretation, but this would take us very far afield indeed to pursue. Another interesting and important direction that this work ultimately points to is calculations (!) of torsions of families of manifolds (i.e. fiber bundles); see [BiL].

1.3. Handlebody theory

The transition from the (simple) homotopy theory of CW complexes (which we have seen connects quite nicely to the algebra of chain complexes) to the geometry of manifolds is effected with the aid of handlebody theory. This technique was first developed by Smale in connection with his proof of the high dimensional generalized Poincaré conjecture (see 1.7). While CW complexes are unions of cells, manifolds are unions of handles. In an n-manifold an **i-handle** is something isomorphic to $D^i \times D^{n-i}$ attached along $S^{i-1} \times D^{n-i}$. The pair (D^i, S^{i-1}) is called the **core**, and (D^{n-i}, S^{n-i-1}) is called the **cocore**. $S^{i-1} \times D^{n-i}$ is called the **attaching region** (see fig. 3).

PROPOSITION. *Every manifold can be described as a handlebody; i.e., by a series of handle attachments (to $\emptyset$ or to its boundary $\times I$).*

In the PL case this comes from thickening (i.e. taking an appropriate regular neighborhood of the simplices in) a triangulation. In the smooth case, one starts with a Morse function (these are generic), and a study of the change in the topology at a critical point gives the decomposition (see [Mi2]). In the topological category, this is only true for dimension other than four.

The point is that there is a handle calculus. One tries to do moves that simplify a handle decomposition. Here is the first:

PROPOSITION (REORDERING). *One can arrange that the handles are attached in ascending order.*

One uses general position to make the core of a lower handle miss the cocore of a higher handle, and then slips the lower handle down.

Next, one has ways of eliminating handles of consecutive dimensions that intersect appropriately.

PROPOSITION. *If two handles of consecutive dimension have cores and cocores that intersect in just one point, then they can be canceled.*

One sees that their union is a ball attached along a face.

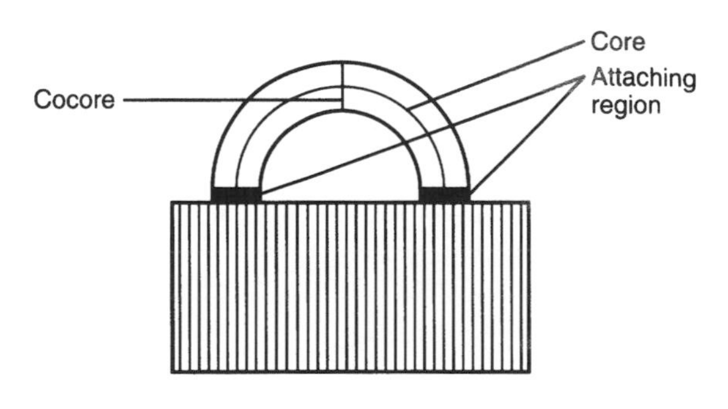

Figure 3. Handle attachment

PROPOSITION (WHITNEY LEMMA). *In dimension at least five, disks' rel boundary can be moved, in a simply connected space, to intersect in the intersection number of points.*

This is the key point where high and low dimensions differ. The proof goes like this. If there are too many intersection points, then there is some pair where they have opposite signs.[2] Connect these points on both disks by arcs, so that one has a circle containing these points. This circle bounds a 2-disk whose interior is disjoint from the others (since dimension is at least five). Pushing one disk past the other using this D^2 is the relevant move. See figure 4.

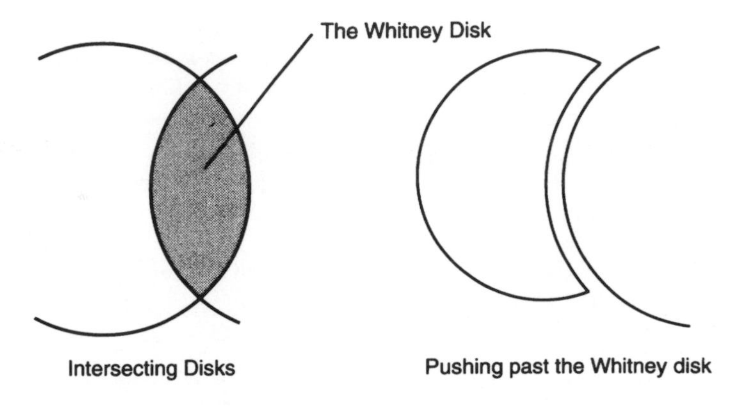

Figure 4. The Whitney trick

This proposition translates algebra into geometry and enables one to effectively manipulate handlebody decompositions of manifolds.

There are $\mathbb{Z}\pi$ **intersection numbers** that can be used for the nonsimply connected generalization of the above.[3] Choosing a base point, and a path from it to each cell, one then travels in each cell to an intersection point. This gives a way to attach a group element (in addition to the sign given by orientation conventions in the usual intersection number) to each intersection: travel to the intersection point along the first cell, and back through the second.

Of course, intersections are closely related to boundary maps in chain complexes. If one takes a cell complex and thickens it to a handlebody,

[2]Signs are given by comparing the orientation given locally by considering the (first disk, second disk) and comparing that to the whole manifold.

[3]These are also relevant to the description of the surgery obstructions in the next chapter.

the boundary maps are the intersections of the cores of the $i+1$ handles with the cocores of the i-handles.[4]

This dictionary enables one to go from CW theorems to appropriately analogous manifold theorems. The next two sections describe examples of this.

1.4. Completing noncompact manifolds

The analogue of making an infinite CW complex finite (up to homotopy) is making a noncompact manifold the interior of a compact manifold with boundary (which is, of course, homotopically a finite complex, and geometrically, a finite handlebody). The results on this problem are due, in the case of simply connected ends (see below), to Browder, Livesay, and Levine [BLL] and, in general, to Siebenmann [Si1].

In order to state the basic results we must first have in place the analogue of finite homological type.

This condition must exclude the possibility of Jacob's ladder, as shown in figure 5, or complements of solenoidal constructions (which cannot always be seen so directly by examining homology: the one pictured in figure 6 has the integral homology of a circle; however, it does have an infinitely generated fundamental group).

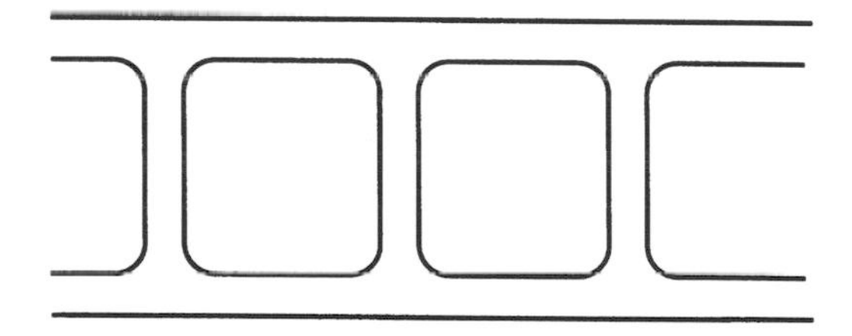

Figure 5. Jacob's ladder extends infinitely in both directions

The basic condition, **tameness**, is that the system of complements of compacta that exhaust the manifold satisfy the Mittag-Leffler condition on homotopy groups. That is, let K_i be a nested sequence of compact

[4]This has an interpretation, in Morse theoretic terminology, in terms of stable and unstable manifolds associated to the gradient flow of a Morse function. The complex obtained in this way, while quite classic and already in the work of Smale, has been rediscovered and reinterpreted by Witten [Wi]. This line of development has extended the ideas of Morse theory to allow one to define chain complexes in infinite dimensional settings, most prominently in the work of Floer on instanton homology and symplectic geometry (see e.g. [Fl]), where gradient flows make sense, as do intersections, but the other terms in the theory do not.

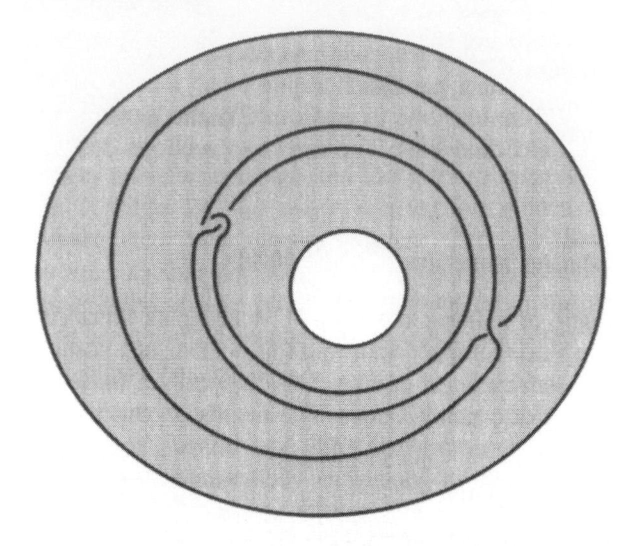

Figure 6. Embed a solid torus in another as the regular neighborhood of the displayed curve. Iterate this procedure, and take the intersection, which is a connected compact space. The complement is an example of a wild end.

sets that fill the manifold W. Then we can consider the sequence of complements

$$W - K_1 \supset W - K_2 \supset W - K_3 \supset \ldots \supset W - K_i$$

and their induced maps on homotopy groups. To begin with, there is the notion of the **ends** of W, which is an element of the inverse limit of the induced sequence on components. (Thus $\mathbb{R}$ has two ends and $\mathbb{R}^2$ has one. The universal cover of the figure eight (or a genus two handlebody) has infinitely many ends.) In particular, we assume that there are a finite number of ends.

It is now not hard to restrict attention to an individual end. We assume in tameness that in $\pi_j(W - K_i)$ the images of all of the later complements, which are a decreasing subsequence of subgroups, ultimately stabilize. The mother of all examples of the failure of this condition is the sequence $\mathbb{Z} \leftarrow \mathbb{Z} \leftarrow \mathbb{Z} \leftarrow \mathbb{Z} \leftarrow \ldots$ where each arrow is multiplication by 2.

Fundamental group tameness implies that there is a sequence of compacta which exhaust the manifold whose complements map by inclusion to one another, inducing an isomorphism on fundamental groups. This common group π is called the **fundamental group of the end**, $\pi_1(\epsilon)$. (The complement of a compactum will be called a **neighborhood of the end.**) One can readily see that it is well defined, and that if the manifold were compactifiable as a manifold with boundary, the fundamental group of the boundary would be isomorphic to this group. Having fundamental group tameness, one can ask for the Mittag-Leffler condition for homology with $\mathbb{Z}\pi_1(\epsilon)$ coefficients, and that the inverse limit be finitely generated. (In the presence of the fundamental group condition, this homological Mittag-Leffler condition is equivalent to the homotopical one.) If this holds, we say that the **end** (which is just an element of the inverse limit of the components of complements of compacta) is **tame**.

For simplicity, we shall assume that our manifold W has one end. N is said to be a **neighborhood of the end** if it is the complement of a compact subset of W. Suppose that N is a one-neighborhood of the end, i.e., that N is connected and that $\pi_1(\Sigma) \to \pi_1 N$ is an isomorphism. Tameness implies that such a neighborhood exists. We define the Siebenmann end obstruction via

$$s(\epsilon) = w(N) \in \tilde{K}_0(\mathbb{Z}\pi_1(\epsilon)).$$

THEOREM (SIEBENMANN). *If ϵ is a tame end, then $s(\epsilon)$ is independent of choices, and if the dimension is at least six, $s(\epsilon)$ vanishes iff ϵ can be completed; i.e. W is the interior of a manifold with boundary such that ϵ is the deleted neighborhood of a (compact) boundary component.*

The independence of the neighborhood of ϵ used to define s is not that hard to demonstrate. It is similar to the exercise at the end of 1.1. That $s(\epsilon)$ must vanish if W can be completed is clear. We will discuss the hard half of the proof a bit in the next section after dealing with the h-cobordism theorem.

For many groups π, the quotient $\tilde{K}_0(\mathbb{Z}\pi)$ is known to be 0; as for the Whitehead group, no torsion-free counterexample is known. For manifolds whose ends have such fundamental group, Siebenmann's theorem takes all the fear out of the noncompactness! Moreover, even if this is nontrivial, so that the obstruction to it being compactified can be nonzero, the fact that this obstruction is understandable should certainly make one fear these ends that much less.

VAGUE PROBLEM. Is there any kind of theory of nontame ends? A hint might lie in the classic literature on the tameness of embeddings of Cantor sets. . . .

1.5. The h-cobordism theorem

The h-cobordism theorem recognizes products. It is due to Smale in the simply connected case and to Barden-Mazur-Stallings in general.

DEFINITION. *An* h-**cobordism** *consists of a manifold triple $(W; M, M')$ such that $M \cup M'$ is the boundary of W, and the inclusions $M, M' \subset W$ are (both) homotopy equivalences. (If the inclusion maps are equivalences* *on* π_1, *then one only needs that* $H_*(W, M; \mathbb{Z}\pi_1) = 0$ *by duality and the Hurewicz-Whitehead theorem.)*

THEOREM. *If M is a manifold of dimension at least five,*

$$\tau : \{\textit{cobordisms with one boundary component } M\} \to Wh(\pi_1)$$

is a 1-1 correspondence. (τ is the torsion of the inclusion map.) In particular, an h-cobordism is a product iff $\tau(W, M) = 0$.

Before discussing the proof I would like to point out that the h-cobordism theorem is really a measure of the nonuniqueness of solutions to the "end completion problem" studied in the previous section. If one has a boundary, then according to the collaring theorem, there is a way to push the boundary into the manifold. If one had two boundaries for the same open manifold, then the region between one boundary and the push-in of the other boundary is an h-cobordism. If τ vanishes, then the two manifolds with boundary are isomorphic. Conversely, we will see soon that for an h-cobordism $(W; M, M')$, $W - M' \cong M \times [0, 1)$, so that if one glues any h-cobordism onto a completion, one obtains a new completion.

One proves the theorem by a handlebody variant of the proof of Whitehead's simplicity criterion. As in 1.3 one finds a handlebody structure such that all the handles appear in the order of their index. Then one does a handle trade (which is a version of rolling) which exchanges a lowest i-handle for an $i + 2$ handle. After doing this enough, one is in a situation where there are only handles of two consecutive dimensions. The matrix representing the boundary map, or, if you prefer, intersections of higher cores with lower cocores, is invertible by virtue of the original space being an h-cobordism. If this matrix is a product of elementary matrices one can construct a diffeomorphism to the product. Elementary matrices correspond to sequences of row operations which can be geometrically mimicked by corresponding handle slides of one handle over another. By this sequence of operations one ends up with a geometric situation which corresponds algebraically to the identity matrix. Handle calculus enables one then to isotop the handles so that all the handles intersect precisely one lower handle in precisely one point, and therefore all the handles are removable; i.e. one has the desired product structure.

The realization of torsions is a handle version of the cellular construction of torsions of homotopy equivalences, an exercise in 1.2.

We conclude this section with a brief discussion of the proof of Siebenmann's end theorem. The idea is to show that in every (one) neighborhood of the end there is a codimension one submanifold M, such that the inclusion of M into the neighborhood of ϵ bounded by E is a homotopy equivalence.

If we can do this, then we have been successful according to two different arguments. The first goes like this. Take a system of smaller and smaller neighborhoods of ϵ and find a sequence $[M_i]$ in each one. We can arrange that $M_i \cap M_j \neq \emptyset$ for $i \neq j$. Note that it is possible to glue any h-cobordism inside $M \times I$ (glue something with torsion $-\tau$ onto a τ h-cobordism to obtain what is, by the h-cobordism theorem, a product). Therefore, we can modify the sequence so that the h-cobordism between M_i and M_{i+1} has $\tau = 0$. Therefore, the region at ∞ bounded by M is an infinite union of cylinders $M \times [i, i+1]$, so that W has a neighborhood of $\infty M \times [0, \infty)$ which we can easily compactify.

In light of Siebenmann's theorem, we have proven the following:

COLLAR DETECTION THEOREM. *If a noncompact manifold W with a tame end having $\pi_1(\Sigma) = \pi_1 W$ is a deformation retract of its boundary, then it is a half open collar.*

This can also be proven directly by a technique called engulfing (due to Stallings). This is nowadays viewed as a method for turning homotopies into isotopies, although originally it was viewed as a way of enclosing contractible sets in balls. (This explains the terminology: the balls expand to engulf the contractible set.) We shall sketch it for the case of trying to prove that a homotopy sphere is a sphere. (We will also give a proof of this from the h-cobordism theorem in 1.7.) The method is not that hard. Suppose X is a subset of dimension k in a k-connected n-manifold, $n - k \geq 4$, and we would like to enclose X in a ball. If cX embedded, then a regular neighborhood of cX would be a ball, and X would certainly be contained in it. Stallings's trick is this. Since X is nullhomotopic, we can extend the given embedding of X to a map of cX. Put this map into general position (generic self-intersection) and consider the self-intersections of cX and everything from there to the cone point (along a join line). This is a low dimensional set, so it can be contained in a ball. cX collapses to this set, and the map is an embedding on this piece, so one can extend the engulfing set one cell at a time as one expands the set. A little more work does the case of codimension three. Now, by engulfing a skeleton and a dual skeleton, one sees that the homotopy sphere is a union of two balls, and from that, it is not too terribly hard to show (if one knows the Schoenflies theorem, as Stallings did, thanks to Morton Brown) that the manifold is topologically a sphere.

To return to Siebenmann's end theorem, one has seen that the difficulty is finding candidate boundaries within ends. To simplify matters, suppose that the relative homology of the end rel boundary is just in some low dimensions; then Wall's work from 1.1 shows that the homotopy type of the end can be obtained by gluing in a finite complex (of low dimension) homotopically iff $s(\epsilon) = 0$. Since it is low dimensional, it is possible to geometrically embed these extra cells. The boundary of a regular neighborhood is actually the candidate boundary, as excision for $H_*(\quad; \mathbb{Z}\pi)$ shows, and the argument is complete.

The general case is done by some more careful handle and embedding arguments, which we will not go through here.

1.5.A. Proper h-cobordism theorem

One can certainly be interested in understanding when a noncompact manifold with boundary is a product. Siebenmann also solved this problem, and we record the solution here for later use, assuming that all ends are tame. (In [Si2] he also discusses some aspects of the nontame case, but the results are a bit harder to state and involve an interesting $\lim^1$ phenomenon involving the lack of Mittag-Leffler on K-groups of neighborhoods of ∞.)

Recall that a map is **proper** if the inverse image of every compact set is compact. A proper map $f : X \to Y$ is a **proper homotopy equivalence** if there is a proper map $g : Y \to X$ such that all compositions are properly homotopic to the identity. We say that (W, M, M') is a proper h-cobordism if each inclusion is a proper homotopy equivalence.

THEOREM. *If* $\dim M \geq 5$, *there is a 1-1 isomorphism*

$$\tau : \{\textit{proper h-cobordisms with one boundary } M\} \to Wh^p(M),$$

where the proper Whitehead group fits into an exact sequence

$$Wh\big(\mathbb{Z}\pi_1(\epsilon)\big) \to Wh\big(\mathbb{Z}\pi_1(M)\big) \to Wh^p(M) \to \tilde{K}_0\big(\mathbb{Z}\pi_1(\epsilon)\big) \to \tilde{K}_0\big(\mathbb{Z}\pi_1(M)\big).$$

That is, the proper Whitehead group is a relative algebraic K-group in the sense of Bass [Ba]. The interpretation of the sequence is as follows. Suppose the manifold at the "bottom of the h-cobordism" is the interior of a manifold with boundary[5]. Then the map to $\tilde{K}_0\big(\mathbb{Z}\pi_1(\epsilon)\big)$ is just the obstruction to putting a boundary on the h-cobordism. It vanishes in $\tilde{K}_0\big(\mathbb{Z}\pi_1(X)\big)$ because the end obstruction has homotopy invariance properties, and to compute over $\pi_1(X)$ one can take advantage of the fact that the bottom is a finite complex. If one can put a boundary on the h-cobordism, then everything is finite, and one can define the torsion of the inclusion of the bottom in the whole thing, as in the compact case. (It is now compact!) This will only be well defined up to the image of $Wh\big(\mathbb{Z}\pi_1(\epsilon)\big)$ in light of our remarks above on uniqueness of boundaries of noncompact manifolds.

The proof of exactness makes use of either the argument given to prove the collar detection theorem or the "Eilenberg swindle," discussed

[5]This is just a convenience, to avoid discussion of "relative finiteness obstructions" of infinite complexes.

in some detail in Part II (see e.g. 5.3). (An algebraic version of this trick is behind the exercise in 1.1 killing the Grothendieck group of infinitely generated projective modules.)

1.6. Some useful formulae

The first formulae that we shall describe are for the finiteness obstructions and torsions of spaces and maps that are obtained by gluing together other spaces and maps.

THEOREM (SUM FORMULAE). *One has the following:*

$$w(X \cup_Y Z) = w(X) + w(Z) - w(Y)$$

where one uses the maps induced by inclusions to get all elements to lie in the same group.

If one has $f : (X, Y) \to (X', Y')$ and $h : (Z, Y) \to (Z', Y')$ homotopy equivalences of pairs, and $f|_y = h_y = g : Y \to Y'$, then

$$\tau(f \cup_g h) = \tau(f) + \tau(h) - \tau(g).$$

The proof is parallel to that of the analogous result for Euler characteristic χ.

EXERCISE. Deduce the result for Euler characteristics from the theorem. Actually, the geometric construction of Ferry described below that relates finiteness obstructions to torsion can be used to deduce the first formula from the second.

EXERCISE. Deduce the additivity of torsions under composition from the sum formula. (Hint: Write down some mapping cylinders.)

The formula for products is also quite simple and is a consequence of the obvious cell structure on a product of CW complexes.

THEOREM (PRODUCT FORMULA). *For products of spaces and maps one has*

$$w(X \times Y) = w(X)\chi(Y) + w(Y)\chi(X),$$

$$\tau(f \times g) = \tau(f)\chi(Y) + \tau(g)\chi(X).$$

Much more subtle is the nature of these obstructions on taking bundles. There the most that can be said is that there is some sort of **transfer** homomorphism defined in some algebraicization of the situation that goes from the K-theory of the base to that of the total space and takes the invariant of the base to that of the total space. The study of such transfers in different settings has been vigorous; in general more than the Euler characteristic of the fiber is relevant.

A special case of the above formulae is where $Y = S^1$, so that $\chi(Y) = 0$. In that case, crossing with Y kills all finiteness obstructions and torsions. This will play an important role for us innumerable times later, so it is worth a closer look.

Let us return to the situation in 1.1. If X is a homologically finite space, so that the Wall obstruction is defined, then, as we discussed, X is finitely dominated, which means there is a finite complex K and maps $i : X \to K$ and $r : K \to X$ such that ri is homotopic to the identity. Of course, K is far from well defined.

EXERCISE/PROPOSITION (MATHER'S TRICK). *The mapping torus $T(ir)$ of $ir : K \to K$ (obtained by identifying $(k, 0)$ with $(irk, 1)$ in the cylinder $K \times [0, 1]$) is homotopy equivalent to $X \times S^1$.*

This gives us a geometric reason why the finiteness obstruction vanishes when crossing with S^1.

The finite dominations occur geometrically with a certain frequency (i.e., not just as an existential fact proven homologically) and variants of this trick are quite useful.

EXERCISE (MAZUR'S THEOREM). By crossing with S^1 and using the exercise in 1.2, show that any tangential homotopy equivalence between manifolds is homotopic to a PL homeomorphism after crossing both manifolds with an appropriate dimensional Euclidean space.

Ferry [Fe2] has observed the strong fact that not only is the homotopy type of the mapping torus determined from that of X, but actually this construction gives a *canonical simple homotopy type* for $X \times S^1$. Ferry then considers the torsion of the map

$$T(ir) \to X \times S^1 \to X \times S^1 \to T(ir)$$

where the middle map is induced by flipping the circle coordinate. This is an element of $Wh(\mathbb{Z} \times \pi)$, which determines the finiteness obstruction. (Of course, it would be a simple homotopy equivalence if X were homotopically a finite complex.)

There is an important, loosely related, decomposition formula, of Bass, Heller, and Swan [BHS][6], which we will also need many times and describe more fully later:

$$Wh(\mathbb{Z} \times \pi) \cong Wh(\pi) \times \tilde{K}_0(\mathbb{Z}\pi) \times \text{ Nils.}$$

We do not at this point have to discuss the Nils. Ferry shows that the torsion above lies in the $\tilde{K}_0(\mathbb{Z}\pi)$ piece and vanishes iff X is homotopy finite, i.e. is a geometric description of the finiteness obstruction. This,

[6]Bass, in his book, takes this formula as defining the negative K-groups, by insisting that it remain valid for K_0, then K_{-1}, K_{-2}, etc.

and the BHS formula, enable one to very often deal with $\tilde{K}_0$ issues by replacing them, if convenient, by Wh issues, which have a rather different geometric meaning.

Thus far, all of the formulae described make sense and are valid for polyhedra (and even ANRs; see [Ch1]). The next formulae regard duality and are truly manifold phenomena, and follow from an examination of the dual triangulation of a PL manifold. (See [KS] for the topological case.)

One way to be led to such formulae is to consider an h-cobordism $(W; M, M')$. Observe that $W = M \times I \iff \tau W = M' \times I$. In other words, because of the h-cobordism theorem, $\tau(W, M) = 0 \iff (W, M') = 0$. This suggests that there is a formula for $\tau(W, M')$ in terms of $\tau(W, M)$.

THEOREM (DUALITY FORMULAE). *If W is an n-dimensional h-cobordism, and $w : \pi \to \{\pm 1\}$ is the orientation character, then one obtains an anti-involution on $\mathbb{Z}\pi$ by sending a group element g to wg^{-1} and hence an involution $*$ on $Wh(\pi)$. One has (Milnor duality formula)*

$$\tau(W, M') = (-1)^{n-1}\tau(W, M)^*.$$

If $f : M' \to M$ is a homotopy equivalence between closed n-manifolds, then

$$\tau(f) = (-1)^{n-1}\tau(f)^*$$

(see fig. 7).

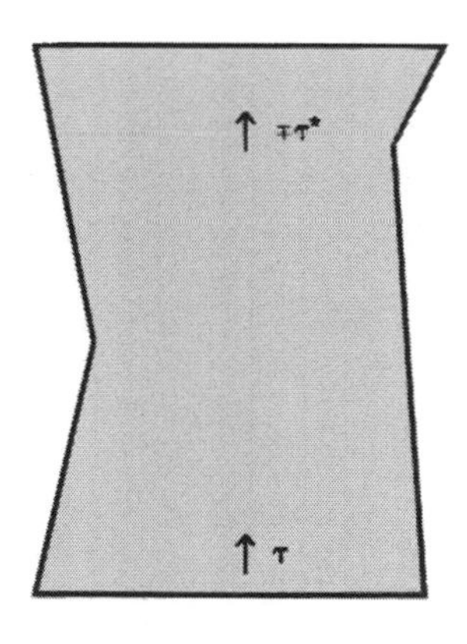

Figure 7. Milnor duality for an h-cobordism

EXERCISE. Using the Milnor duality formula and the composition formula for homotopy equivalences, find the formula for the torsion of the homotopy equivalence between the ends of an h-cobordism. Using this, prove that no two distinct three dimensional lens spaces with fundamental group $\mathbb{Z}_7$ are h-cobordant.

This method works for all primes with odd class number and in all dimensions if we work smoothly (although, naturally enough, this requires some more number theoretical input). A different method is required to prove the general (and nonsmooth) non-h-cobordism result. See [AB] (and also 4.7 and 13.4 below).

1.7. Some applications

In this section we will use without proof the algebraic facts that $Wh(0) = Wh(\mathbb{Z}) = 0$, which are not that hard. Our first application, Smale's, was the reason the h-cobordism theorem was ever considered!

THEOREM (GENERALIZED POINCARÉ CONJECTURE). *If Σ^n, $n \geq 6$, is a homotopy n-sphere, then Σ is PL homeomorphic to the sphere.*

Remove the interiors of two small nonoverlapping balls. By a Mayer-Vietoris sequence argument and excision, one homologically has an h-cobordism, and general position or Van Kampen's theorem provides us with the fundamental group data necessary. Now the h-cobordism theorem implies that this region is an annulus. If one glues an annulus to a ball, one obtains a bigger ball, so that we obtain Σ as the union of two balls glued together along their boundaries.

LEMMA (ALEXANDER TRICK). *Any (PL) homeomorphism of the sphere extends to a (PL) homeomorphism of the ball.*

Just radially extend.

The Alexander trick enables one to construct a PL homeomorphism now from Σ to the sphere.

The radial extension of a diffeomorphism is not smooth at the origin unless the diffeomorphism is an orthogonal transformation. In fact, the lemma is false, and the generalized Poincaré conjecture fails smoothly. These "counterexamples" are the famous exotic differential structures on the sphere. See [KM].

The above argument does not work for $n = 5$, but one can do this case using material from the next chapter.

Now let us turn to knot theory. An embedding is standard if there is an isomorphism of the ambient manifold sending the submanifold to some standard embedding of that submanifold. Subspheres have standard equatorial embeddings, and the next theorem characterizes this embedding among all embeddings.

THEOREM (LEVINE'S CRITERION). *A locally flat embedding of $S^{n-2} \subset S^n$ is standard, $n \geq 6$, iff the complement is a homotopy circle.*

(ZEEMAN UNKNOTTING). *A locally flat embedding of $S^{n-c} \subset S^n$ is always standard, $n \geq 6, c \geq 3$.*

Actually, Levine's criterion is true smoothly [Lv1] as well, but Zeeman's theorem is very false smoothly [Ha, Lv2]. The proof of both results in the PL category is similar, so we only do Zeeman unknotting. This follows quite directly from the following result:

THEOREM (CONCORDANCE IMPLIES ISOTOPY). *Any (smooth, PL, or topological) embedding $M^m \times I \subset W^w \times I$, $w - m \geq 3$, is equivalent to a product embedding $(M \subset W) \times I$.*

(Again, the smooth version of Zeeman unknotting fails because one cannot cone in the smooth category.)

One constructs the equivalence first on a regular neighborhood V of $M \times I$ and then in the complement of a yet larger neighborhood, and finally patches these together over the region in between. We have an isomorphism of the submanifold to $M \times I$, which we can extend to a regular neighborhood–in the smooth case, by the tubular neighborhood theorem and the fact that bundles on a cylinder are determined by what they are on an end, and by some similar principle in the other categories.[7] The complement has the same fundamental group as W, and excision with the Hurewicz-Whitehead theorem shows the complement is an h-cobordism. The sum formula for τ shows that $\tau = 0$, so it is a product. The region between two regular neighborhoods is always annular, so all that remains is extending a homeomorphism of $\partial V \times I \times \{0\} \cup \partial V \times \{0\} \times I \cup \partial V \times I \times \{1\}$ to $\partial V \times \{x\}$ (analogous to the Alexander lemma). But this is obviously possible. ($\partial V \times I \times I \cong [\partial V \times I \times \{0\} \cup \partial V \times \{0\} \times I \cup \partial V \times I \times \{1\}] \times I$.) (see fig. 8.)

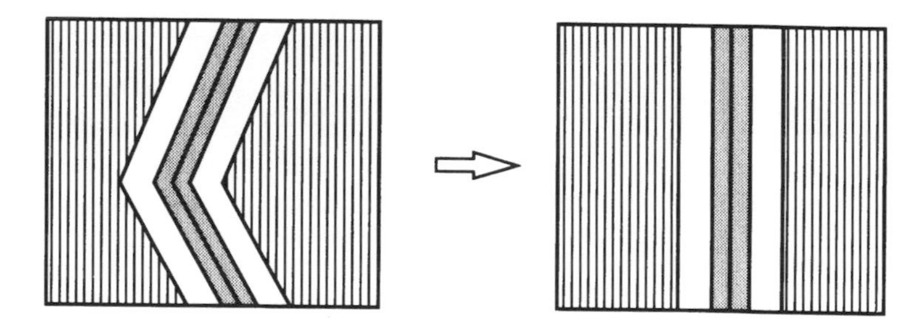

Figure 8. Proof of concordance implies isotopy. Build PL homeomorphism from shaded region to shaded region using the h-cobordism theorem. Finally, do the unshaded annular region directly, by hand.

[7] In the PL case, this is the notion of Block bundle. For the topological case, see [KS] and 6.3.

We close this section with the construction of some interesting group actions on the sphere. It is much more primitive than the results that will be available in the next chapter but is nonetheless striking as an application of algebra to the production of interesting geometric examples.

EXAMPLE Let $p \geq 5$ be a prime, and let $i \geq 2$; then there are infinitely many inequivalent free $\mathbb{Z}_p$ actions on S^{2i+1}.

Start with a lens space (1.2.A). We will construct infinitely many $\tau \in Wh(\mathbb{Z}_p)$ such that the other end of the h-cobordism with torsion τ is not diffeomorphic to the lens space. (We leave the fact that they are not equivalent to each other to the reader.)

Firstly, the elements. There is a pullback (Rim) square of rings:

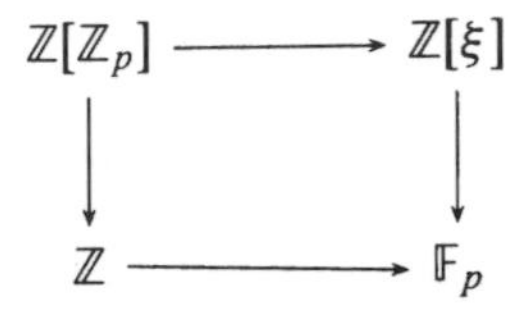

where ξ is a primitive p-th root of unity, and $\mathbb{F}_p$ is the field w+ith p elements. From this it follows that the $(p-1)$st power of any unit in $\mathbb{Z}[\xi]$ comes back from a unit in $\mathbb{Z}[\mathbb{Z}_p]$. The corresponding 1×1 matrix is our desired element. An example of a unit in $\mathbb{Z}[\xi]$ is $(\xi + \xi^{-1})^r$. Observe that this unit is invariant under the duality $*$, which in $\mathbb{Z}[\xi]$ corresponds to complex conjugation.

Now why are the ends of these h-cohordisms different?[8] One has $(W; L, L')$. The map $L' \to L$ is the composition of the retraction of W to L with the inclusion of L' into W. Using the composition formula and the duality formula one finds the torsion of this map is τ^2, which is nontrivial, and therefore not homotopic to a diffeomorphism.

EXERCISE. Use Reidemeister torsion (1.2.A) to distinguish these spaces.

EXERCISE. Prove the result of [RoS] classifying smooth semifree G actions on the disk with fixed set a subdisk in terms of the Whitehead group of $\pi_0(G)$. (A group action is semifree if any point fixed by a nontrivial element of the group is fixed by all elements of the group.) When is the product of two semifree actions linear?

1.8. Notes

All of the constructions of this chapter take place most easily in the PL category, but they also make sense in the smooth category. The results are the same in the topological category, but requirc much more work. Topological invariance of torsion and the finiteness of ANRs were first

[8] I recognize that this is the solution to one of the earlier exercises.

proven by Hilbert manifold techniques by Chapman and West respectively. See [Ch1] (and also [Ch5]) for an exposition. Another proof can be obtained using controlled methods that will be discussed in chapter 9. For manifolds, the missing material can be found in [KS]. This includes the topological h-cobordism theorem and other analogues of facts that are much simpler for the PL category. Later, when we get to stratified spaces, the differences between the topological and PL categories will become much more pronounced. For manifolds, there are almost no differences in high dimensions. In dimension four the celebrated work of Freedman [FrQ] and Donaldson shows very striking differences.

A general reference for much of the material in this chapter is [Luck1]. For the finiteness obstruction there is a book [V] that also devotes a certain amount of space to actual calculations. Cohen's book [Co] is a beautiful treatment of simple homotopy theory and also contains proofs of some of the formulae we use. Milnor's expository article [Mi3] is very valuable. The novice should beware that Milnor believed (but corrected in proof) that $Wh(\pi)$ is torsion free for π finite; this is not the case, and it affects some of the statements that are proven by calculation. However, this forces the reader to really understand what is going on. The paper also contains the duality formula (due to him) for h-cobordisms. Jonathan Rosenberg has recently written an introduction to algebraic K-theory that contains a great deal of material of use to readers trying to gain hands-on familiarity with the algebra (and geometry) discussed here. It will appear as a Springer book. Also, Steve Ferry has recently written notes which contain much related geometric information.

A final general reference on torsion and on the h-cobordism theorem is [KMdR].

A useful reference for PL topology, which contains a proof of the h-cobordism theorem, is [RS4]. Siebenmann's thesis was never published. An account is contained in [Ke] with an unnecessary Noetherian hypothesis on $\mathbb{Z}\pi_1(\epsilon)$ and can also be found, in general, by specializing [Q2]. Infinite simple homotopy theory (the proper theory) can be found in [Si2].

Engulfing was invented by Stallings and remains a powerful tool. For a more recent, very valuable extension, see [Ch2].

The sum and product formulae are due to [KwSz]. (See also [Ge1] for the parallel finiteness calculation.)

The proof of Mazur's theorem, outlined in exercises, I discovered as a graduate student. See [Maz] for the original proof. Milnor [Mi5] used this to disprove the hauptvermutung for polyhedra[9]; I will give the argument in 5.3.

[9]This was the conjecture that homeomorphic polyhedra are PL homeomorphic.

For the precise, subtle connection between Ferry's embedding of $\tilde{K}_0(\mathbb{Z}\pi)$ in $Wh(\pi \times \mathbb{Z})$ and the Bass-Heller-Swan splitting, see [Ra6].

The unknotting theorems and "concordance implies isotopy" theorem can be found in the book [Hud] with the original proof using an intricate construction called "sunny collapsing". The proof given here (see also [RS4]) relies on the h-cobordism theorem. However, sunny collapsing remains a useful technique; see [G2].

I believe that the kind of examples of nonlinear actions on the sphere coming out of torsion were first constructed in [Mi10]. (His application was to show that the sphere admitted infinitely many conjugacy classes of free cyclic group actions except for a few small orders.) A complete topological and PL classification appears in [BPW] and [Wa1] and will be described below in chapter 2 and reinvestigated in Part III. The results on Whitehead groups used in our construction are quite primitive; the reader should consult [O1] for the great strides that have been made in calculation.

2 Surgery Theory

This chapter is devoted to developing the theory of surgery as it was classically understood, say in [Br1] and [Wa1]. The main result asserts that the manifolds within a simple homotopy type are determined by tangential data and invariants related to the quadratic form theory associated to the integral group ring of the fundamental group.

The first two appendices (2.4.A and 2.4.B) describe variants of the general theory that would confuse a first reading but are both of intrinsic interest and necessary for section 5.2. Appendices 2.5.A–C describe some of the calculations of the classifying spaces that enter into the theory and make the theory more concrete. (One can actually do examples!)

Interestingly enough, the calculations for the topological and PL cases have much simpler form. Appendix 2.5.C, on F/O, necessary for the smooth theory, is included only for reasons of completeness. We have not developed any theory of smooth stratified spaces for conceptual geometrical reasons that lie deeper than the computational difficulties of the homotopy theory involved in smooth topology (which are themselves quite hard; these computational difficulties begin with that holy grail, the stable homotopy groups of spheres).

The main result necessary from chapter 1 for an understanding of this chapter is the h-cobordism theorem (see 1.5). For the reader interested in only simply connected manifolds this says that every h-cobordism is a product (in high dimensions). Such a reader can safely ignore all remarks involving algebraic K-theory with no great loss. Furthermore, many of the applications of chapter 4 are available to readers of only this chapter.

2.1. Poincaré duality

As this chapter studies the existence and classification of manifold structures within a given (simple) homotopy type it behooves us to study the most obvious homotopical restriction on manifolds: Poincaré duality. For notational simplicity we will ignore orientations. (Orientations enter in describing an anti-involution on $\mathbb{Z}\pi$ necessary for describing duality relations.)

A finite[1] complex X is said to be an n-dimensional **Poincaré complex** if there is a class $[X] \in H_n(X)$ such that $\cap[X] : H^i \to H_{n-i}$ is an isomorphism, with $\mathbb{Z}\pi$ coefficients ($\pi = \pi_1 X$). Note that homology with $\mathbb{Z}\pi$ coefficients is ordinary homology of the universal cover viewed as a $\mathbb{Z}\pi$ module via the covering translations, while cohomology means cohomology with compact supports. (One could alternatively use ordinary cohomology and Borel-Moore homology, i.e., homology with closed supports (see e.g. [Bo]), and would not get a different concept.) This is precisely what one would get by defining cohomology as the homology of the dual chain complex.

Manifolds are Poincaré complexes with this definition, but they actually satisfy a stronger form of Poincaré duality on the chain level. Let C_* and C^* denote the cellular chain complex and cochain complex respectively. Both of these have natural bases: by cells and the dual basis to the cells respectively. Using a chain approximation, one can define a chain **duality map** $\cap[X] : C_* \to C^{n-*}$. M is said to be a **simple Poincaré complex** if this map is a simple chain homotopy equivalence.

EXERCISE. Note that simplicity of a Poincaré complex is not necessarily a homotopy invariant. It is a simple homotopy invariant. Show that the torsion of the duality map satisfies duality $\tau = (-1)^n \tau^*$. If $Y \to X$ is a homotopy equivalence with torsion σ, the difference between the torsions of the duality maps of X and Y is given by $\sigma + (-1)^n \sigma^*$. Observe that this gives an obstruction, called the **simplicity obstruction**, to a Poincaré complex being homotopy equivalent to a manifold. See 2.4.A and the following exercise for more on this.

EXERCISE. Find a cobordism with fundamental group $\mathbb{Z}_p$ between the lens spaces analyzed in the problem at the end of 1.6. (This can be done explicitly, with considerable effort. It is easier to use bordism theory: in this situation, the relevant group of bordism classes can be computed to vanish.) Glue the ends together by a homotopy equivalence. Show that the resulting Poincaré complex is not homotopy equivalent to a simple one using the obstruction from the previous exercise.

REMARK/PROBLEM. A natural class of Poincaré spaces consists of the finite (dimensional) H-spaces, according to Browder [Br4]. It is not known whether these are finite complexes, and assuming they are, whether their simplicity obstructions are zero.

In light of the previous exercises, the setting of simple Poincaré complexes is most appropriate to the question of whether or not a polyhedron is simple homotopy equivalent to a manifold.

---

[1] Browder has shown that finite domination follows from duality if one knows that the fundamental group is finitely presented.

There are also relative notions of all this. (Y, X) is said to be a **Poincaré pair** if there is a class relating absolute homology of Y to the relative cohomology of the pair, and X is also Poincaré. If X is disconnected, then there are also various Lefshetz dualities relating the homology relative to some components of X to the cohomology relative to the remaining ones. Also, if the fundamental group of X injects into that of Y, then the Poincaré duality for X is automatic.

And, of course, there are obvious notions of simple Poincaré pairs and the like. If one were interested (and there are some good reasons to be, although many of them are now historical), then one could also study Poincaré cobordism, simple Poincaré cobordism, etc.

Many of the more familiar signature type invariants of manifolds can be defined using only the underlying Poincaré structure. The most simple, the ordinary signature studied by Thom and Hirzebruch, is defined from the middle dimensional cohomology of a $4k$ dimensional Poincaré space. Poincaré duality implies that the inner product on the free part[2] of this cohomology gives a symmetric unimodular quadratic form, so that one can diagonalize. The signature is the difference between the number of positive and negative eigenvalues.

For manifolds this is multiplicative in finite sheeted coverings, as a consequence of the Hirzebruch signature formula in the smooth case (see [MS] and below), but for Poincaré spaces this fails [Wa5]. This gives what might be the simplest examples of Poincaré complexes not homotopy equivalent to manifolds. (An exercise in 2.4 describes how to do this explicitly, using little machinery.) It is remarkable that one uses an invariant defined by the Poincaré duality, the homotopical form of the resemblance to a manifold, to contradict the possibility of the space being a manifold.

The G-signature of [AS, pt. III] is also defined if one has a G action on a Poincaré space.

There are also deeper invariants of quadratic forms that give invariants of Poincaré spaces extending the idea of these previous examples. For instance, the paper [Ms1] uses Fredholm representations for infinite groups to define a collection of signatures that are also defined in the generality of Poincaré spaces. (Similarly, the "signatures with coefficients in an almost flat bundle" defined by Connes, Gromov, and Moscovici [CoGM] are defined for all Poincaré complexes. Similarly, all methods for dealing with the Novikov conjecture (4.6.A and chapter 14) implicitly give additional invariants of Poincaré complexes.)

In general, it is a very deep problem to try to understand which bordism invariants of manifolds extend to be bordism invariant of Poincaré spaces.

---

[2]The free part of an abelian group is its quotient by its torsion subgroup.

2.2. Spivak fibration

In the spirit of trying to make Poincaré spaces more like manifolds, we shall next describe the type of bundle theory that Poincaré spaces have. It is a homotopical version of usual bundle theory.

DEFINITION. *Let X be a CW complex. A* **spherical fibration** ξ *over X is a Serre fibration $E \to X$, with homotopy fiber homotopy equivalent to a given sphere. Two of these are* **fiber homotopy equivalent** *if there is a homotopy equivalence of total spaces which homotopy commutes with the "projection".*

We will sometimes refer to maps which are not themselves fibrations as spherical fibrations if they become such after being turned into Serre fibrations by the usual path space construction [Sp].

There is a process for stabilizing spherical fibrations: the (iterated) fiberwise suspension or equivalently the fiberwise join with a trivial bundle. We will use the same symbol for a spherical fibration and its stabilization, and usually identify them. Stable homotopy equivalence classes of spherical fibrations over X form a group under join. (The reader should observe that join is the analogue of Whitney sum for bundles. Given an orthogonal vector bundle, one obtains a spherical fiber space by looking at the unit sphere bundle. Then the underlying sphere bundle for a Whitney sum is the join of the underlying spherical fiber spaces.) This group is sometimes indicated by $KSph$.

PROPOSITION. *There is a space BF such that $KSph(X) \cong [X, BF]$. $\pi_i(BF) = \lim \pi_{i+k-1}(S^k)$. In other words, the homotopy groups of BF are the stable homotopy groups of spheres.*

The existence of BF was established in [Sts] and can be proven using the Brown representation theorem [Brne, Sp], which gives a general criterion for when a homotopy functor can be represented as maps into some (classifying) space. The statement about its homotopy groups can be seen by thinking about the clutching maps that describe spherical fibrations over a sphere. They are maps from S^{i-1} to the autohomotopy equivalences of some sphere. On taking adjoints, one gets the desired result.

We have mentioned the map that gives the underlying spherical fibration to any orthogonal bundle. Of course, orthogonal bundles are given by $KO(X) \cong [X : BO]$ and this forgetful map is equivalent to a map

$$J : BO \to BF.$$

This is the J-homomorphism long studied by homotopy theorists, as it can be used to give the first systematic infinite family of nontrivial elements in the stable homotopy groups of spheres. Geometrically it has an interpretation in terms of the Pontrjagin-Thom construction of stable homotopy as bordism of framed manifolds. (See [Th[3], Mi7].) A map from a sphere into O corresponds to a change of framing of the trivial bundle over that sphere. In other words, one has a framed manifold, whose underlying manifold is a sphere and whose cobordism class is the desired element of stable homotopy.

There are analogues of J for the other categories PL and Top; they are just forgetful maps $BCat \to BF$.

These homotopical ideas are the ingredients for the primary obstruction to making a Poincaré space into a manifold.

DEFINITION PROPOSITION EXERCISE. *If X is a Poincaré space, then if one embeds X in a high dimensional Euclidean space, the map from the boundary of a regular neighborhood to X is a spherical fibration. The stable spherical fibration is well defined up to fiber homotopy equivalence. This stable spherical fibration is called the* **Spivak fibration**.

Hint: Use 1.2 (and 1.6 with the $\times S^1$ trick to handle silly torsion difficulties).

REMARK. This proposition includes the theorem of Atiyah [A3] that the underlying spherical fibration of the normal bundle to a manifold is a homotopy invariant.

EXERCISE. Show the converse (see [Br1, Ra2, pt. II]; this is due to Quinn (unpublished)) of the above proposition, namely that if the homotopy fiber of the map from the boundary of a regular neighborhood to X is a sphere, then X is a Poincaré complex. As a corollary deduce that if a finite complex has some cover which is a Poincaré space, then the complex is a Poincaré space. (This is not true for simple Poincaré complexes; cf. e.g. [CW6].) More difficult, show that if one has a fibration of finitely dominated (e.g. finite) complexes, $F \to E \to B$, then E is Poincaré iff F and B are [Q8]. (Hint: The homotopy fiber relevant for E is the join of those for F and B.)

Recall that the **Thom space** of a vector bundle over a compact base is the one point compactification of the total space. Given a spherical fibration, one can form its Thom space as well. It is the mapping cone of the projection map. (This is the cylinder of the projection with the total space identified with a base point.)

PROPOSITION. *The top homology class of the Thom space of the Spivak fibration is in the image of the Hurewicz homomorphism.*

---

[3]This breathtaking paper of Thom introduces transversality, uses it to reduce the calculation of bordism groups to stable homotopy theory, and then does those calculations for unoriented bordism and rationally for oriented bordism. It was one of my greatest pleasures to read the original years after I had learned its content from other sources.

Note the top homology class is the image under the Thom isomorphism (valid for oriented spherical fibrations by the same proof as for oriented bundles) of the fundamental class of the Poincaré space. It can be viewed as the fundamental class of the regular neighborhood of X. This is a codimension zero submanifold of the sphere. The map from the sphere to the Thom space which takes the fundamental class of the sphere to the top class is simply the collapse map that maps the complement of the regular neighborhood to the base point. This homotopy class, by definition of the Hurewicz homomorphism, works.

REMARK. The proposition actually characterizes homotopically the Spivak fibration [Spi]. However, the preimage of the fundamental class under Hurewicz is not well defined. There is an ambiguity stemming from fiber autohomotopy equivalences.

At this point we have a well-defined element $Sp(X) \in KSph(X)$. We shall close this section with:

PROPOSITION. A necessary condition for X to be homotopy equivalent to a Cat manifold is that $Sp(X) \in Im\ KCat(X)$.

The stable Cat normal bundle of an embedding of the corresponding manifold in Euclidean space would be the lift.

2.3. Reducibility and normal invariants

In this section we will elucidate the geometric meaning of the lifts of Spivak fibration.

DEFINITION. *A spherical fibration will be said to be* **Cat reducible** *if there is a Cat bundle ξ which is fiber homotopy equivalent to it. ξ, with its identification with the given fibration, will be called a reduction of the spherical fibration.*

To rephrase matters we have a diagram

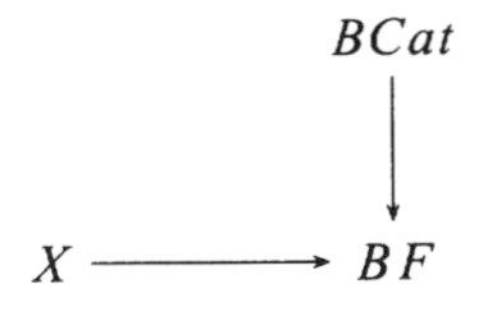

and we are concerned with lifts. Given one reduction, the set of all reductions is in a 1-1 correspondence with $[X : F/Cat]$. (F/Cat is the homotopy fiber of the natural map $BCat \to BF$.)

An important case where one starts off with a reduction is where X is a manifold, and one reduction is given by its stable normal bundle. Another interpretation of $[X : F/Cat]$ is that it is given by Cat bundles with a fiber homotopy equivalence to a trivial bundle.

Now for the geometry:

DEFINITION (BROWDER). *A* **(degree one) (Cat) normal invariant**[4] *for a Poincaré complex X consists of a degree one map from a Cat manifold* $f : M \to X$*, a Cat bundle* ξ *over X, and a stable trivialization of the sum* $\tau_M \oplus f^* \xi$*, where* τ_M *is the tangent bundle of M. A* **normal cobordism** *is the same sort of object over* $X \times I$*.*

We will see in the next section that normal invariants are the kind of objects on which we can perform surgery.

PROPOSITION (SULLIVAN). *Normal invariants up to normal cobordism are naturally in a 1-1 correspondence with reductions of the Spivak fibration. Hence for a Cat manifold M, they are naturally in a 1-1 correspondence with* $[M : F/Cat]$.

The proof is transversality. For instance, to go from a reduction to a normal invariant, one takes the map from a sphere to the Thom space of the reduction ξ and takes the transverse inverse image of the 0-section. The fact about Hurewicz homomorphism makes this a degree one map, and transversality gives the bundle data. In the reverse direction, one observes that the bundle ξ must be equivalent to the Spivak fibration in light of the latter's homotopy characterization.

For a more direct (but more or less equivalent) proof of the 1-1 correspondence in the manifold case, see [RSu].

2.4. The surgery exact sequence

In this section I would like to explain two things. The first is how one goes about answering the question of when a degree one normal invariant is normally cobordant to a (simple) homotopy equivalence. This involves a process called surgery, which might be obstructed, and leads, as in the h-cobordism theorem, to an obstruction group that depends on the fundamental group (and the dimension mod 4 and orientation character). The second topic is how one goes from this obstruction theory to a classification of manifolds within a given simple homotopy type (the surgery exact sequence).

We consider a degree one normal map $f : M \to X$ (we'll suppress the bundle data). How does one improve the map f? We consider the process of **surgery**, first introduced in [Mi1] and then developed in [KM, Br1, No, Su1, Wa1]. It is an analogue of the process considered in 1.1 where we improved maps between CW complexes by attaching cells to the domain. Suppose that f is already an isomorphism on π_j for $j \leq i-1$.

[4]Unless otherwise stated all normal invariants will be degree one. In chapter 6 we will have some use for normal invariants of other degree. However, they do not play any role in our treatment of surgery to a homotopy equivalence.

Then one can find singular spheres in M (i.e. just maps from spheres into M) that represent the kernel of f_{i*}.

EXERCISE. Show that any degree one map induces a surjection on fundamental group. (Hint: Otherwise, factor the map through an intermediate cover.)

EXERCISE. As an application of Poincaré duality, show that the induced map on homology for any degree one map is split surjective. Therefore, the same is true by Hurewicz for the first homotopy group where the map is not an isomorphism.

Suppose that one can represent these singular spheres by disjoint embedded framed subspheres (i.e. subspheres with trivialized normal bundle). Then $(M \times I) \cup (D^{i+1} \times D^{n-i})$ will be a cobordism, and one can extend the map f and all the bundle data. If i is below the middle dimension, then general position produces the subspheres. The bundle data give the necessary trivializations because the normal bundle of M is pulled back from X; therefore, the restriction of the bundle to any sphere in M which goes nullhomotopically to X must be trivial. Fairly direct Mayer-Vietoris calculations give the fact that the homology kernel (and therefore the homotopy kernel) is killed. (See Fig. 9.)

What happens as we get to the middle dimension?

In even dimensions one runs into two problems: firstly, one must produce these disjoint embedded spheres with trivial normal bundles, and secondly, one must guarantee that surgery on them improves f. (Actually, the first problem is really two: we must embed the spheres, and then we must make them disjoint.) In the odd dimensional case, one only has the second problem. The difficulties are all quantitatively measured in Grothendieck groups of symmetric or antisymmetric *quadratic* forms over $\mathbb{Z}\pi$ for the even dimensional case, and in automorphisms of such forms in the harder, odd dimensional case. For even dimensions, one

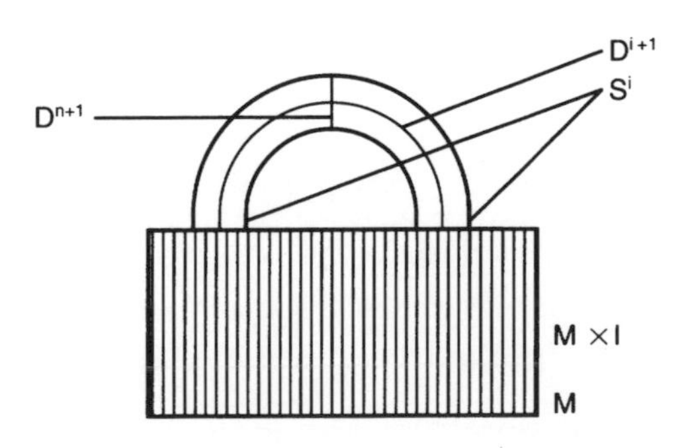

Figure 9. Surgering an i-sphere in M.

takes framed immersed subspheres and measures intersections and self-intersections[5] using the $\mathbb{Z}\pi$ intersection number, and in high dimensions, these numbers measure precisely the minimum number of intersections and self-intersections as in the proof of the h-cobordism theorem. If the associated quadratic form is *hyperbolic*, i.e. has the form

$$\begin{pmatrix} 0 & I \\ \pm I & 0 \end{pmatrix},$$

then surgery will kill the kernel. (The $\pm$ is determined by the dimension mod 4.) Thus one gets groups that depend on the dimension mod 4, π, and the orientation character (to describe the type of symmetry) and that measure the difficulty in surgery.

The odd dimensional case is much more complicated and is based on automorphisms of quadratic forms or, more nicely, by thinking of pairs of self-annihilating subspaces of a hyperbolic form. The chapter explaining how this goes is one of the hardest in [Wa1] and represents one of the book's central achievements. (One could give a very clumsy approach to the problem by crossing with a circle, solving an even dimensional problem, and using Farrell's theorem 4.6 to remove the circle. A beginner could live with this approach, but, philosophically, it is terrible.) We will not discuss how this is done here[6]. To summarize, we have the following critical theorem:

THEOREM ([Wa1]). *There are (covariantly functorial abelian) groups*[7] $L_n(\mathbb{Z}\pi, w)$ *that depend on* n mod 4, *the ring* $\mathbb{Z}\pi$, *and an orientation character* $w : \pi \to \mathbb{Z}_2$ *such that a degree one normal invariant* $f : M^n \to X$, $n \geq 5$, *is normally cobordant to a (simple) homotopy equivalence iff an obstruction* $\Theta(f) \in L_n(\mathbb{Z}\pi, w)$ *vanishes, where* π *is the fundamental group of* X *and* w *is the orientation character.*

When there will be no confusion, we will ignore the orientation character and just write $L_n(\mathbb{Z}\pi, w)$ as $L_n(\pi)$.

The four dimensional periodicity in L-theory is induced geometrically by $\times \mathbb{C}P^2$. (In general $\times K$, for K a simply connected manifold with signature 1, induces a periodicity [Wa1, Mo, Ra2].)

Before continuing, we should discuss manifolds with boundary a bit. For manifolds with boundary there are two obvious questions: (1) how to

---

[5]The intersections are basically homological in nature, but self-intersections are more subtle. These are actually affected by the stable framing given as part of the definition of normal invariant.

[6]The algebraic theory of surgery, discussed below, gives a uniform treatment of the obstruction groups in odd and even dimensions.

[7]The groups are slightly different for the problems of homotopy equivalence and simple homotopy equivalence. See appendix 2.4.A.

classify manifold pairs homotopy equivalent to a given one, and (2) how to classify manifold pairs homotopy equivalent to a given one where the boundary is already given as a fixed homotopy equivalence (say a Cat isomorphism) to the original boundary.

EXERCISE. Formulate the Poincaré versions of these questions and their normal invariant theories.

The second problem is the easier one to handle in our framework. Since the boundary is fixed throughout the entire surgical procedure, the algebra is absolutely unchanged.

FIRST ADDENDUM (TO SURGERY THEOREM). *The same obstruction measures the obstruction for surgering degree one normal invariants rel boundary. This obstruction is natural with respect to codimension zero inclusions.*

The second statement means that if we include a codimension zero submanifold in a larger manifold and glue a homotopy equivalence to the complement to the rel ∂ normal invariant, then the surgery obstruction of the larger object is the image of the surgery obstruction of the submanifold under the map induced by inclusion.

For not rel ∂ problems there still is an obstruction theory, but it is rather more complicated to algebraicize. Still, one can get quite a bit of information by just thinking about what a relative group should look like.

SECOND ADDENDUM (TO SURGERY THEOREM). *The obstruction to surgering a degree one normal invariant (of dimension ≥ 6) not rel boundary lies in a group $L_n(\pi, \pi')$ (with the obvious notation; π' will be a groupoid if ∂ is disconnected; for a disjoint union, L is additive). It fits into the obvious exact sequence of a pair (∂-map given by taking the surgery obstruction of the restriction to the boundary).*

$$\ldots \to L_n(\pi') \to L_n(\pi) \to L_n(\pi, \pi') \to L_{n-1}(\pi') \to L_{n-1}(\pi) \to \ldots$$

We mention a very important special case, whose proof is nicely presented in Wall's book:

$\pi - \pi$ THEOREM ([Wal, THEOREM 3.3]). *If $(X, \partial X)$ is a Poincaré pair with $\pi_1(\partial X) \to \pi_1 X$ an isomorphism, then any degree one normal invariant with target $(X, \partial X)$ can be surgered to a homotopy equivalence.*

This theorem would appear entirely natural if one thought of the notation $L_n(\pi, \pi)$ as being some sort of relative group, and then "excision" would show that this group is 0. This is not a good way to think about things, not the least because the labeling of surgery groups is always done by listing only the fundamental groups of the parts of the manifold to be surgered: if we were working relative to the boundary, we would be using the absolute group. (Also, excision is a complicated topic in surgery theory.)

But this group does vanish. One way to think about the theorem is this Surgery theory is cobordism invariant (the typical example of a surgery obstruction is a signature), so any $\pi = \pi'$ problem has vanishing obstruction, as far as the boundary part is concerned. Now, having solved the boundary problem, we would like to solve the absolute (rel boundary) problem that now remains. This is, indeed, obstructed, but we can kill this absolute obstruction by changing the solution to the boundary problem (see the Wall realization theorem, below).

It also asserts that in some sense the obstruction to surgery is cobordism invariant (if one keeps track of the fundamental group). If a surgery problem is cobordant to one that is solved, i.e. is already a (simple) homotopy equivalence, then it is solvable by surgering, relative to the already solved boundary component, the whole cobordism to a homotopy equivalence.

Part of the $\pi - \pi$ theorem's fundamental nature can be seen from the following exercises. If you get stuck, you can see them all done simultaneously in [Wal, chapter 9].

EXERCISE. Using the $\pi - \pi$ theorem, show that if $L(K)$ is the obstruction to doing surgery with target K, then $L(K)$ just depends on the dimension of K, its fundamental group, and its orientation character. It might help to use the following:

EXERCISE. Using the $\pi - \pi$ theorem, show that every surgery obstruction that occurs for any n-dimensional space with fundamental group π occurs as a rel ∂ problem on a regular neighborhood of a 2-complex.

EXERCISE. As a consequence of the previous exercise, prove the

WALL REALIZATION THEOREM. *Let X be a given $n-1$ dimensional manifold, and let $\alpha \in L_n(\pi)$ be a given element. Then there is an n-manifold M with two boundary components, $X \cup X'$, a map $M \to X \times I$ which is a Cat isomorphism on one component and a (simple) homotopy equivalence on the other, and whose rel ∂ surgery obstruction is α.*

It is possible to prove the realization theorem directly by a construction called **plumbing**. To realize an even dimensional $= 2k$ dimensional obstruction, represented by a given $j \times j$ matrix over $\mathbb{Z}\pi$, one starts with $M \times I$, where M is an arbitrary $2k-1$ dimensional manifold with fundamental group π. We will attach to $M \times I$ k-handles with self-intersections and which intersect each other according to the matrix. Figure 10 shows how to realize the 1×1 matrix (2) (a self-intersection issue) in the simply connected case. (See figure 10.)

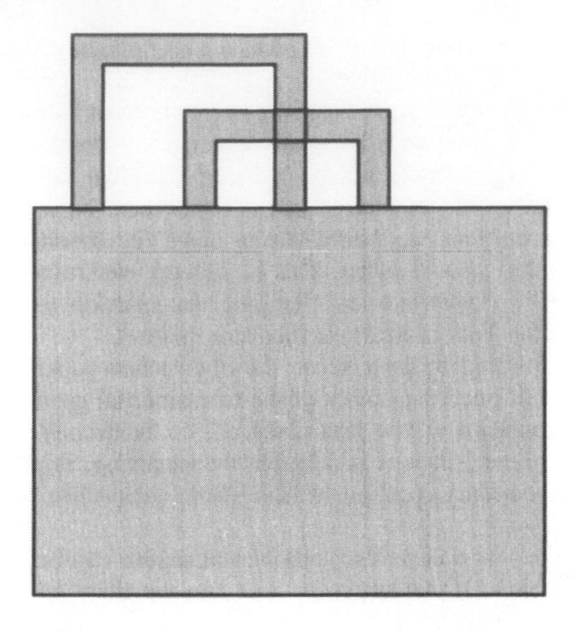

Figure 10. Wall realization of the hyperbolic form

To realize intersections between different handles (the off-diagonal entries), one has to make identifications of the k-handles $D^k \times D^k$ with $D^k \times D^k$ interchanging coordinates; see figure 11.

Unimodularity of the matrix is what forces the result of these surgeries to be simple homotopy equivalent to M. The surgery obstruction of the collapse map to M is rigged to be exactly the given matrix.

EXERCISE. Using the previous exercises and the $\pi - \pi$ theorem, prove the exact sequence for a pair in L-theory.

EXERCISE. Glue the two ends of the result of a Wall realization together by the simple homotopy equivalence, to obtain a closed simple Poincaré complex. This complex has a degree one normal invariant, with the surgery obstruction the given one. Can you use this to give simple Poincaré complexes not homotopy equivalent to manifolds? (Hint: Let π be a finite group and use a quadratic form which violates the multiplicativity of signatures as discussed in 2.1.)

Actually, Wall first does the absolute theorem and then the $\pi - \pi$ theorem (this is unobstructed, so there is no need to develop the algebraic machinery of the closed case), and then uses a variant of the above outline to prove the existence of relative groups and their exact sequence. In fact, one might say that *the $\pi - \pi$ theorem is equivalent to the existence*

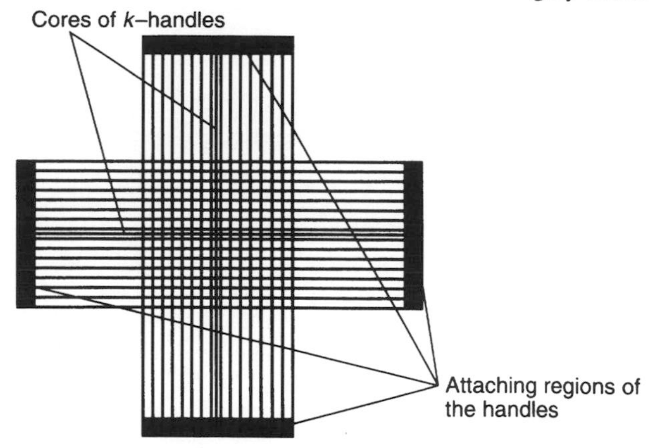

Figure 11. Plumbing. We glue two copies of k-handles together by interchanging core and cocore directions.

of a surgery theory (where the obstructions might not be computable in practice).

In any case, we are almost ready for the surgery exact sequence. First we define the object computed by the surgery sequence:

DEFINITION. *Let M be a Cat manifold. $S^{Cat}(M)$, the* **Cat structure set of** *M, consists of pairs (M', f), where M' is a Cat manifold and $f : M' \to M$ is a simple homotopy equivalence. Two objects, (M', f) and (M'', f'), represent the same element iff there is a Cat isomorphism $g : M' \to M''$ such that f is homotopic to $f'g$.*

Notice that a nontrivial element of $S(M)$ is either the result of a funny manifold (i.e. one not isomorphic to M) or the result of a self-homotopy equivalence that is not homotopic to an isomorphism.

Now we come to the central theorem:

THEOREM (THE SURGERY EXACT SEQUENCE). *Let M^n, $n \geq 5$, be a connected oriented Cat manifold with fundamental group, π and orientation character $w : \pi \to \mathbb{Z}_2$. Then there is an exact sequence*

$$\ldots \to [\Sigma M; F/Cat] \to L_{n+1}(\pi, w) \to S^{Cat}(M) \to$$
$$[M; F/Cat] \to L_n(\pi, w).$$

Note the sequence continues indefinitely to the left.

REMARK. The surgery exact sequence, which is actually a consequence of the $\pi - \pi$ theorem, has the remarkable philosophical implication that to study all manifolds with a given fundamental group (and orientation character), it is only necessary to study one with that fundamental group (and orientation character) in every congruence class mod 4 of dimensions.

EXERCISE. Formulate this theorem for Poincaré spaces using the Spivak fibration.

EXERCISE.[8] Making use of the fact that surgery obstructions are the same in all categories, prove the main result of smoothing, triangulation theory: the Cat structures on a high dimensional Cat manifold (when this makes sense) are in a 1-1 correspondence with lifts of the Cat normal bundle to $BCat$.

EXERCISE. Formulate both versions of a surgery exact sequence for manifolds with boundary. One of them uses the groups $L(\pi, \pi')$ discussed above.

To prove the surgery exact sequence is not hard given the other basic geometric theorems that we've already discussed. All the maps are already defined. The boundary map is produced using the Wall realization theorem.

EXERCISE. Show that the $X' \to X$ produced by that theorem is independent of the normal cobordism M[9]. (M, of course, is not itself well defined.) Composites are tautologously 0 ($[\Sigma X; F/Cat]$ consists of normal invariants for $X \times I$ that are Cat isomorphisms on both ends).

"Kernel equals image" for $[X; F/Cat]$ is the surgery theorem. If something in $S^{Cat}(M)$ goes to 0 in $[X; F/Cat]$, one has a normal cobordism, and one can try to surger to a simple homotopy equivalence to $M \times I$, giving an obstruction in $L_{n+1}(\pi)$. If an element of $L_{n+1}(\pi)$ goes to 0 in the structure set, then it represents a surgery problem where both ends are X, so that one gets a rel ∂ surgery problem on $X \times I$, lifting back to $[\Sigma X; F/Cat]$. For future reference, we record the surgery groups of the trivial group:

PROPOSITION. *We have*

$$L_n(e) = \begin{cases} 0 & \\ \mathbb{Z}_2 & \\ 0 & \\ \mathbb{Z} & \end{cases}, \quad n \equiv \begin{matrix} 1 \\ 2 \\ 3 \\ 0 \end{matrix} \quad \text{mod 4 } \textit{respectively.}$$

[8]This exercise is slightly dishonest intellectually. It is only good mathematics for putting smooth structures on PL manifolds. For analyzing topological manifolds it would be very roundabout to first establish surgery theory.

[9]You should make use of the h-cobordism theorem of 1.5.

The $\mathbb{Z}$ is given by (sign M − sign X)/8 for a surgery problem $M \to X$. The $\mathbb{Z}_2$ is given by the **Kervaire-Arf invariant**, which is an invariant (the only one of Witt type!) of quadratic forms over $\mathbb{F}_2$. See e.g. [Br1, RSu] for a description.

2.4.A. The Rothenberg sequences

In the previous section we were pretty cavalier about whether we were surgering to achieve a homotopy equivalence or a simple homotopy equivalence. There are theories for both, with L-groups denoted $L_*^h(\pi)$ and $L_*^s(\pi)$ respectively. (There are similar gadgets for pairs, etc., and one can have a more severe decoration on the boundary than on the interior, etc.)

The superscripts are often called **decorations**. One can actually decorate using an involution invariant (see 1.6) subgroup of the Whitehead group to study the obstruction to surgering to achieve a homotopy equivalence with torsion in that subgroup. (These were first introduced by Cappell in [Ca4].) We will later have situations where we decorate using other algebraic K-groups.

There are also correspondingly different structure sets $S^s(M)$ and $S^h(M)$; the former was defined in the text, and the latter has objects that are homotopy equivalences to M, and the equivalence relation allows h-cobordisms in place of homeomorphisms.

EXERCISE. Prove the surgery exact sequence for $S^h(M)$. Why do the equivalence relations change automatically when the objects do?

Obviously the relationship between $L^s(\pi)$ and $L^h(\pi)$ is governed by $Wh(\pi)$. (After all, if $Wh(\pi) = 0$, then homotopy equivalence is the same as simple homotopy equivalence, so the different versions of surgery theory must coincide.) The precise formula is given by the Rothenberg sequences. Consider, now, to what extent the torsion of a homotopy equivalence is unchanged by h-cobordisms.

Using the Milnor duality formula, and the fact that the torsion of a composition is the sum of the torsions, one quickly sees that the torsion is changed by $\tau \pm \tau^*$, where τ is the torsion of the h-cobordism. One is then led to the group

$$\{\tau = \pm\tau^*\}/\{\sigma \pm \sigma^*\}$$

depending on the dimension as in Milnor duality. The torsion lies in the numerator because of the duality relationship that holds for a torsion of a homotopy equivalence between closed objects satisfies. (This should seem familiar to the reader who has been assiduously working out the exercises.) This group is called the **Tate cohomology** of $\mathbb{Z}_2$ acting on the module $Wh(\mathbb{Z}\pi)$. We shall denote it by $H^d(\mathbb{Z}_2; Wh(\mathbb{Z}\pi))$. A little geometric work then leads to the Rothenberg sequence

$$\ldots \to H^{d+1}\big(\mathbb{Z}_2; Wh(\mathbb{Z}\pi)\big) \to S^s(M) \to S^h(M) \to H^d\big(\mathbb{Z}_2; Wh(\mathbb{Z}\pi)\big),$$

and similarly, in L-theory,

$$\ldots \to H^{d+1}\big(\mathbb{Z}_2; Wh(\mathbb{Z}\pi)\big) \to L^s(\mathbb{Z}\pi) \to L^h(\mathbb{Z}\pi) \to H^d\big(\mathbb{Z}_2; Wh(\mathbb{Z}\pi)\big).$$

EXERCISE. Rigorously prove these sequences using the h-cobordism theorem and the $\pi - \pi$ theorems for s- and h-surgery.

EXERCISE. Work out a Rothenberg sequence for manifolds with boundary. Explain why $L^s(\pi,\pi) \cong L^h(\pi,\pi) \cong 0$ (i.e. the $\pi - \pi$ theorem is true for both simple homotopy and homotopy surgery) yet not every h-cobordism of a $\pi - \pi$ manifold is trivial.

2.4.B. Proper surgery

Everything that we have done so far for compact manifolds and manifolds with boundaries can be done for noncompact manifolds with tame ends (1.4). The reader probably realizes that noncompact manifolds are analogous to manifolds with boundary, not rel ∂.

One can formulate the notion of a proper Poincaré pair and a simple proper Poincaré pair using locally finite homology and the Whitehead group that enters the proper h-cobordism theorem (1.5.A). The normal invariants are entirely unchanged.

There is a $\pi - \pi$ theorem for proper surgery. One needs to have the ends be $\pi - \pi$ as well. Thus a whole surgery theory exists. The L-groups are functors of (dimension, orientation character, and) the proper fundamental group data of the manifold.

Recall that if X is compact, then $Wh^p(X \times \mathbb{R}) \cong \tilde{K}_0(\mathbb{Z}\pi)$. This leads to an essential role for projective modules in proper surgery. For instance, we have a Rothenberg sequence

$$\ldots \to H^{d+1}\big(\mathbb{Z}_2; \tilde{K}_0(\mathbb{Z}\pi)\big) \to L^s(X \times \mathbb{R}) \to L^h(X \times \mathbb{R}) \to H^d\big(\mathbb{Z}_2; \tilde{K}_0(\mathbb{Z}\pi)\big).$$

Now Siebenmann's theorem (1.4) identifies

$$L^s(X \times \mathbb{R}) \cong L^h(X) \cong L^h(\mathbb{Z}\pi),$$

because with the proper simplicity one can put a boundary on these open manifolds. Thus, we get a sequence

$$\dots \to H^{d+1}\big(\mathbb{Z}_2; \tilde{K}_0(\mathbb{Z}\pi)\big) \to L^h(\mathbb{Z}\pi) \to L^h(X \times \mathbb{R}) \to H^d\big(\mathbb{Z}_2; \tilde{K}_0(\mathbb{Z}\pi)\big).$$

The group $L^h(X \times \mathbb{R})$ is often written as $L^p(\mathbb{Z}\pi)$ (called the **projective L-group**). Algebraically, it is defined the same way $L^h(\mathbb{Z}\pi)$ is, except projective modules are used in place of free ones. (See [No, Ra4] for this algebra.) Notice that the above sequence is very similar to the $L^s \to L^h$ Rothenberg sequence, with $\tilde{K}_0$ replacing Wh. It is often referred to as the Rothenberg-Ranicki sequence.

One can also interpret the projective L-groups as measuring the obstruction to taking a finitely dominated Poincaré complex and making it a manifold after $\times S^1$. See [PR]. This is closely related to the Bass-Heller-Swan formula and Farrell's fibering theorem (see 4.5). Clearly, it measures the obstruction for surgery after taking the product $\times \mathbb{R}^1$.

Unfortunately it is, in general, difficult to describe a proper L-group in terms of absolute L-groups of the interior and of the end. One would want to have L^h in the interior and L^p on the ends, but these do not combine to give the L-theory decorated by the relative K-theory.

2.5. F/Top and the characteristic variety theorem

All of the previous material isn't worth a whole lot unless one can compute F/Cat and the various L-groups. The remainder of this chapter is devoted to F/Cat for the various Cats. Top has the easiest answer, so we will start with it. L-groups can be studied either geometrically or algebraically (and in some cases analytically!) and will be returned to in chapter 4.

Firstly, we can easily compute the homotopy groups (at least in high dimensions) of F/Top from the generalized Poincaré conjecture (1.7) and the surgery exact sequence. The GPC asserts that $S(D^n \operatorname{rel} \partial) = 0$ for $n \geq 5$. Plugging this into the surgery exact sequence one discovers that

$$\pi_i(F/Top) = L_i(e) = 0,\ \mathbb{Z}_2,\ 0,\ \mathbb{Z} \text{ for } i \equiv 1, 2, 3, 0 \bmod 4 \text{ respectively.}$$

Special arguments are needed to verify that this continues through low dimensions. (Except $i = 0$. F/Top is connected.)

Knowing the homotopy groups of F/Top is a good first step for computing what $[X : F/Top]$ is. For instance, since F/Top is an H-space, general nonsense[10] implies that

$$[X : F/Top] \otimes \mathbb{Q} \cong \oplus H^{4i}(X; \mathbb{Q}).$$

(This holds for F/PL and F/O as well.)

However, we can go further and describe the integral homotopy type of F/Top explicitly. Before doing this I would like to say a few words regarding **localization**.

Let S be a set of prime numbers. By $\mathbb{Z}_{(S)}$ we mean the ring of fractions whose denominators are relatively prime to the elements of S. For an abelian group, A, a good way to concentrate on information only involving the primes of S is to localize by setting $A_{(S)} = A \otimes \mathbb{Z}_{(S)}$. One way to say this is to observe that if $(t, S) = 1$, $\times t$ is an isomorphism on $\mathbb{Z}_{(S)}$. Furthermore, any map from A to an abelian group satisfying this property factors through $A_{(S)}$. (One way to see this is that by hypothesis, for each t, one can complete the diagram:

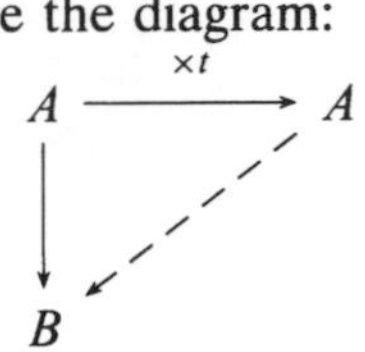

in a unique way. Consequently, the map factors through the direct limit, which is $A_{(S)}$.)

The same can be done for H-spaces. (Actually, one can localize simply connected spaces in general but H-spaces suffice for our purpose.) If H is an H-space, then one can produce a localization $H_{(S)}$ which localizes at S all mapping spaces $[-, H]$, in particular homotopy groups. It also localizes homology and cohomology, as one can argue using some spectral sequence arguments. One construction of the localization is again as a direct limit of power ($\times t = t$-th power of an additive group) maps.

Thus, we shall describe the diagram

$$\begin{array}{ccc} F/Top & \longrightarrow & F/Top_{(2)} \\ \downarrow & & \downarrow \\ F/Top[1/2] & \longrightarrow & F/Top \otimes \mathbb{Q} \end{array}$$

[10]In short the argument goes like this. The Milnor-Moore theorem on the structure of Hopf algebras implies that the rational cohomology of F/Top is a graded polynomial algebra on the generators of rational homotopy (since it is an H-space, via Whitney sum of bundles). The map to a product of Eilenberg-MacLane spaces given by these primitive cohomology classes (think of cohomology as maps to an Eilenberg-MacLane space) gives an isomorphism on rational cohomology. Therefore, by the mod C Hurewicz theorem, for a finite complex X the induced map of mapping groups $[X : -]$ is a rational isomorphism. If we knew that F/Top was an infinite loop space (which it is), then one could argue that the Atiyah-Hirzebruch spectral sequence for the F/Top cohomology theory rationally degenerates.

This will then describe for us for any finite complex X the abelian group $[X : F/Top]$ as the pullback of $[X : F/Top_{(2)}]$ and $[X : F/Top[1/2]]$ over $[X : F/Top \otimes \mathbb{Q}] = \oplus H^{4i}(X; \mathbb{Q})$.

THEOREM (KIRBY-SIEBENMANN, SULLIVAN). *There are homotopy equivalences*

$$F/Top_{(2)} \cong \prod K(\mathbb{Z}_2, 4i-2) \times K(\mathbb{Z}, 4i) \quad i \geq 1$$
$$F/Top[1/2] \cong BO[1/2]$$

where $K(A, n)$ is the Eilenberg-MacLane space that classifies $H^n(-; A)$. The map to rational cohomology is the obvious one for 2-localization and is the Pontrjagin character (Chern character of the complexification) for the 1/2-localization.

In other words,

$$[X : F/Top]_{(2)} \cong \prod H^{4i-2}(X; \mathbb{Z}_2) \times H^{4i}(X; \mathbb{Z}_{(2)}) \quad i \geq 1$$
$$[X : F/Top][1/2] \cong KO^0(X)[1/2]$$

and there is a short exact sequence

$$0 \to [X : F/Top] \to$$
$$KO^0(X)[1/2] \times \bigoplus H^{4i-2}(X; \mathbb{Z}_2) \times H^{4i}(X; \mathbb{Z}_{(2)}) \to$$
$$\bigoplus H^{4i}(X; \mathbb{Q}) \to 0$$

REMARK. The proof is not that hard for the reader who believes everything said till now, but it would take us a bit afield to go through the details. Suffice it to say that the $\pi - \pi$ theorem (or rather its consequence, bordism invariance of surgery obstructions) enables one to build a map

$$\Omega_n(F/Top) \to L_n(e)$$

by viewing the left (bordism of manifolds with maps into F/Top) as bordism of normal invariants, of which one can take the surgery obstruction. One uses the work of Thom on bordism [Th] and of Connor and Floyd [CF] relating bordism to K-theory to produce maps which the above calculation of homotopy groups can show to be homotopy equivalences[11].

Sullivan likes to give his theorem a more geometric sound. He asserts that given a manifold M there is a **characteristic variety** X in M that

[11]Here are some of the details. At 2, Thom shows bordism is a product of Eilenberg-MacLane spectra. Therefore the homomorphism actually produces classes in $H^i(F/Top; L_i(e))$ which give the decomposition. For the argument away from 2 one needs to observe that both sides are modules over a smooth oriented bordism ring $\Omega(*)$, the left-hand side by $*$, and the right by taking the product with signature. A variant of the main theorem of [CF] implies that one then gets a map $KO^0(F/Top) \otimes \mathbb{Z}[1/2] \to \mathbb{Z}[1/2]$, which then gives, by a universal coefficient sequence, a map $F/Top \to BO[1/2]$.

consists of ordinary manifolds and $\mathbb{Z}_n$-manifolds (I'll get to those in a moment), such that a normal invariant is trivial iff the surgery obstructions on all of these subobjects are trivial.

A $\mathbb{Z}_n$-manifold is a manifold with n diffeomorphic boundary components, all glued together. The reader should verify that signature mod n is a cobordism invariant of $\mathbb{Z}_n$-manifolds. (Observe that if M is a manifold, then nM bounds $M\times$ the cone on n points, so one loses the multiples of n.) Similarly one can define mod n surgery obstructions for normal invariants of these. (This is a special case of the general process for defining surgery obstructions for singular spaces that will be considered in Part II.) Sullivan's theorem specifies such a subcollection of M, so that the transverse inverse image for a normal map gives a collection of $\mathbb{Z}_n$-surgery problems, whose obstructions vanish iff the normal invariant is trivial. The collection is referred to as the characteristic variety.

Almost always it is the homotopical form of the calculation of F/Top that is useful, but the geometric form is handy from time to time. The replacement theorem in 13.5 was first proven using the geometric version, although the argument presented there does not use the characteristic variety theorem at all!

2.5.A. The signature operator and Sullivan orientations

The discussion away from the prime 2 of F/Top has an important complement.

THEOREM (SULLIVAN ORIENTATIONS). *There is a class $\Delta(M) \in KO_n(M^n) \otimes \mathbb{Z}[1/2]$ which is an orientation. If $f : W \to M$ is a degree one normal invariant, then the normal invariant, localized away from 2, of f is $\Delta(M)/f_*\Delta(W)$.*

This class Δ can be produced in two ways. The first way is due to Sullivan, and the second is to apply index theoretic ideas.

The first approach can be implemented in two ways. One can argue homotopy theoretically using MSPL, the classifying space for PL bordism, in place of F/Top in the previous chapter.

More geometrically, embed M in a Euclidean space with codimension $4r$ and regular neighborhood $(N, \partial N)$. We shall describe an element of $KO^{4r}(N, \partial N)$ which corresponds under (Spanier-Whitehead) duality (cf. [Ad2]) to $\Delta(M)$. As in 2.5 it is only necessary to give a pair of homomorphisms

$$\begin{array}{ccc} \Omega(N, \partial N) & \longrightarrow & \mathbb{Z} \\ \downarrow & & \downarrow \\ \Omega(N, \partial N; \mathbb{Q}/\mathbb{Z}[1/2]) & \longrightarrow & \mathbb{Q}/\mathbb{Z}[1/2] \end{array}$$

with the right properties with respect to taking products with closed oriented smooth manifolds. For the first map, one takes a bordism class in $(N, \partial N)$ and takes the intersection with X and computes the signature. The second map goes the same way using $\mathbb{Z}_n$-manifolds to represent the bordism with coefficients and taking mod n signatures of such spaces. (Note that $\mathbb{Q}[1/2]$ is the direct limit of $\mathbb{Z}_n$ for odd n.)

The second way to define Δ involves two ideas. The first is that Atiyah [A4] has shown how to assign to every elliptic operator a K-homology class. (Subsequently Kasparov [Kas1] and Brown, Douglas, and Fillmore [BDF] gave more complete functional analytic descriptions of K-homology in these terms; see also [BD] for an entire description of K-homology in terms of indices.) Then, to get Δ, in the smooth case one can apply this to the classical signature operator [AS, pt. III], and for the PL and topological cases, one can use the operator of [Te] and [SuT] (cf. [Hil]).

Atiyah's idea is simply this: the index with coefficients in bundles gives a pairing

$$K^0(X) \otimes Ell(X) \to \mathbb{Z}$$

which is almost, but not quite, enough to describe an element of $K_0(X)$. However, on inflating this with families parametrized by a space Y one gets a pairing

$$K^0(X \times Y) \otimes Ell(X) \to K^0(Y),$$

which as one varies Y does suffice for giving the element. One should think of the use of general Y's as like inflating the bordism cycles to bordism of $\mathbb{Z}_n$-manifolds to get a hold of torsion phenomena.

I will not review here the definitions of the various types of signature operators. One can see [Ros1] for an exposition of some of this circle of ideas and in particular an account of analytic proofs of Sullivan's theorems.

In [RsW4] we show that the K-theory class at 2 of the signature operator is the image under a certain natural transformation from homology to K-theory of a topological characteristic class of the tangent bundle of M.

2.5.B. Rochlin's theorem and F/PL

Even before the celebrated work of Donaldson, it was known that the theory of smooth four dimensional manifolds was possessed with some oddity. Almost all of the then known peculiarities stemmed from one theorem:

ROCHLIN'S THEOREM. *If M is a closed smooth spin 4-manifold, then* $16 | \operatorname{sign}(M)$.

To put things in perspective slightly, realize first, that we can arrange for M to be simply connected up to cobordism by some elementary surgeries. In that case, M spin boils down to the assertion that $x \cup x \equiv \mathrm{mod}\, 2$ for all $x \in H^2(M; \mathbb{Z})$. Van der Blij's lemma (see [Se1]) says that the signature of a unimodular quadratic form with evens along the diagonal (when described as a matrix) is always divisible by 8. Rochlin's theorem gives an extra divisibility by 2.

One way to prove Rochlin's theorem is to use the fact [AH2] that the $\hat{A}$-genus of any spin manifold of dimension $4 \bmod 8$ is even. (Nowadays, we would attribute this to the fact that in these dimensions the complex spinor representations are quaternionic, so that the kernels and cokernels of the Dirac operator are of even complex dimension and therefore have even index.) Now, for any 4-manifold, the index theorem implies that $\mathrm{sign} = 8 \times (A - \mathrm{genus})$.

16 is the best possible. The famous Kummer surface has signature 16. (Topologically, the Kummer surface is obtained by taking the quotient of T^4 by the involution that flips all coordinates. This quotient has 16 singular points. The links of these singular points are copies of $\mathbb{R}P^3$. $\mathbb{R}P^3$ is the boundary of the unit tangent disk bundle to S^2, so we remove neighborhoods of the singularity and glue in 16 copies of this unit tangent disk bundle.)

Rochlin's theorem has profound impact on high dimensional manifolds as well. First of all, below dimension eight, every PL manifold has a smooth structure (unique till dimension seven). Secondly, all of the analysis of 2.5 works for F/PL just as well as it did for F/Top. Even their homotopy groups are abstractly isomorphic. Things only deviate in dimension 4. Consider now

$$\Omega_4(F/PL) \to L_4(e) = \mathbb{Z}.$$

In the topological case this was onto, and responsible for, a cohomology class in $H^4(F/Top; \mathbb{Z}_{(2)})$ which evaluated 1 on the image of the Hurewicz homomorphism. In the PL case, Rochlin's theorem shows that this is impossible. On the image of the Hurewicz homomorphism, we are computing the simply connected surgery obstruction of a degree one normal invariant with target a sphere; the domain is a PL spin manifold, so by Rochlin's theorem this has signature divisible by 16, and hence the surgery obstruction is even.

One does get all multiples of 2 using appropriate maps from connected sums of the Kummer surface to S^4.

The upshot of all this is that $F/PL \to F/Top$ is an isomorphism on homotopy groups except in dimension four, where both groups are $\mathbb{Z}$ and the map is multiplication by 2. Consequently, $Top/PL \cong K(\mathbb{Z}_2, 3)$. This proves (and then displays $Top/PL \cong K(\mathbb{Z}_2, 3)$) there is an obstruction (called the **Kirby-Siebenmann invariant**) in $H^4(M; \mathbb{Z}_2)$ associated to any topological manifold, which in high dimensions vanishes iff the manifold has a PL triangulation; the triangulations are classified by $H^3(M; \mathbb{Z}_2)$.

Indeed, to be more honest about the logical development of the whole subject, one should realize that all of the surgery theory described must first be worked out for the PL case, with Rochlin's theorem entering to produce a low dimensional peculiarity in the homotopy type of F/PL.

Topological manifolds are much harder to analyze. A major step was Novikov's theorem on the topological invariance of rational Pontrjagin classes. (We will explain an approach to this using more modern generalizations of surgery theory in chapter 9.) Precise topological classificatory information under some hypotheses was then discovered by Sullivan and Lashof-Rothenberg.

The real breakthrough which enabled all the further complete analyses was Kirby's torus trick [Ki], which reduced the isotopy classification of homeomorphisms of Euclidean space to the problem of classifying PL manifolds homotopy equivalent to the torus (this was solved by Hsiang-Shaneson and Wall; we will discuss this in chapter 4). In fact, the main difficulty that then followed was deciding whether Top/PL is contractible or has one nontrivial group. This and the methods needed for mimicking the usual PL constructions (e.g. transversality, handlebody theory, simple homotopy theory) were developed in [KS].

2.5.C. F/O

This appendix is included for reasons of completeness only. It summarizes the calculation of F/O due to Sullivan. See [MM, MQRT] for more details.

While F/Top is analyzed using its interpretation in terms of normal invariants, F/O is analyzed in terms of a more purely homotopy theoretic framework.

Recall that F/O is the fiber of the classical J-homomorphism $BO \to BF$ and is related to the issue of how much of an orthogonal bundle is invariant under fiber homotopy equivalence. This was the object of Adams's well-known series of papers [Ad] which motivated, stated, and explained the significance of the Adams conjecture, solved by Sullivan [Su3] and Quillen [Ql1] and later given a beautifully simple proof by Becker and Gottlieb [BG].

For convenience we will p-adically complete all spaces (this is an analogue of localization but is a bit more drastic; see [Su3]).

The Adams conjecture is more or less the statement that if ψ^* denotes the Adams operation (see e.g. [A5]), then one can construct a map $BSO[p] \to F/O[p]$ lifting $\psi^k - 1$, i.e. fill in a missing diagonal arrow in the following diagram:

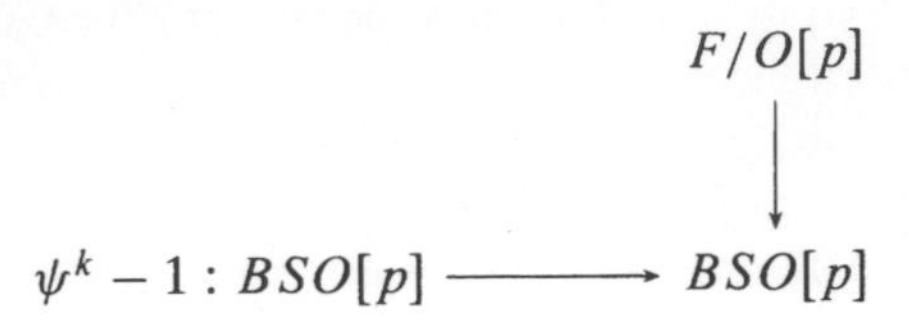

for k a generator of $\mathbb{Z}^*_{p^2}$ for p odd and $k = 3$ for $p = 2$. Let $\alpha : BSO[p] \to F/O[p]$ be a lift. The definition of F/O gives a map the other way (indeed, it is the vertical map in the diagram above!). These enable one to decompose $\Omega\, BF$ as a product of two spaces $Im\ J \times Cok\ J$, whose names describe their homotopy groups. ($Im\ J$ can be viewed as the homotopy fiber of $\psi^k - 1$.)

THEOREM (SULLIVAN). *For each prime,*

$$F/O[p] \cong BSO[p] \times Cok\ Jp.$$

This is, in some ways, a very sad result. The second factor involves the hard to understand part of the homotopy groups of spheres. Maps into it are what smooth normal invariants are about.

EXERCISE. Show that $Cok\ J$ integrally splits off F/O. Warning: There are many subtle and difficult points regarding the H-space and infinite loop space structures on these spaces and the maps between them. We refer the interested reader to [MM, MQRT].

PROBLEM. Is there an infinite loop space structure on F/O which makes the surgery exact sequence into a sequence of abelian groups and homomorphisms? In the topological category, this is possible, as will be explained in the following chapter (3.4).

2.6. Notes

The surgery process was first described by Milnor [Mi1] as a "procedure for killing the homotopy groups of a manifold". In [KM], it was extended to deal with the problem of classifying smooth structures on the sphere. (These authors noted, in particular, how homotopy groups of spheres enter, as in 2.5.C.) The general simply connected theory of surgery on a map was developed by Browder and Novikov; see [Br1]. They were able to apply this idea to classify the smooth manifolds homotopy equivalent to a given simply connected one, up to finite indeterminacy. The formulation in terms of Spivak bundles, etc., is due to Sullivan [Su1,2]. Wall then developed the nonsimply connected theory extensively, and his opus [Wa1] must be studied by all students of high dimensional topology. The survey of Lees [Ls] is a valuable introduction to the surgery theorem.

Both [Br1] and [Wa1] spend some space on Poincaré complexes. Spivak's original paper is not that hard to read, and [Wa5] contains interesting low dimensional information and examples. For some deeper geometric information, see [Lvt1, J1, Q4], and especially [HV]. We will return to more algebraic theory in the next chapter.

The signature-related invariants of Poincaré complexes and algebraic invariants of L-groups do connect quite closely with K-theory of C^*-algebras. The appendix, (2.5.A) on the signature operator and the Sullivan orientation is an example of this. Yet others are scattered throughout the text.

The Rothenberg sequence appears in Shaneson's thesis [Sh1]. Proper surgery appears in [Ta] and also [Mau].

The results on classifying spaces are described in this chapter in the order of their simplicity. More information can be found in [MM, MQRT]. Historically, Sullivan first analyzed F/PL including the low dimensional peculiarities. For the low dimensional part he used the smoothing of low dimensional PL manifolds and the connection between F/O and the J-homomorphism. F/Top came later, after smoothing and triangulation were understood through the work of [KS], which itself was the culmination of a long research program that included, [KM, LR1, Su1, 2]. (The original proof of Rochlin's theorem [Ro] was via the connection to homotopy theory. It was discovered in trying to reconcile an incorrect calculation by Rochlin of the stable 3-stem using the Pontrjagin-Thom construction, with the correct answer provided by the method of killing homotopy groups. The proof described here I first heard from S. Stoltz and is presented in the charming book [LwM].)

The Adams conjecture, as mentioned in the text, is due to Quillen and Sullivan. Quillen's proof is closely related to his calculation of the algebraic K-theory of finite fields. One can also try to read [Su4] for an inspirational "big picture". The method of Becker and Gottlieb is much more elementary and is based on a transfer associated to principal compact Lie group fibrations and an appropriate "splitting principle".

The method of proving Sullivan's result on orientations from the work of Sullivan and Teleman was the joint work of Rosenberg and myself. The result at 2 settled an enigma that bothered us of how operator theorists can produce a topological invariant unknown to topologists (i.e., the signature operator at 2). It wasn't unknown. The invariant is essentially the class defined in [MoS]. The map from homology to K-theory is best understood using intersection homology. We will return to the connection between intersection homology and classifying spaces in Part III.

3 Spacification and Functoriality

This chapter deals with a more formal view of the material of the previous chapter. We shall show how to "spacify" the results of the previous chapter so that all long exact sequences become the long exact sequences of fibrations. The various terms have to do with "blocked" versions of the problems of surgery theory. (To have a problem blocked over X is a PL analogue of the smooth notion of a family of problems parametrized over X.)

We shall also see that in the topological case we can make the surgery exact sequence a sequence of abelian groups and homomorphisms, having fourfold periodicity (akin to Bott periodicity). Finally, we shall apply this to see that surgery theory is actually a covariant functor from spaces to abelian groups (or, better yet, to spectra). The push forward is rather similar to Atiyah-Singer's topological index.

These developments are rather technical but seem to provide the correct framework for the formulation of the classification theorems in chapter 6.

The reader who feels entitled to some applications immediately can go straight to chapter 4; it is a convenience, which is not absolutely necessary, to take with you the belief that G/Top has an infinite loop space structure which makes the surgery map a homomorphism. (Actually, the group structure is the one implicit in the calculation of F/Top in 2.5; the addition on $[X; F/Top]$ is given by the addition of the functions on the characteristic variety of X associated to the elements.)

3.1. Spacification

In this section we describe a formal process that makes spaces out of the terms in the surgery exact sequence and is such that the various long exact sequences become exact sequences on homotopy groups of fibrations. The next section gives a geometric interpretation of this, which we will use many times later in this text.

Other functors, such as the Whitehead groups, also have "spacified versions", but for them the spectra are more sophisticated than the ones we deal with here, and it is not a simple formal process that constructs them. If one spacified using the technique of this chapter, one would get the wrong answers for topological questions (but the correct ones for certain PL applications!).

The formal apparatus for "spacification" is Δ-sets. A very useful explanation of the use of simplicial techniques can be found in [May]. A suitable general reference for Δ-sets, and their homotopy theory, is [RS1]. The general idea is that we describe a space by its simplices and how they're fitted together. Then one can take a geometric realization of this space and interpret combinatorially its homotopy groups (under decent conditions). We will be content to explain how this works in some examples.

To get the idea, let's start with a very easy example from first-year topology. Suppose that we define a space by having vertices be points of a fixed space X. Edges will be paths in X: they connect vertices. Faces will be maps of the 2-simplex into X and the boundary map is the usual one. If one geometrically realizes this, one gets the singular complex of X. Its homotopy groups, given by the combinatorial data of a map of a simplex into this complex, *assemble* into an element of the homotopy of X (because the simplices that make up a sphere will fit together and one will get a map of a sphere into X from these data) and in fact the natural map from the singular complex into X is a weak homotopy equivalence. In this example, one sees that for a reasonable theory one needs to have a subdivision property (so that one can realize the homotopies that one expects to find). This is called the Kan condition.

A more typical example for us goes like this. Let the vertices of $\underline{\Omega}_i$ be closed Cat manifolds of dimension i. Edges should be cobordisms. A simplex Δ should correspond to manifolds with multifaceted boundary, decomposed with the same combinatorics of Δ. In this example one can readily see (using transversality) that $\pi_j(\underline{\Omega}) \cong$ the bordism group of $i+j$ dimensional Cat manifolds.

Now we can spacify just about any situation where one has suitable relative theories and gluing operations. (I will not axiomatize this, but the method is formal enough.) (In bordism, the bordism of manifolds with some fixed lower dimensional boundary pattern is either empty or closed bordism of the same dimension by a "difference construction".)

For instance, for a manifold M, to spacify the structure set, $\mathbf{S(M)}$, one considers a simplex Δ to be a map of a space W whose boundary is decomposed according to the combinational data of $\partial\Delta$, as in the bordism example, to $M \times \Delta$, which is a (decorated appropriately, see 2.4 and 2.4.A) homotopy equivalence when restricted to each subsimplex. The h-cobordism theorem identifies $\pi_0(\mathbf{S(M)}) \cong S(M)$. More generally,

$$\pi_j(\mathbf{S(M)}) \cong S(M \times \Delta^i, \operatorname{rel} \partial).$$

Even stronger, there is a loop space statement

$$\mathbf{S}(M \times \Delta^i, \text{ rel}) \cong \Omega^i \mathbf{S}(\mathbf{M}).$$

(The fact that one sees group structures on $S(M \times \Delta^i, \text{ rel } \partial)$, $i \geq 1$, via the isomorphism to $\pi_i(\mathbf{S}(\mathbf{M}))$ is something one can see directly geometrically by gluing together two elements of $S(M \times \Delta^i, \text{ rel } \partial)$ along a face in their boundary (and reconfiguring the boundary).)

EXERCISE. Show that in all categories, if one uses this addition of structures, the surgery exact sequence for $S(M \times \Delta^i, \text{ rel } \partial)$, $i \geq 0$, becomes an exact sequence of groups and homomorphisms.

The normal invariants can also be spacified; we denote this space by **NI(M)**. A simplex Δ corresponds to a normal invariant (decomposed as usual) of $M \times \Delta$. One has a homotopy equivalence

$$\mathbf{NI(M)} \cong \mathbf{Maps}[M : F/Cat],$$

using the material of 2.3 (i.e. the proof that normal cobordism classes of maps into M are given by the homotopy classes of maps into F/Cat).

It remains to spacify the surgery obstructions. The key to doing this is finding a bordism interpretation of surgery. This can be done using the $\pi - \pi$ theorem and realization. Wall does this in chapter 9 of his book [Wa1] (surgeons always know what people are talking about when they mention chapter 9).

Basically, $L(\pi)$ will be made up out of all surgery problems with fundamental group π. Since we only have realization when we move on to manifolds with boundary, rel ∂ (and furthermore, since we'd like to be able to map manifolds with boundary into the surgery groups), these must be included among the 0-simplices. However, because of the $\pi - \pi$ theorem, if a map is cobordant to a homotopy equivalence (even changing what the target of the normal invariant is during the cobordism), then it is normally cobordant, in the conventional sense, to a homotopy equivalence.

More explicitly, we define $\mathbf{L}_i(\boldsymbol{\pi})$[1] as follows. A 0-simplex consists of a degree one normal map $W, \partial W \to M, \partial M$[2] with an isomorphism[3] $\pi_1 M \to \pi$, such that the map $\partial W \to \partial M$ is already a homotopy equivalence. A 1-simplex is a cobordism of such objects. The boundary of the

[1]We are ignoring orientations as usual.

[2]In the range it does not matter if we use Cat manifolds or Poincaré spaces, as the reader can see by going through the ensuing arguments.

[3]One only needs a homomorphism; see [Wa1, chapter 9]. That is the difference between his version of "restricted" and "unrestricted" objects. With the latter one can define a group structure just by disjoint union and have the empty set as a base point. We shall not bother with this because the reader can verify that these spaces are their own fourth loop spaces, so there is a group structure on components given this way.

cobordism will have three pieces: a top, a bottom, and the side. The top and bottom will be rel ∂ surgery problems, and the side will be a homotopy equivalence between manifolds with (usually disconnected) boundary. i-simplices are similar things.

By crossing with I and "repaving the boundary" one sees that all homotopy equivalences can be connected[4] to the base point. Furthermore, the $\pi - \pi$ theorem shows that taking the surgery obstruction provides an injective map to $L_i(\pi)$. Surjectivity is given by realization.[5]

EXERCISE. Build an assembly map $\mathbf{Map}[\Delta^j, \partial : \mathbf{L}_i(\boldsymbol{\pi})] \to \mathbf{L}_{i+j}(\boldsymbol{\pi})$, and show that it is a homotopy equivalence. Show that L-spaces actually form a spectrum. ($\times$ with $\mathbb{C}P^2$). $\Omega^i \mathbf{L}_j \cong \mathbf{L}_{i+j}$. (See [Ad2] for a basic reference on spectra; [Ad3] provides an informal introduction. For a modern version, see [LMS].)

Combining all the above spaces one obtains, for M an i-manifold, a sequence readily seen to be a fibration (up to homotopy):

$$\mathbf{S(M^i)} \to \mathbf{NI(M)} \; (\cong \mathbf{Maps}[M : F/Cat]) \to \mathbf{L}_i(\boldsymbol{\pi})$$

for $i \geq 5$ as usual. This method of describing surgery groups makes dealing with relative groups and the like simple; they are cobordism groups of relative objects or equivalently homotopy groups of homotopy fibers.

3.2. Blocked surgery

Now we give a geometric interpretation of the structure and surgery spaces of the previous section.

Let F be a manifold or polyhedron or stratified space. An F **block** bundle over a polyhedron[6] X consists of the following. Over vertices of X one has F's. Over 1-simplices one has $F \times I$'s. Over a simplex Δ one has $F \times \Delta$'s. This differs from the notion of a bundle in that one doesn't have a projection map to Δ such that each inverse image is a copy of F, etc.

Doesn't the above definition sound a lot like a Δ-set? Thus one forms the associated Δ-set $\mathbf{B\widetilde{Cat}(F)}$. Vertices are Cat manifolds isomorphic to F, edges are manifolds isomorphic to $F \times I$, etc. An F block bundle is then given by a homotopy class of maps $X \to \mathbf{B\widetilde{Cat}(F)}$. (The reason for

[4] See previous note.

[5] If one defines the L-group formally by this cobordism group, realization is automatic.

[6] In some places it is convenient to have topological block bundles over nonpolyhedra. This is an awkward notion, but it can be made clear (if messily) for locally triangulable spaces (like manifolds).

the notation is that $\Omega\widetilde{\mathbf{BCat}}(\mathbf{M})$ has as components the Cat autoisomorphisms of F up to *pseudoisotopy*.)[7]

A primary reason for the interest in block bundles is that they play the role in the PL category of fiber bundles in the smooth category. For instance, while the smooth tubular neighborhood theorem identifies vector bundles with the germ neighborhoods of a given manifold in higher dimensional manifolds (i.e., every embedding has a neighborhood which is diffeomorphic to the total space of a vector bundle, and the vector bundle is unique up to isomorphism), in the PL case, neighborhoods of locally flat submanifolds look like the mapping cylinder of a "projection" of block bundles whose fibers are spheres. (See [RS2] for the general theory and for many specific calculations. The reader should, as an exercise, re-prove as many of the main results of that paper as she can by direct blocked surgery arguments.)

One can also form a similar Δ-set $\mathbf{BF}(\mathbf{M})$ out of M-fibrations (i.e., the homotopy notion). What we called BF before is just the limit under suspension of $\mathbf{BF}(\mathbf{S^i})$. A little thought shows that (with the s-decoration) one has for a manifold, M, a fibration (except that not all components of $\mathbf{S}(\mathbf{M})$ actually occur in the fiber)

$$\mathbf{S}(\mathbf{M}) \to \widetilde{\mathbf{BCat}}(\mathbf{M}) \to \mathbf{BF}(\mathbf{M}).$$

This enables one to study the M block bundles surgery theoretically.

Let us apply this to a specific problem (which we will have need for in chapter 8). Suppose that we have a map $E \to B$. When is it homotopic to a block fibration? We answer this with a theorem whose meaning will be explained immediately below:

THEOREM (See [Q1, BLR, LvR, CW8]). *Suppose that* $\dim E - \dim B \geq 5$;[8] *then a map* $f : E \to B$ *is homotopic to a block fibration iff*

(1) *the homotopy fiber F is homotopy equivalent to a finite complex*[9] *so that E is then simple homotopy equivalent to the total space of a fibration with finite fiber,*
and
(2) *an associated obstruction element vanishes; this obstruction is an element of the set of components of the fiber of an "assembly map"*

$$A\ : \mathbf{Sections}\ (E(\mathbf{L}(\mathbf{F}) \downarrow B)) \to \mathbf{L}(\mathbf{E})$$

---

[7] Autoisomorphisms are isotopic if they can be connected by a 1-parameter family of such; they are pseudoisotopic if there is an autoisomorphism of $F \times I$ which restricts on the boundary components to the given ones.

[8] One can do something in lower dimensions, ad hoc, sometimes. For instance, think about the case where $\dim E = \dim B$.

[9] Recall from the exercise in 2.2 that this implies that F is a Poincaré space.

For condition (1) the reader should note that any fibration over a finite base with finite fiber has a natural simple homotopy type. E is only homotopy equivalent to this canonical simple type, and we have the torsion of this comparison to contend with as a first obstruction.

From the F fibration over B, one obtains an associated fibration with **L(F)** fiber by taking L-spaces fiberwise. The sections of this fibration are referred to in condition (2). Over each simplex Δ in B one takes the transverse inverse image and tries to do surgery to make the map (simplexwise) simple homotopy equivalent to the corresponding $\Delta \times F$. This gives a section of the associated bundle, but on assembly, it is trivial, because when we glue these pieces together (i.e., *assemble* them) we get the global manifold E, which is homotopy equivalent to the total space of the associated fibration, i.e., is a solved surgery problem!

If this obstruction is trivial, then we can do blocked surgeries to a blocked fibration and surger the normal cobordism into an s-cobordism. These together provide the homotopy to a solution.

EXERCISE. Show that elements of π_1 of the fiber of the assembly map are the different block bundles fiber homotopy equivalent to $E \to B$.

EXERCISE. Re-prove the above theorem by comparing $S(E)$ and the classification of F block bundles. Note the **exponential** or **adjoint isomorphism NI(E)** $\cong$ Sect(E(**NI(F)**) $\downarrow B$)), as one sees geometrically or via the isomorphisms of both **NI(E)** and Sect(E(**NI(F)** $\downarrow B$)) to function spaces.

EXERCISE. If one spacifies $Wh(\pi)$ using the ideas of 3.1, what does one get? Construct an involution on this space. What does it do on homotopy groups? Can you see a Rothenberg fibration?

REMARK. This spacification of Wh is not really that useful. In particular, it is not the one that arises in pseudoisotopy theory [Wald 1] or even in our later work on stratified spaces. Much more useful and interesting is the version of Wh that arises as the fiber of the assembly map in Quillen's (or Waldhausen's) K- (or A-) theory.

3.3. Algebraic theory of surgery

If one examines things carefully enough, one discovers that many (though not all) of the constructions we've discussed so far in surgery do not really depend that much on the geometry of manifolds and Poincaré complexes, but rather on the algebra of their underlying chain complexes with the Poincaré duality map. This is analogous to some of our discussions in K-theory, e.g. the purely chain complex version of the finiteness obstruction (1.1).

I do not want to develop this algebraic point of view extensively, but at several junctures it will be very convenient to have. For one thing, it turns out that algebraically, surgery is exactly a cobordism group (of a less cumbersome type than the (Wall) chapter 9 sort we've used before), so that one can manipulate elements using a more geometric language. For another, it provides a smaller and simpler model than Poincaré bordism and allows one to do certain arguments.

Here is the basic idea.[10] An algebraic Poincaré complex over a ring R, with an anti-involution $-$, consists of a chain complex C^* and a chain map

$$\varphi : C^{n-*} \to C_*,$$

which has some good properties. First of all, we want φ to be a chain homotopy equivalence. Secondly, we'd like φ to be ϵ-symmetric. Consider φ^*. It also gives a map $C^{n-*} \to C_*$. Ideally, $\varphi^* = \epsilon\, \varphi$. (The ϵ comes from the fact that we occasionally would like to deal with skew-symmetric forms and the like.) That is, however, too much to ask for, in general. Instead one requires φ^* to be chain homotopic to $\epsilon\, \varphi$.

The issue is that now by dualizing again one gets another chain homotopy from φ^* to $\epsilon\, \varphi$. So one has to assume that these are chain homotopic. And then we have a symmetry condition on this homotopy. Etc.

All of this can be summarized concisely algebraically as a map from the standard resolution of $\mathbb{Z}$ over $\mathbb{Z}[\mathbb{Z}_2]$ into $\mathrm{Hom}(C^*, C_*)$ of degree n.

(Actually, we have explained the idea of a **symmetric algebraic Poincaré complex**. Ranicki [Ra1] also defines the notion of a **quadratic APC**, which is a more refined thing. It captures the more subtle self-intersections in the definition of surgery obstructions.)

Clearly this definition is closely modeled on the idea of a Poincaré complex (2.1). One can define similar algebraic notions related to the idea of a Poincaré pair. Consequently there are ideas of cobordism of algebraic Poincaré complexes and the like.

DEFINITION. *$L^n(R)$ is the cobordism group of symmetric algebraic Poincaré complexes. $L_n(R)$ is the cobordism group of quadratic algebraic Poincaré complexes.*

Ranicki shows that for $R = \mathbb{Z}\pi$, $L_n(R) \cong L_n(\mathbb{Z}\pi)$ in the sense of the previous chapter; i.e. that the above is a good cobordism theoretic algebraic description of the surgery obstruction groups.

REMARK. The reader used to the K-theory literature should not make the mistake of associating variance by subscript and superscript. Both functors are covariant.

[10] Actually, for many purposes the more recently defined **visible APCs** of Weiss [Ws] are even more useful. The fiber of the visible assembly map (see below) is isomorphic to the fiber of the quadratic assembly map. See the notes for more discussion.

REMARK. It is possible to decorate L-groups of this sort. One can insist that the C's be based and that the torsion of ω live in some subgroup of K_1. Similarly, one can use projective modules only living in some subgroup of K_0.

From a Poincaré complex one can derive from its chain complex an example of an SAPC. Degree one normal maps naturally gain QAPC structures on their mapping cones (the framing giving rise to the quadratic refinement).

There is a forgetful map from QAPC to SAPC. Under this map the obstruction for a normal map becomes the difference of the SAPC's range and domain.

Finally, if $1/2 \in R$ or for an arbitrary R if we $\otimes \mathbb{Z}[1/2]$, there is no difference between $L^n(R)$ and $L_n(R)$. This means that away from the prime 2 surgery obstructions are differences of intrinsic invariants, namely the SAPC associated to the manifolds. These invariants are sometimes referred to as the **underlying algebraic complex**, the **Mischenko-Witt element**, or the **((Mischenko-)Ranicki) symmetric signature** of X. They are closely related to the class of the signature operator on M, for M a manifold, in the K-theory of the C^*-algebra of the fundamental group [KaM1, Kas2]. We have seen for Whitehead torsion that it is only occasionally possible to obtain this invariant of a map as the difference between intrinsic invariants of spaces (see 1.2.A); in surgery, away from 2, we always can.

One can generalize a bit further and define what a (Q, S)APC n-ad is. It is something modeled on the n-simplex Δ. Using these complexes, one defines algebraic spectra $\mathbf{L^n(R)}$ and $\mathbf{L_n(R)}$ precisely by analogy to the bordism space $\underline{\Omega}_i^{Cat}$ defined above. There are maps, for instance,

$$\underline{\Omega}_i^{Cat} \to \mathbf{L}^i(\mathbb{Z})$$

defined by taking the underlying chain complexes of manifolds. Or, more generally,

$$B\pi_+ \wedge \underline{\Omega}_i^{Cat} \to \mathbf{L}^i(\mathbb{Z}\boldsymbol{\pi}),$$

inducing the map $\pi_*(B\pi_+ \wedge \underline{\Omega}_i^{Cat}) = \Omega_*^{Cat}(B\pi) \to L^i(\mathbb{Z}\pi)$ and sending a cobordism class to Its symmetric signature.

The reader should be warned that for Ranicki's groups the right-hand side is not 4-periodic in general. ($\pi = \mathbb{Z}_2$ is an example.)

Since the symmetric L-theory resembles manifolds, one can take products and develop a pairing

$$L^n(R) \otimes L_m(S) \to L_{n+m}(R \otimes S)$$

analogous to taking a surgery problem over $S(=\mathbb{Z}\pi)$ and taking the product with a manifold over $R(=\mathbb{Z}\pi')$ to get a surgery problem over $R\otimes S(=\mathbb{Z}[\pi\times\pi'])$.

This can be jazzed up further to give a pairing of spectra. Using the element $\mathbb{C}P^2$ in $L^4(\mathbb{Z})$, one then obtains a map of spectra $\mathbf{L}_m(S)\to\mathbf{L}_{m+4}(S)$, which is a homotopy equivalence. Since $\mathbf{L}_{m+4}(S)\cong\Omega^4\mathbf{L}_m(S)$, these spectra have fourfold periodicity.

Using, for instance, the calculation

$$L^n(\mathbb{Z})=\mathbb{Z},\mathbb{Z}_2,0,0 \text{ for } n\equiv 0,1,2,3 \bmod 4 \text{ respectively},$$

one can see that the effect of taking the product of a surgery obstruction by a simply connected manifold only depends on its symmetric signature in these small groups. The $\mathbb{Z}$ is just the ordinary signature, and the $\mathbb{Z}_2$ is the DeRham invariant which counts the number mod 2 of $\mathbb{Z}_2$ summands in the $2i$-th homology of the $4i+1$ manifold. This product formula was first proven by Morgan by an elaborate geometric argument.

The algebraic theory of surgery is wonderful also for its ease in setting up exact sequences for the analysis of different L-groups. For instance, Rothenberg sequences have as relative terms APCs with boundary having a simple structure and with interior not having a simple structure (which can then be identified with the Tate cohomology of 2.4.A). Also, one can prove general localization theorems for L-theory analogous to the ones in algebraic K-theory [B1a, Ql2]. A remarkable consequence of L-theory that will sometimes be of use to us is due to Ranicki [Ra3]:

PROPOSITION. *If $\mathbb{Z}\subset R\subset\mathbb{Q}$, then the map $L_*(R\pi)\to L^*(\mathbb{Q}\pi)$ is an isomorphism away from 2.*

REMARK. We have concentrated here on the classification theory of manifold structures on a Poincaré complex. Ranicki [Ra1] has also applied these ideas to describe a *total surgery obstruction.*[11] It lies in a group composed naturally of the components of a deloop (or equivalently, third loop) of the structure space. If X is a polyhedron, one can describe the total surgery obstruction as the obstruction to nullcobording the local obstructions to Poincaré duality (the homology of the chain map corresponding to capping with the orientation class), all the time keeping track that the global "assembled" version is trivial, as X satisfies Poincaré duality. (See also [LvR].)

[11]We are essentially ignoring a single $\mathbb{Z}$ here that comes from the difference between F/Top and $\mathbf{L}(e)$. As explained in 9.4.C this is accounted for by homology manifolds. Another way to get this all right is to work relative to a given manifold structure on a top cell.

EXERCISE. Using this description, show that a polyhedral homology manifold X is canonically simple homotopy equivalent to a topological manifold. (With more effort, show that the map can be taken to be a resolution, i.e., $\varphi : M \to X$ such that φ maps $\varphi^{-1}\mathbb{O} \to \mathbb{O}$ by a proper homotopy equivalence for each open subset $\mathbb{O}$ of X.) This result is due to Galewski and Stern [GaS].

REMARK. We have only defined here the "connective L-spectra". One can (and it's quite useful) invert the $\times\mathbb{C}P^2$ periodicity to build variants with homotopy in all dimensions, positive and negative. For most geometric purposes, these are indistinguishable.

3.3.A. The structure of L-spectra

We have seen that the L-spectra arise naturally in the geometric problem of blocked surgery. They will occur many times in the sequel. The following is a useful characterization. (We place it here since it applies to the algebraic spectra as well.)

THEOREM (See [TW]). *At the prime 2 all L-spectra are products of Eilenberg-MacLane spectra. Away from 2 they are all products of (loops of) BO with coefficients.*

L is given an infinite loop space structure by crossing with $\mathbb{C}P^2$. This is similar to the method for producing the infinite loop space structure on BU, Bott periodicity, given in [A1] where one uses the Dolbeault complex on $\mathbb{C}P^2$.

The way one introduces A coefficients into a Ω-spectrum is by taking the function space $[M(A) : -]$, where $M(A)$ is the Moore space with coefficients A.

All L-spectra are modules over $\mathbf{L}^*(\mathbb{Z})$ by $\otimes$. Note that $\mathbf{L}^*(\mathbb{Z})$ is itself a module spectrum over MSO (the smooth bordism spectrum), which is Eilenberg-MacLane at 2. Therefore, the first part follows from general nonsense. The statement away from 2 follows from the method of Sullivan described in 2.5.

Also, the statement of Sullivan's geometric characteristic variety theorem can be rephrased as

PROPOSITION. *A map of $\mathbf{L}^*(\mathbb{Z})$-module spectra is nullhomotopic iff it is zero on all homotopy groups with coefficients $\pi_i(\ \ ;\mathbb{Z}_k)$.*

Homotopy groups with coefficients are defined using maps of Moore spaces. They relate to usual homotopy groups via universal coefficient theorems and Bockstein exact sequences.

In addition to the calculability of L-spectra, we will also see that the fibrations that arise geometrically (as in Part II) tend to be *flat* (by which I mean pulled back from fibrations over an aspherical space) and are often trivial, even when the original block fibration they come from is not. An example of this arises in the next section in proving the Thom isomorphism for structures.

3.4. Applications to manifold surgery

The previous sections of this chapter were more algebraic and homotopical than is my wont, but we shall now see that there are a number of useful implications of these theories.

PROPOSITION. *There is a homotopy equivalence* $\mathbb{Z} \times F/Top \cong \mathbf{L}_0(e)$.

Let us not worry about low dimensions. The map from F/Top to $\mathbf{L}_0(e)$ is given by viewing a simplex in the former as a normal invariant and then mapping into the target by taking the associated surgery obstruction.

Slightly differently, one can consider the fibration for $\mathbf{S}(\Delta \text{ rel } \partial)$, which is contractible by the h-cobordism theorem, and then deloop, since both sides are their own fourth loop spaces (up to components) by 2.5 and 3.1.

Using this proposition one can give F/Top a new H-space structure, as the 0 component of $\mathbf{L}_0(e)$.

THEOREM. *With this H-space structure, the surgery fibration is a map of infinite loop spaces:*

$$\mathbf{S}(\mathbf{M}^i) \to \mathbf{Maps}\,[M : F/Top] \to \mathbf{L}_i(\pi).$$

In particular, the homotopy groups gain abelian group structures, and the maps are homomorphisms.

The point is that $\mathbf{Maps}\,[M : F/Top]$ are now part of $\mathbf{Maps}\,[M : \mathbf{L}_0(e)]$ and that the assembly map

$$\mathbf{Maps}\,[M^i : \mathbf{L}_*(\mathbf{e})] \to \mathbf{L}_{*+i}(\pi)$$

is naturally an infinite loop map using the infinite loop structure from 3.1. The fiber $\mathbf{S}(\mathbf{M}^i)$ then inherits an infinite loop structure.

EXERCISE. (For those who read 2.5.A.) Show that there is a unique H-space structure on F/PL so that the map F/PL to F/Top is an H-map. In this structure, the PL surgery exact sequence is an exact sequence of abelian groups and homomorphisms. This structure comes from an infinite loop space structure, and the map is actually an infinite loop map.

Now, two of the three terms in the fibrations are (almost) their own fourth loop space. This should translate into a statement for the third, the structure set.

SIEBENMANN PERIODICITY. *There is an exact sequence of abelian groups*

$$0 \to S(M) \to S(M \times D^4 \operatorname{rel} \partial) \to \mathbb{Z}.$$

The map to $\mathbb{Z}$ vanishes if M is a manifold with boundary being studied rel ∂.

EXERCISE. Prove this.

REMARK. These periodicities are geometric forms of Bott periodicity away from the prime 2.

Actually, we now know [BFMW] that the components of the fiber of the assembly map correspond to s-cobordism classes of ANR homology manifolds simple homotopy equivalent to X. If we replace manifolds by ANR homology manifolds, we then get periodicity exactly. The $\mathbb{Z}$ element is a measure of the local structure of the homology manifold. When we work rel ∂, the ∂ value determines the local structure. In chapter 9, we will discuss this further.

Just as Bott periodicity generalizes to a Thom isomorphism in K-theory, so too does Siebenmann periodicity in the theory of structures.

THEOREM. *If $E(D^{4k} \downarrow M)$ is an orientable block fibration over M, then there is a periodicity map $S(M) \to S(E \operatorname{rel} \partial)$ which fits in an exact sequence as above.*

We shall ignore all issues involving the extra $\mathbb{Z}$ that perturb the periodicity. The reader can hopefully trace this on his or her own. That is, we'll identify F/Top with $\mathbf{L}(e)$. Alternatively, we'll be working with homology manifolds and leave the question of manifoldness as a final issue that the interested reader can check for.

To prove the isomorphism theorem, we need to see an isomorphism

$$[E \operatorname{rel} \partial : \mathbf{L}_0(e)] \leftarrow [M; \mathbf{L}_0(e)]$$

which spacifies well and commutes with the assembly maps and the isomorphism $L_i(M) \cong L_{i+4k}(E \operatorname{rel} \partial)$. We will only produce the isomorphism; the details of all the commutativities are a bit of a nuisance.

Well, from 3.2 we have

$$\mathbf{NI}(E \operatorname{rel} \partial) \cong \operatorname{Sect}(E(\mathbf{NI}(D^{4k} \operatorname{rel} \partial)) \downarrow M).$$

But, there is the canonical surgery equivalence (requiring an orientation character preserving map to the trivial group, which exists because the bundle is oriented)

$$\mathbf{NI}(D^{4k} \operatorname{rel} \partial) \to \mathbf{L}_{4k}(\mathbf{e})$$

with which we can make the identifications

$$\operatorname{Sect}(E(\mathbf{NI}(D^{4k} \operatorname{rel} \partial) \downarrow M)) \cong \mathbf{Maps}[M; \mathbf{L}_{4k}(\mathbf{e})] \cong [M; \mathbf{L}_0(e)].$$

REMARK. Thom isomorphisms are usually associated to Thom classes. (See [Ad2], but the reader should expect to do some work to connect the viewpoint there with the multiple Poincaré duality and deformation viewpoint we use. The reader can compare the discussion in [Kas2, RsW2] that does something similar for topological K-theory.) Given an oriented topological n-manifold, there is a Thom class $\Delta(M) \in H_n(M; \mathbf{L}^*(\mathbb{Z}))$ which induces the above Thom isomorphism theorem. It is produced from the map of spectra from bordism to SAPC discussed above. Away from 2 it is the Sullivan orientation from 2.5.

REMARK. This class was first defined in [Ra1]. He calls it the symmetric L-theory orientation of a manifold. (For homology manifolds, it is not integrally an orientation, even in $\mathbb{Z}[1/2]$, unless they are resolvable.) I prefer to call it the signature class, because it is a refinement of the class of the signature operator for manifolds, and so too is its generalization in 6.1 when we restrict attention to certain natural classes of stratified spaces, as the reader of chapters 12 and 13 will see.

Just as Atiyah and Singer use the Thom isomorphism in K-theory (in [AS, pt. I]) to define the topological index of an elliptic operator, which is a wrong-way map in K-cohomology, we can use the above theorem to define a *covariant functor structure on structure groups*.[12]

DEFINITION. *Let X be a finite CW complex. If its dimension is i, we define,*[13] *for large $j (\geq 2i + 3)$, $S_j(X) = S$ (regular neighborhood of $X \subset \mathbb{R}^j$ rel ∂). Then for smaller values of j one can extend using periodicity. If X is not finite, we just define $S(X) = \lim S(K)$ over K finite, $K \to X$.*

Note that if M is an oriented i-manifold, then we have an isomorphism $S(M) \cong S_i(M)$.

Now if $f : X \to Y$ is an orientation true[14] map, then we define the **induced map** $S_j(X) \to S_j(Y)$ as follows. For j sufficiently large we embed the regular neighborhood of X into that of Y as a codimension zero submanifold (by thickening an embedding of X into a regular neighborhood of Y). Now one can extend any structure in $S_j(X)$ by the identity map on the complement to give an element of $S_j(Y)$.

Applied to $S(M \times D3) \cong NI(M)$ one obtains a covariant functor structure on a group that had previously been given a contravariant one from its structure as the homotopy classes of maps. Using this, one realizes that what is actually natural is the following version of the structure

[12] Since we are dealing with topological structures we have already promoted the sets to be groups.

[13] Here we are ignoring orientation. Strictly speaking one should use a product of an even dimensional real projective space and Euclidean space to capture the theory with $X \to B\mathbb{Z}_2$.

[14] I.e., commutes with w_1.

sequence:

$$\ldots \to L_{n+1}(\pi) \to S(M) \to H_n(M; \mathbf{L}(e)) \to L_n(\pi).$$

or, spacily,[15]

$$\mathbf{S}(M) \to \mathbf{H}_n(M; \mathbf{L}(e)) \to \mathbf{L}_n(\pi).$$

Functoriality then shows that we have a commutative diagram from a map $M \to B\pi$:

$$\begin{array}{ccc} H_n(M; \mathbf{L}(e)) & \longrightarrow & \mathbf{L}_n(\pi) \\ \downarrow & & \downarrow \\ H_n(B\pi; \mathbf{L}(e)) & \longleftarrow & \mathbf{L}_n(\pi), \end{array}$$

with which we can compute the surgery obstruction. The map on the bottom line is often called the **assembly map**. We will discuss more algebraic definitions of it in chapter 9; for now it suffices to think of it as Poincaré dual to the assembly provided by assembling simplices.

For instance, immediately from this diagram we see that for π finite, the surgery obstruction for a degree one map between closed manifolds in $L_n(\pi)$ rationally lies in $L_n(e)$. Using the result that $L_n(\pi)$ has no odd torsion, one can quickly reduce calculations to the case where π is a 2-group; see [Wa6]. Indeed, for finite groups the map $H_n(B\pi; L(e)) \to L_n^h(\pi)$ is very well understood [HMTW] and one can then view the maps in the surgery exact sequence as computed. (With the "s-decoration" the present state of computation is much worse.)

For infinite groups the issues involved in calculating the assembly map are rather different and more closely tied to geometry than to algebra. See 4.5.A and chapter 14.

3.5. Notes

Spacification was first done in the simply connected case by Casson for purposes of understanding the hauptvermutung (see [Ar]). Quinn, in his thesis [Q1], extended this to the nonsimply connected case. In his Cambridge thesis Ranicki also produced these spectra algebraically by a complicated method; the version in [Ra1,4] is as transparent as the geometric version using his cobordism of algebraic Poincaré complex perspective. The symmetric L-spectrum was also constructed in [Ms2].

---

[15] Note that by this point, we have tacitly inverted the periodicity maps so that the L-spectra are now nonconnective. As far as π_i, $i \geq 0$, statements are concerned, all that might happen is an extra $\mathbb{Z}$ in $\pi_0(S(M))$.

The primacy of the $\pi - \pi$ theorem is, of course, observed in [Wa1]. Shaneson, in his thesis [Sh1], seems to have been the first to realize that this implies that one really has to understand only one manifold very well to understand all manifolds with that fundamental group. Cappell and Shaneson [CS1] also built their theory of homological surgery on the $\pi - \pi$ theorem.

The subject of blocked surgery was first studied by Casson [Cs] as a first step to the problem of fibering one manifold over another. Quinn points out in his thesis that the surgery spaces have some relevance to this problem. [BLR] develops the theory in detail (e.g., for instance pointing out that one is really dealing with sections of a bundle rather than a mapping space) but only in special cases, presumably restricting attention to the only cases that seemed calculable. Levitt also used blocked ideas in several of his papers. The joint paper with Ranicki, [LvR], is particularly elegant. The formulation I have chosen was probably first written down in [CW8] (although lectured on many years earlier) and is used there to show that one can construct PL group actions on manifolds with a plethora of possible fixed sets.

The algebraic theory of surgery, as presented here, is due to Ranicki, who has proved much more. Such fundamental topics as localization, assembly, arithmetic squares, etc. are best dealt with in this framework. That material is well covered in his book [Ra4]. A more primitive formulation is due to Mischenko [Ms2] that only works away from the prime 2. (Mischenko used symmetric duality maps, not chain homotopy symmetry as we described in the body of the text.) Weiss [Ws] has provided a very interesting refinement of the symmetric signature that still suffices for product formulae and has much better assembly properties. In his theory, the total surgery obstruction is the obstruction to decomposing the visible symmetric signature of X over X, i.e., pulling it back to $H_{*-1}(X; \mathbf{L}^*(\mathbb{Z}))$. The fiber of Weiss's visible assembly map is the same as the fiber of the quadratic assembly map.

There are many versions of the assembly map. There have been important applications of it to giving formulae for surgery obstructions and the symmetric signature in [TW, Wa5]. The structure of L-spectra was observed by both Jones and Taylor-Williams.

The result in the exercise on F/PL was first observed in work of J.P. May [MQRT].

Siebenmann periodicity appears first, as does the group structure on structure sets, in [KS]. [Ni] is also a useful reference for this and many of the other concerns of this chapter. A geometric description, which we will have use for later, appears in [CW3].

One can obtain information on the surgery obstructions on closed manifolds directly by bordism methods. This is done in [Wa1]. However, the assembly point of view has been indispensable for more recent results.

(In fact it has been the need for solid descriptions for computational purposes that has led many authors to give versions of the assembly map; see [Ra5] and the references included there.)

The functoriality of surgery is a formal consequence of [Ra1]. The geometric form given here I developed jointly with Cappell for the purpose of computing some homomorphisms of structure sets. The version given here appears in [Wei4] as well.

4 Applications

The reader who has made it to this point deserves more than just a promise that good things lie ahead and that everything just studied is to be soon generalized and extended. It is time to learn new things about manifolds.

There have been many different applications of surgery to date, and the issue of which ones to choose is a difficult and personal one. I have tried to pick ones that are inherently striking, exhibit a useful technique or unexpected phenomenon, or have served as motivation for other developments.

On the other hand, I have not done any actual *calculations* of surgery groups and obstructions in these pages. For finite groups, this has a well-developed, highly complex algebraic theory that every surgeon must at some point come to terms with, but is outside the scope of these notes. I refer the reader to the lectures of Hambleton [Hm] and the references there for an in-depth study of this important topic. When necessary in the following pages, I will invoke a calculation, but here we'll stress how far it is possible to go on geometric insight with a minimal amount of calculation.

The calculations that arise for infinite torsion-free groups have been attacked by methods that are, for the most part, "geometric". (The idea that one solves a geometric problem by reducing it to algebra is far too facile. Often the algebraic problems that arise are quite eccentric from the algebraic point of view, for instance, when is $\mathbb{Z}\pi$ a decent, Noetherian ring?) We will see some of this here and in the applications in chapter 9. Infinite groups with torsion really seem to mix the phenomena of both, and we will analyze this, in some cases, using the geometry of certain natural stratified spaces in chapter 14. Farrell and Jones have recently developed some powerful new methods for this problem as well [FJ5].

Some of the applications here are almost trivial given all the machinery that we have developed; some are more challenging. In some cases, I will give inefficient arguments for why something is true if I feel that there is a pedagogical benefit in doing this. I hope that the next pages bring the reader as much joy as their contents brought me.

4.1. Homotopy CP^n's

Everything done in this section follows more easily from the material of the next. We know enough to compute with the surgery exact sequence and see anything there is by plugging in $M = CP^n$. However, we will be more ad hoc and geometrical using the $\pi - \pi$ theorem and calculation of the simply connected surgery obstruction groups.

First, recognition.

PROPOSITION. *A simply connected CW complex has the homotopy type of CP^n iff its cohomology algebra is a truncated polynomial algebra on one two dimensional generator $\mathbb{Z}[c]/c^{n+1} = 0$. There are two homotopy classes of homotopy equivalences to CP^n, which differ by composition with complex conjugation.*

The two dimensional cohomology class of a generator (there are two) gives a map to CP^∞ (= $K(\mathbb{Z}, 2)$, the classifying space for 2nd integral cohomology). By the cellular approximation theorem one can arrange for the image to lie in CP^n. Moreover, a little obstruction theory shows that the homotopy class of such a compression is unique. This map is a homotopy equivalence by the Whitehead theorem.

For another interpretation:

PROPOSITION. *The quotient spaces of free S^1 actions on S^{2n+1} have the homotopy type of CP^n. The actions are conjugate iff there is an isomorphism of the quotients preserving the two dimensional cohomology class.*

One sees that the quotient is of the correct type by a Gysin sequence and the previous proposition.

In the smooth and PL categories, the quotient is automatically a manifold. In the topological situation, this is not the case; see e.g. [Bi] or [Dvr] for many examples. Nonetheless, it will follow from later discussions (in 13.2) that these exotic actions can all be *deformed* into nice locally smooth (i.e. with manifold quotient) actions.

Let us recall what CP^n looks like. One has a decomposition of

$$S^{2n+1} = (S^{2n-1} \times D^2) \cup (S^1 \times D^{2n})$$

which preserves the group action. This describes CP^n as the union of the total space of a D^2 Hopf bundle over CP^{n-1} and a cell D^{2n}. This decomposition can be thought of as the compactification of affine space as projective space where one glues in a projective space at ∞.

We also have copies of CP^i, $i \le n$, sitting "equatorially" in CP^n.

Now let W be a homotopy CP^n and $f : W \to CP^n$ a homotopy equivalence. By making f transverse to these subprojective spaces, we can define the **splitting invariants** as follows:

$$s_i(W) = \text{surgery obstruction of}$$
$$(f|_{f^{-1}CP^i} : f^{-1}CP^i \to CP^i) \in L_{2i}(e),\ i \leq n-1.$$

Recall that these groups are $\mathbb{Z}$ or $\mathbb{Z}_2$ according to whether i is even or odd.

THEOREM. *The set of splitting invariants determines a topological homotopy CP^n for $n \geq 3$. Furthermore, all values of these invariants arise.*

REMARKS. Freedman [Fr2] has extended this to $n = 2$. We will see in 4.3 that $s_i(W) = 0$ iff we can make f transverse to CP^i with the inverse image a homotopy CP^i. For $i = n - 1$ this means that W is the Thom space of the Hopf bundle over a lower dimensional homotopy complex projective space. Equivalently, the circle action on the sphere is a join of an action on a lower dimensional sphere and the obvious action on the circle. This *desuspension*, when it exists, is unique by the above theorem.

This theorem can be viewed as just the statement that the union of the subprojective spaces is the characteristic variety.

For all values of $i \neq n-1$, if one relaxes the condition that f be transverse, we will see in 4.4 that one can always arrange for $f^{-1}CP^i = CP^i$. For $i = n-1$, the transversality is (up to homotopy rel f^{-1} when it's a submanifold) automatic. (This is related to chapter 14's material, but can be done by hand using the spacified version of 4.6.)

EXERCISE. Show that the even splitting invariants determine and are determined by the Pontrjagin classes of W.

One can prove the above theorem as follows. If one punctures (W and) CP^n by removing a small open ball, then one sees that $S(CP^n - \text{ball}) \cong [E; F/Top] \cong [CP^{n-1}; F/Top]$ (where the union is the total space of the 2-disk bundle over CP^{n-1}), which is detected by these invariants. (Just compute, using obstruction theory or 2.5.)

By the GPC (1.7) one knows that the boundary is automatically a sphere and one can therefore glue the missing ball back in. (There is only one way to glue the ball back in because of the Alexander trick.)

Actually, to do this calculation, one does not need the homotopy type of F/Top, only its homotopy groups. Rothenberg, early in the history of surgery, before Sullivan's classification of normal invariants, did the whole calculation in that way. He realized explicitly the splitting invariants and used the geometry of the previous remark to show they characterized.

To realize all of these invariants by hand we need to know about **Milnor** and **Kervaire manifolds**. These are PL manifolds $M^i(\alpha) \to S^i$ for each $\alpha \in L_i(e)$ realizing the surgery obstruction α. These exist for $i \geq 5$.[1] One way to see them is to apply the Wall realization theorem to S^{i-1} and cap off the other end using the GPC. Kervaire ($i \equiv 2 \bmod 4$) and Milnor ($i \equiv 0 \bmod 4$) constructed these manifolds by hand, using plumbing, and Wall's proof of his realization theorem (which we outlined in chapter 2) is an extension of their construction. These were (for suitable α and i) the original nonsmoothable manifolds.

EXERCISE. Knowing topological invariance of rational Pontrjagin classes, show that $M^8(1)$ is not smoothable. (Hint: Use the Hirzebruch signature formula to show that p_2 would not be an integer.)

REMARK. If f is a (weighted) homogenous polynomial of several complex variables with the origin as an isolated singularity, then the intersection of the unit ball with the ϵ-level set is often a Kervaire or Milnor manifold. (They are called Brieskorn varieties.) See [Mi8] for more information, such as specific calculations for specific polynomials. Often the Brieskorn description of an exotic sphere enables one to understand it better, such as by putting a group action on it or embedding it somewhere.

Suppose now that we've realized a sequence of splitting invariants for $i \leq n$. To do $n+1$, connect sum this lower realization with a Kervaire or Milnor manifold as appropriate to realize the splitting invariant. Then the total space of the Hopf bundle mapping to the CP^{n+1} ball is a $\pi - \pi$ problem, so that one can surger this to a homotopy equivalence of pairs. Then cone the boundary (which, as a homotopy sphere, is a genuine sphere).

EXERCISE. Do the PL classification.

EXERCISE. Show that the map $W \to W$ that sends $c \to -c$ on H^2 is homotopic to a homeomorphism.

PROBLEM. (Fairly hard at this point, but return to it from time to time.) In which cases can you make this homeomorphism an involution?

REMARK (FOR A SECOND READING?). The reason such involutions are still somewhat more mysterious than many others is because (1) the fixed set is in the middle dimension, so the gap hypothesis does not hold, and (2) the action is not homologically trivial, so the only extant techniques that apply without the gap hypothesis do not apply in this case. However,

---

[1] We have tacitly used their existence in our description of the homotopy types of F/Top and F/PL. We gave cohomology classes to describe a homotopy equivalence (at 2), but to verify that these do provide such an equivalence we need to compute that they are an isomorphism at 2. The Kervaire and Milnor manifolds correspond to elements of homotopy that generate.

one can classify, using the material of chapter 14, the actions that have the isovariant homotopy type of the linear action.

EXERCISE. Work out the parallel theory of homotopy quaternionic projective spaces.

REMARK. I should mention that Brumfiel has done some analysis of the smooth case. As far as I know, the only copies of this manuscript can be found in the Fine Hall Library (Princcton).

4.2. Simply connected manifolds

The general case of simply connected manifolds is no different than that of complex projective space.

We start with existence.

THEOREM. *A simply connected Poincaré space is homotopy equivalent to a topological manifold iff its Spivak bundle has a topological reduction.*

REMARK. The reader should verify this for $n \leq 3$ by classifying simply connected Poincaré complexes. For $n = 4$ this follows from Freedman's work. The proof that follows also works for PL for $n \geq 5$. This result is false for the smooth case (Exercise: Using the material on G/O work out which Milnor manifolds are homotopy equivalent to smooth manifolds) and for PL for $n = 4$ (Donaldson).

Suppose $n \geq 5$. Then the issue is whether the surgery obstruction is 0. If it isn't, correct for this by taking the connected sum with a Kervaire or Milnor (KM) manifold.

COROLLARY. *Every simply connected finite H-space is homotopy equivalent to a closed manifold. Every connected finite H-space has a finite cover homotopy equivalent to a closed manifold.*

REMARK. It is not known whether one really has to take finite covers. It is also not known whether or not the manifold can always be taken smooth (or parallelizable). For more information, see [CW6,7]. However, even without taking finite covers, the result is "locally" true [Wei9].

To prove this in the simply connected case, one notes that Browder [Br4] has shown that finite H-spaces are Poincaré spaces and that their Spivak bundles are trivial, so the result follows. If the fundamental group is torsion free, then it is an elementary exercise in covering space theory[2] to show that one can split off a torus. The general result follows.

[2] First split off a single circle. Observe that the circle is a homotopy retract of the H-space, and argue that such are split factors of the H-space. Alternatively, realize that every space with a map to a circle is homotopy equivalent to the mapping torus of the monodromy on the associated infinite cyclic cover. Argue that if this is a connected cover of an H-space, the monodromy is homotopic to the identity.

Classification is not that hard. We will not bother to make it more explicit than the following:

THEOREM. *If M is a simply connected manifold, then $S(M) \cong [M\text{-point}; F/Top]$. Equivalently, it is the kernel of the surjection $[M : F/Top] \to L(e)$.*

Away from 2 the manifolds are classified by the ratios of their Sullivan orientations. The rational L-classes can be varied at will (within a $\mathbb{Z}$-lattice in cohomology) subject only to the Hirzebruch signature formula.

The second formulation follows from the first. Equivalently, one uses Kervaire and Milnor manifolds to show that the action of L_{n+1} is trivial (i.e. each element of L_{n+1} arises as a cylinder connected sum with a KM manifold).

4.3. Browder's splitting theorem

The results of this section are true in all categories.

The splitting problem is the following. One is given $f : M' \to M$, a (simple) homotopy equivalence, and $N \subset M$, a submanifold. The goal is to homotop f so that the transverse inverse image, N', of N is homotopy equivalent to N. If this is possible, we say that f is **splittable along** N.

In codimension one, if the normal bundle is trivial, the answer is close to always; see [Ca1]. With nontrivial normal bundle there is a more complicated, beautiful theory; see [CS2]. In codimension two the problem is quite subtle, see [CS1] for some discussion. In higher codimensions one has:

THEOREM (BROWDER [Br2]). *Suppose that $f : M' \to M$ is a simple homotopy equivalence, and $N \subset M$ is a submanifold of codimension ≥ 3 and dimension ≥ 5. Then f is splittable iff the surgery obstruction of $f^{-1}(N) \to N$ in $L_n(\mathbb{Z}\pi_1 N)$ is trivial.*

The proof is quite simple, instructive, and summarized in figures 12 and 13 (sometimes called the top hat trick for a reason the reader who draws a picture of the range of the following surgery problem will see).

First one surgers abstractly N' assuming the vanishing of the obstruction. One can thicken the normal cobordism of N' by the normal bundle of N in M and glue it onto a copy of $M' \times I$. Then one has the picture in figure 13 and works rel $M \times 0 \cup$ a neighborhood of $N' \times I$ to surger the expanded cobordism. We succeed using the $\pi - \pi$ theorem (the codimension hypothesis is used here to see that we are in a $\pi - \pi$ situation). The target is just a redrawing of $M \times I$ so the domain is now an h-cobordism. However, since the fundamental group of the complement is π, we can glue an h-cobordism with the negative of the torsion

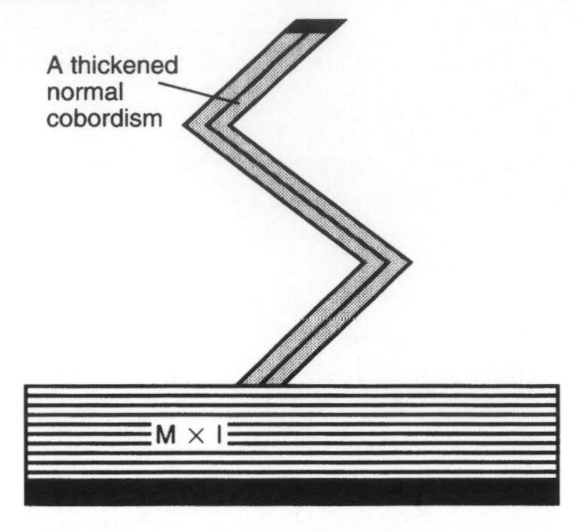

Figure 12. Normal cobordism extension. One first surgers the submanifold to obtain a homotopy equivalence and then does surgery relative to the darkened regions (which is possible by the $\pi - \pi$ theorem) to produce an s-cobordism.

Figure 13.

onto the complement. The s-cobordism theorem then asserts that this is a homotopy to a patently split homotopy equivalence. QED.

Note that one does not really need M and N to be manifolds. All one needs is some Poincaré structure for the surgery and bundle data normal to N for the transversality.

EXERCISE. Prove that one can always split along a codimension one submanifold V which divides the ambient manifold into two components W_i if the map $V \to W_1$ induces an isomorphism on fundamental groups.

REMARK. The splitting problem without this hypothesis is analyzed in [Ca1-4]. We will soon see applications of this to certain types of topological rigidity.

EXERCISE. Using Browder's splitting theorem interpret the splitting invariants for homotopy CP^n geometrically. (The only thing remaining to see is how to split along the codimension two subprojective space. Reprove Browder's theorem in this special codimension two case.)

While you're at it, prove that there is a unique embedding of CP^n in CP^{n+1} in the usual homology class.[3] (Hint: This is not a surgery exercise!)

4.4. Embedding theory

The main result of this section is the reduction of the question of the existence of PL or topological locally flat embedding to homotopy theory. This is due to Browder, Casson, Haefliger, Sullivan, and Wall (cf. [Wal]). Here we will present more or less the classical proof. It's possible to give a more efficient proof using blocked surgery (it's the first exercise). In chapter 11, we will present yet another approach and give a significant generalization of this theorem outside the realm of manifolds.

DEFINITION. *A Poincaré embedding of X in Y consists of a spherical fibration $\xi \downarrow X$, a Poincaré pair (Z, E), and homotopy equivalences between E and the total space of ξ, and of Y with the union of the mapping cylinder of ξ and Z along E.*

THEOREM. *$M^m \subset N^n$ (locally flatly), $m - n \geq 3$, iff M Poincaré embeds in N. More precisely, every Poincaré embedding comes from a geometric embedding.*

EXERCISE. Prove the theorem as follows. Codimension one split along the sphere bundle of ξ (exercise from the last section). This provides a codimension zero submanifold of N homotopy equivalent to M. Observe that there is no obstruction to block fibering this over M (i.e., observe that the obstruction space to block fibering is contractible because of the $\pi - \pi$ theorem). The fiber is a disk. Now, M embeds in any block disk bundle over itself.

EXERCISE. Formulate a relative version of the theorem. Using the concordance implies isotopy theorem (see 1.7) show that there is a unique embedding realizing a given Poincaré embedding.

EXERCISE. Deduce from the embedding theorem that F_q/PL_q stabilizes for $q > 2$. Prove this directly by relating this space to $\mathbf{S}(S^q)$ and using blocked surgery to show that that space is independent of q (for $q > 2$).

EXERCISE ([STANKO]). Show that every (i.e., even wild) topological embedding in codimension at least three gives rise to a Poincaré embedding.

---

[3] I've always loved this fact. If one takes the connected sum of the usual embedding with a knotted sphere pair, and there are infinitely many of these, the projective space manages to unknot itself. Perhaps one should view this as a flexibility result rather than a rigidity one!

(For locally flat embeddings, this is clear; in general the local neighborhood system is homologically alright by Alexander duality applied locally, and then one must apply a + construction; see [Ad3].) Deduce that if a manifold embeds in any way in codimension at least three, it embeds locally flatly. Do this argument carefully with respect to an open cover to prove that every topological embedding can be approximated by a locally flat embedding, which is unique up to small isotopy. (This is a taming theorem.)

REMARK. The PL codimension two situation is very different [CS1,4] with respect to existence, uniqueness, and approximation. The topological situation is very poorly understood at present, although on some evidence (related to the material in chapters 9 and 10) I have conjectured that the above theorem is true (nonlocally flatly) although "concordance implies isotopy" fails. In other words, in a nonlocally flat concordance sense, codimension two is identical to all higher codimensions topologically.

As for codimension one, Cappell's splitting theorem implies that often for a Poincaré embedding there is a unique embedding of some unique manifold simple homotopy equivalent to M in W. (See 11.5.) In particular, M itself will not usually realize the Poincaré embedding. After all, for M' homotopy equivalent to M, M Poincaré embeds in $M' \times I$, but will embed there iff M is h-cobordant to M'.

Let us now (anticlimactically?) sketch the classical proof of this theorem. We build an analogue of classical surgery. A normal invariant for a Poincaré embedding is a degree one map of a manifold pair realizing the various stable and unstable bundle data of the embedding. One identifies the aggregate of these with $[Y : F/Cat] \times [X : F_c/Cat_c]$, where BF_c and $BCat_c$ denote the classifying space of $c - 1$ dimensional spherical fibrations and c-dimensional Cat (block) bundles, respectively.

The surgery obstructions for this theory are just the product $L(Y) \times L(X)$. This is essentially the content of the Browder splitting theorem, which asserts that surgery can be done on a submanifold ambiently if it can be done at all. (The fact about surgery being possible on the ambient manifold is just $\pi - \pi$; the submanifold is too small to prevent any of the necessary surgeries.)

Consequently, the key issue is to see that $[X : F_c/Cat_c] \to [X : F/Cat]$ is an isomorphism for all X (a fact which was already sketched once in the exercises).

As in the case of manifolds, all we have to do is understand one situation well enough to get this homotopical information. In this case it is the Zeeman unknotting theorem (1.7) that saves us. Zeeman unknotting, for S^i in S^{i+c}, when interpreted in the present context, boils down to an isomorphism of homotopy groups $\pi_i(F_c/Cat_c) \to \pi_i(F/Cat)$ for Cat $=$ PL, Top and $c \geq 3$ (or if plugged into the original surgery exact sequence gives the isomorphism $\pi_i(F_c/Cat_c) \to L_i(e)$, which amounts to the same thing). This says that $F_c/Cat_c \to F/Cat$ is a weak homotopy equivalence, so that for any X, $[X : F_c/Cat_c] \to [X : F/Cat]$ is an isomorphism, and the proof is complete.

4.5. Extension of group actions

The material in this section is much more recent than any of the other results in this chapter. Nonetheless I include it because it is a pleasant geometric result whose statement involves no surgery or, seemingly, classification theory, but its proof, while surgical, involves no calculation. Also, because the details of homotopy theory of classifying spaces do not enter, the results are equally true in all categories. On the other hand, we are making real use of the generality implicit in allowing arbitrary Poincaré spaces as targets; our targets are constructed as homotopy pullbacks. The results here are, however, only the tip of an iceberg. The most recent survey of the developments that followed this material is the now hopelessly outdated [Wei2].

THEOREM (ASSADI-BROWDER (unpublished),[4] WEINBERGER [Wei9, pt. I] **(Extension across homology collars)**. *Let $(W; M, M')$ be a simply connected $\mathbb{Z}_{(g)}$-homology h-cobordism.*[5] *Suppose that a group G of order g acts freely on M and that the action is trivial (i.e. each element induces the identity) on $H_*(M; \mathbb{Z}[1/g])$. Then there is an extension of the action to such an action on W iff*

$$\sum(-1)^i\,[H_i(W, M; \mathbb{Z})] = 0 \in \tilde{K}_0(\mathbb{Z}G)$$

where we are following the exercise in 1.1 in viewing every finite module of order prime to G as giving an element of $\tilde{K}_0(\mathbb{Z}G)$. If an action exists, it is unique up to an element of $Wh(G)$.

(One can change an action by $Wh(G)$ by gluing an h-cobordism onto the quotient of one of the ends.)

We remind the reader that for cyclic groups the $\tilde{K}_0$ condition of the theorem is automatic. This theorem answers a problem in [J2].

REMARK. The philosophical context of this theorem is given by Smith theory, which deduces $\mathbb{Z}_{(g)}$-homological results from a group action. (See e.g. [Bre] for a textbook reference.) The above theorem is a first step in systematically deducing geometric converses. In that regard, and some others, it is like the h-cobordism theorem. The analogue of surgery in this program can be found in [CW4, DW]. The extension problem in the nonfree case is more complex. We refer the reader to [CW5] for a

[4]See also [AV] for a nonsimply connected extension.

[5]In other words, both inclusions are homology equivalences with coefficients in $\mathbb{Z}_{(g)}$, i.e. localized at the primes dividing g.

beginning. Hopefully the stratified surgery exact sequence will have more to say about this in the future.[6]

To prove the theorem, we merely have to construct a Poincaré pair (to be the putative quotient) (2.1), determine when it has a vanishing Wall obstruction (1.1), construct a normal invariant (2.3), do surgery (this is easy–the simple connectivity shows that we're in a $\pi - \pi$ situation) to obtain a manifold, and finally arrange that the universal cover of this manifold be W.

The first step, constructing the putative quotient, is a little delicate. It uses a functorial extension of the localization theory we described in 2.5. Basically, the quotient for W looks like W away from g and M/G at g, mixed together by the rational homotopy data given by the inclusion. (This construction only works nicely if the homological triviality is assumed.) (To build the other boundary piece for our pair, do the same for M'.) One checks that one gets a Poincaré pair.

The finiteness is dealt with using the exercises of 1.1.

The existence of normal invariants is guaranteed by general nonsense. One is trying to extend the normal invariant we already have on M/G. The obstruction to doing this can be computed (by localization theory) one prime at a time. For the primes dividing the order of G, the obstruction group is 0 because of our homology collar condition. For the others, one compares to W, which, being a manifold, has a normal invariant!

Finally, a little homological argument and the Atiyah-Hirzebruch spectral sequence show that $[W/M : F/Cat] \leftarrow [(W/G)/(M/G) : F/Cat]$ is an isomorphism. Since we are $\pi - \pi$, the transfer from structures on the quotient to structures on the manifold is an isomorphism, so we now have existence.

We also have uniqueness, but it is important to realize that we've been using h-surgery (2.4.A), so the uniqueness is only up to h-cobordism. Hence the $Wh(G)$ ambiguity.

The interested reader can consult the original source [Wei9, pt. I] for the details of all of these arguments.

4.6. Farrell fibering and Shaneson's formula

Farrell's theorem describes when a high dimensional manifold fibers over the circle.[7] We state informally the result [Fa1]:

[6] Actually, the characteristic class material of chapter 13 implies a major extension of the results of [CW5] for even order groups (where one is entitled to invert 2 by the condition that we are dealing with a homology collar).

[7] In dimension three, this question is answered by Stallings [Sta2]. Above dimension five, the case of simply connected fibers was analyzed by Browder and Levine [BLe]. For some remarks and counterexamples in the low dimensional case, see [Wei1]. In dimension five, the result is correct if one uses the notion of "approximate fibration" to replace fiber bundle. This is conceivably true even in dimension four.

THEOREM. *A map $M \to S^1$ is homotopic to a fibration iff the infinite cyclic cover associated to the map (pulling back $\mathbb{R} \to S^1$) is homotopy equivalent to a finite complex and an element of $Wh(\pi)$ vanishes.*

We sketch the proof from [Fa2]. One sees that the infinite cyclic cover has a tame end, and if the Siebenmann obstruction vanishes, one can produce a boundary, B, which one pushes into the interior. This should be a single fiber. Unfortunately, the covering translate may intersect B. (If not, the region between B and TB would be an h-cobordism, and if the torsion vanished, M would then be described as a mapping torus.) By compactness, there is an n such that both $T^n B \cap B = \emptyset$ and $T^{n+1} B \cap B = \emptyset$. One glues the regions bounded by B and $T^n B$ and by B and $T^{n+1} B$ together along B. Hopefully, as before, the resulting h-cobordism is trivial. One can now glue $T^{n+1} B$ to $T^n B$ and produce something h-cobordant to the original manifold. If enough Whitehead obstructions vanish, the original manifold can be fibered over S^1 with fiber B.

With more work involving the Bass-Heller-Swan splitting (see 4.5 below) one can identify the obstructions with the torsion of the homotopy equivalence to the space $\mathbb{R}^1 \times_T B$, i.e., the mapping torus of the self-homotopy equivalence of B obtained via the generator of the group of covering translations. (See also [Si5, Ra6].)

Shaneson's thesis[8] takes off from this starting point and yields a calculation of surgery groups. For simplicity we work topologically.

THEOREM. $\mathbf{S}^s(\mathbf{M} \times \mathbf{S}^1) \cong \mathbf{S}^s(M \times I \operatorname{rel} \partial) \times \mathbf{S}^h(M)$, *and similarly in L-theory,* $\mathbf{L}^s_{\mathbf{n+1}}(\mathbb{Z} \times \pi) \cong \mathbf{L}^s_{\mathbf{n+1}}(\pi) \times \mathbf{L}^h_{\mathbf{n}}(\pi)$.

In general, one gets a fibration for the structures on a manifold which fibers over S^1.

The map $\mathbf{S}^s(M \times S^1) \to \mathbf{S}^h(M)$ is obtained by looking at the fiber produced by Farrell's fibration theorem. (One can spacify Farrell's theorem, because it has a relative version for manifolds with boundary, whose boundaries are already fibered over the circle.) By $\times S^1$, there is a section of this map.

If this map to $\mathbf{S}^\mathbf{h}(\mathbf{M})$ vanishes on some element,[9] one cuts open the manifold along this fiber and glues on both boundary components the h-cobordism from this fiber to M, yielding an element in $S^s(M \times$

[8] The periodicity and functoriality were unavailable at the time, so more complicated arguments were employed. One first proved the result for L-groups using, basically, chapter 9 of [Wa1] and Farrell's thesis, and then one used the surgery exact sequence for computing. In fact, one of Shaneson's key ideas was avoiding low dimensional problems by using the periodicity of L-groups.

[9] I am confusing groups and spaces here. The reader should get used to this kind of going back and forth between spaces and elements of their homotopy groups. While perhaps confusing, it is prevalent.

I rel boundary). The reader can see that on gluing the ends together, one is not changing the ambient manifold.

These constructions together prove Shaneson's first result.

The result about L-groups follows by taking the fiber to the obvious splitting of normal invariants. (The splittings are compatible with each other.)

REMARK. Strictly speaking, there should be a dimension assumption (as such is necessary in surgery and in Farrell's theorem) in the theorem, but if we interpret structure sets algebraically, then as a consequence of the theorem for high dimensions (geometrically), one obtains it even in low dimensions. We will thus use Shaneson's formula uncritically in all dimensions ≥ 5.

EXERCISE. By functoriality, we have a map $\mathbf{S}^s(\mathbf{M} \times \mathbf{S}^1) \to \mathbf{S}^s_{n+1}(M) \cong \mathbf{S}^s(\mathbf{M} \times \mathbf{I}\,\mathrm{rel}\,\partial)$. Show that the composition $\mathbf{S}^h(\mathbf{M}) \to \mathbf{S}^s(\mathbf{M} \times \mathbf{S}^1) \to \mathbf{S}^s_{n+1}(\mathbf{M})$ can be nontrivial. (Hint: Work in L-theory and compute using the fact that by crossing with the 2-disk D^2 one obtains a way of killing the composite $\mathbf{L}^h(\mathbf{M}) \to \mathbf{L}^s(\mathbf{M} \times \mathbf{S}^1) \to \mathbf{L}^h_{n+1}(\mathbf{M})$ that does not work in $\mathbf{L}^s_{n+1}(\mathbf{M})$. Trace this back to the Tate cohomology term in the Rothenberg sequence.)

Now, for a torus, $Wh = 0$, so $\mathbf{S}^h(\mathbf{M}) \cong \mathbf{S}^s(\mathbf{M})$, and one reduces the calculation of $\mathbf{S}(\mathbf{T}^k \times \mathbf{D}^n)$, after unraveling the induction, to $\mathbf{S}(\mathbf{D}^{n+k}) = *$. Consequently, homotopy tori are tori if $n + k \geq 5$. The same argument applies to iterated circle bundles (nilmanifolds). We will discuss other aspherical manifolds in chapter 9.

EXERCISE. Show that the obstruction to block fibering over a torus or nilmanifold is governed by Wh obstructions.

EXERCISE (DUE TO RANICKI). Using the proper h-cobordism theorem (1.5.A) and proper surgery (2.4.B) prove the analogues of Shaneson's theorems $\mathbf{S}^h(\mathbf{M} \times \mathbf{S}^1) \cong \mathbf{S}^h(\mathbf{M} \times \mathbf{I}, \mathrm{rel}\,\partial) \times \mathbf{S}^p(\mathbf{M})$ and, similarly in L-theory, $\mathbf{L}^h_{n+1}(\mathbb{Z} \times \pi) \cong \mathbf{L}^h_{n+1}(\pi) \times \mathbf{L}^p_n(\pi)$.

This is a good place to briefly discuss some of the low dimensional points we avoided in discussing F/Top.

EXERCISE (See [Sh1]). Using Shaneson's formula for $\mathbf{L}$ and Rochlin's theorem (2.5.B), compute $S^{PL}(S^3 \times S^1 \times S^1)$.

EXERCISE (See [HsSh, Wa1]). By the same approach, compute $S^{P/PL}(T^n)$.

Using the double suspension theorem (DST) (cf. [Dvr]), which states that the double suspension of a manifold homology n-sphere is a manifold (and hence by GPC the sphere), one can do the following:

Apply the plumbing construction (Wall realization) to the 3-sphere to obtain a *homology* 3-sphere at the "other end". If we take the cone on the boundary components, we get a nonmanifold with just one singular point. However, the DST asserts that after $\times S^1$ this is a topological manifold. (Why?) Using this result do the following:

EXERCISE. Show that the two elements of $S^{PL}(S^3 \times S^1 \times S^1)$ coming from the action of $L_6(\mathbb{Z} \times \mathbb{Z})$ are topologically s-cobordant.

REMARK. As we've mentioned in 2.5.B, to be logically correct, one first calculates F/PL and through much work (the classification of PL tori is one of the ingredients) establishes many of the formal aspects of the topological category. After having done all that, one is still left with an ambiguity of whether Top/PL is contractible or $K(\mathbb{Z}_2, 3)$. The above example is approximately what was involved in settling this (local contractibility of homeomorphism spaces was used rather than DST, which was then only a conjecture). Nowadays one could use Freedman's work to directly produce a 4-manifold, rather than the 5-manifold obtained by crossing with a circle. For the complete development of the foundations of topological manifolds, see [KS].

4.6.A. *The Novikov conjecture*

In this appendix I want to give just an introduction to this important problem. We will on other occasions, especially chapters 9 and 14, return to it, and deepen our understanding. I have written elsewhere a more comprehensive survey of some aspects of this problem [Wei5].

We have seen (4.2) that for simply connected manifolds the characteristic classes of homotopy equivalent manifolds are entirely variable subject only to the Hirzebruch relation.

EXERCISE (MISCHENKO). Using the factorization of the surgery obstruction through the assembly map (3.4), show that the only possible restrictions on the rational characteristic classes of homotopy equivalent manifolds with fundamental group π are describable in terms of $f_*\big(L(M) \cap [M]\big) \in H_*(B\pi; \mathbb{Q})$ where $f : M \to B\pi$ classifies the fundamental group. (Do the same for $\Delta(M)$ in $KO[1/2]$.)

NOVIKOV CONJECTURE. *If $g : M' \to M$ is a homotopy equivalence, then*

$$f_*\big(L(M) \cap [M]\big) = f_* g_*\big(L(M') \cap [M']\big) \in H_*(B\pi; \mathbb{Q}).$$

This element is called the **higher signature** *of M because, by Hirzebruch's formula, when $\pi = e$, this is the usual signature.*

Now, in general it is not possible to expect the integral version of this class to be exactly homotopy invariant. The high dimensional lens spaces

($\times S^1$) are (simple) homotopy equivalent but sometimes have different images for higher signature class in $H_*(B\pi; \mathbb{Z})$. It is possible, however, that the only counterexamples to the integral version involve fundamental groups with torsion.

INTEGRAL NOVIKOV CONJECTURE. *If $g : M' \to M$ is a homotopy equivalence, and π is torsion free, then*

$$f_*(\Delta(M)) = f_*g_*(\Delta(M')) \in H_*(B\pi; \mathbf{L}^*(\mathbb{Z}))$$

where Δ is the class described in 3.3 *(and restricts to the signature operator away from* 2*.)*

EXERCISE. Show that the Novikov conjecture is equivalent to the rational injectivity of the L-theory assembly map. The integral injectivity is also equivalent to the integral Novikov, but the only proof I know of this equivalence relies on [Ws]. However, you should be able to show that integral Novikov follows from the injectivity of assembly.

COROLLARY. *The integral Novikov conjecture is true for the free abelian groups and for fundamental groups of nilmanifolds.*

We showed earlier that the assembly map is an isomorphism, as the structure spaces are contractible. In general, there is the following:

BOREL CONJECTURE. *If $f : M' \to M$ is a homotopy equivalence between aspherical manifolds that is a homeomorphism on the boundary, then f can be homotoped* rel ∂ *to a homeomorphism.*

For the state of the art on this problem, see [FJ1,2,3,4].

If we assume that f is a simple homotopy equivalence, the Borel conjecture would be equivalent to the assertion that the assembly map is an isomorphism. As stated, the conjecture also has the implication that

CONJECTURE. *If M is as above, then $Wh(\pi) = 0$.*

And again, one might expect that one doesn't really need a finite dimensional $K(\pi, 1)$ for these "isomorphism" or "vanishing" conjectures, but rather that for all torsion-free groups the assembly map is an isomorphism.

EXERCISE. Show that integral Novikov implies Borel stably, i.e., that after crossing with Euclidean space M' and M become homeomorphic. (Hint: Use $\pi - \pi$.)

The statement $Wh = 0$ follows from the components of the following conjecture, which is rather parallel to the Borel:

K-THEORY BOREL. *The algebraic K-theory assembly map*[10]

$$H_*(B\pi; \mathbf{K}(\mathbb{Z})) \to K_*(\mathbb{Z}\pi)$$

is an isomorphism for π torsion free.

The Bass-Heller-Swan formula (1.6), generalized to twisted Laurent extensions and higher algebraic K-theory by Quillen [Ql2], establishes this conjecture for the free abelian and fundamental group of nilmanifold cases.

Recently [BHM] have solved the K-theory Novikov (rationally) for all groups with finitely generated homology by a homotopy theoretic adaptation of an idea of [Cn] to Waldhausen's algebraic K-theory of spaces [Wald1]. (This asserts injectivity of the K-theory assembly map.)

In this setting, it is reasonable to replace $\mathbb{Z}$ (in both K and L settings) by other rings. We have seen that even for $\pi = \mathbb{Z}$ this cannot be correct because of the Nil terms in the BHS formula. It now seems possible that the Nils that arise for $\mathbb{Z}$ are the whole difficulty for all torsion-free groups, and that the fibers of the assemblies are functors of this Nil. For some evidence, see [FJ1, pt. II, ConS, G1], but I will not be more specific here.

One can extend this philosophy to other situations where there are assembly maps. A very powerful instance of this is where one uses the K-theory of C^*-algebras. $C^*\pi$ is a certain completion of the complex group ring of π. In the case where π is free abelian, this should be thought of as the space of continuous functions on the torus (via Fourier series). There is a natural map

$$K_*(B\pi) \to K_*(C^*\pi)$$

where one can, as before, make various injectivity and isomorphism conjectures. Actually the $C^*\pi$ version implies rational (in fact, $\mathbb{Z}[1/2]$) Novikov because there is a commutative diagram (defined using the spectral theorem):

$$\begin{array}{ccc} H_*(B\pi; \mathbf{L}(\mathbb{Z})) & \longrightarrow & L_*(\mathbb{Z}\pi) \\ \downarrow & & \downarrow \\ K_*(B\pi) & \longrightarrow & K_*(C^*\pi) \end{array}$$

where the left vertical is a $\mathbb{Z}[1/2]$ injection. Perhaps for this reason it is called in the literature the **strong Novikov conjecture**. (Note that it does not imply the integral Novikov conjecture!) It also has the virtue of applying to elliptic operators other than the signature operator, and enables

[10] Not defined here.

one to prove results parallel to Novikov in other geometric settings. For a prototypical example of this, see [Ros2].

Needless to say, there has been enormous activity on all of these problems. References for proofs of various cases of Novikov by various methods are [Lus] [HsSh] [Ca5,3] [FH2] [Kas2] [KaS] [FRW] [FeW1,2] [Car] [CM] [CoGM]. What happens for groups with torsion will have to wait for chapter 14....

The following are some (numbered) exercises (with references) related to the Novikov conjecture. I have to admit that these exercises are not at all routine (even while admitting that others in this book can also be quite difficult). Almost always the cited references contain additional information.

EXERCISE 1 (KAMINKER-MILLER). Using Ranicki's result on localization (3.4) prove a version of Novikov for rational homology equivalence. See [Wei9, pt. II] for how to deal with maps that are not degree one. (Or do it yourself!)

EXERCISE 2. Cappell [Ca1] proved that one can codimension one split a homotopy equivalence, modulo *Wh*, if the submanifold's fundamental group injects and is "square root closed"; i.e. any element whose square lies in the subgroup lies in the subgroup. Using this (and the technique of Shaneson's thesis), prove a Mayer-Vietoris sequence for L-groups of such amalgamated free products. Apply this to Borel. (See [Ca 5,4].)

REMARK. Waldhausen [Wald3] has proven such a Mayer-Vietoris sequence, modulo Nils, for the K-theory Borel. Cappell has found counterexamples to codimension one splitting when the square root closed condition does not hold. These are based on UNil groups, an analogue of Waldhausen's groups. These examples, and UNil phenomena in general, are among the most mysterious in topology.

EXERCISE 3 (LUSZTIG [LUS]). Let T be the dual torus to $\mathbb{Z}^n$; i.e. $T = \mathrm{Hom}(\mathbb{Z}^n, S^1)$. Thus T parametrizes the flat bundles over a manifold with free abelian fundamental group. If one twists the signature operator by these flat bundles and considers the index of this family [AS, pt. IV] to give an element of $K^*(T)$, then it captures the higher signature of M and it is homotopy invariant.

EXERCISE 4 (GROMOV-LAWSON [GL1]). One of the early applications of the index theorem was to show that $\langle \hat{A}(M), [M] \rangle = 0$ for a spin manifold with positive scalar curvature. (See [AS, pt. III].) The argument, due to Lichnerowicz, is based on the Bochner-Weitzenboch formula $D^*D = \Delta + \kappa/4$ where D is the Dirac operator, Δ is the Laplacian, and κ is the scalar curvature. Ind $D = \langle \hat{A}(M), [M] \rangle\, \kappa$ positive implies that D^* and D have no kernel, so the index vanishes. Apply Lusztig's argument to see that the higher $\hat{A}$-genus vanishes for spin manifolds of positive scalar curvature with free abelian fundamental group. The **Gromov-Lawson (-Rosenberg) conjecture** is the same statement in general and sometimes refers to a converse (for spin manifolds with torsion-free[11] fundamental group) as well.

EXERCISE 5 (CAPPELL [Ca5]). Show that if Novikov holds for a group π, it holds for any group containing π as a subgroup of finite index.

EXERCISE 6 ([Wei9, pt. II]). Suppose that G is a group that acts freely on M, $\pi_1 M \to \pi_1 M/G$ splits, and the action of G induced on twisted homology is trivial. Then Novikov implies that the higher signature of M vanishes. (Hint: Compute this signature directly and from exercise 1.)

In fact, this vanishing is equivalent to the Novikov conjecture.

REMARK. While one cannot see equivalence from the C^*-algebra version, more subtle formulae for the homologically trivial actions with fixed points or other singular data can be derived. This is based on extending a G-signature theorem proof of the simply connected case of the previous exercise. See [RsW5].

PROBLEM. Show that in addition if M is Kaehler and G acts holomorphically preserving the Kaehler class and the fundamental group is free abelian, then the **higher Todd genus** (or even the higher Todd polynomial; see [Hi]) vanishes.

EXERCISE 7 ([RsW3]). Show that if the Borel conjecture holds for ∂W, and (integral) Novikov for W, then any homotopy equivalence of pairs $(V, \partial V) \to (W, \partial W)$ induces an equality between higher signatures of V and W in the relative group homology. The same does not hold just for maps of pairs which are homotopy equivalences. (However, there are some invariance properties even for this more general class of maps: the most obvious of which follow from the fact that the double of such a map is a homotopy equivalence.)

In [Lo], p. 229, Lott considers a type of Novikov conjecture for manifolds with boundary of a special sort. These boundary conditions seem to give a homotopy invariant algebraic Poincaré nullcobordism for the boundary, so that one is lifted from a Novikov conjecture of pairs to an absolute one.

---

[11] Rosenberg has even suggested a converse for groups with torsion, but which is not always as explicit. However, he has worked out explicitly what it means for manifolds with finite fundamental group (and appropriate spin conditions), but it is far from verified. In recent work, he and Stoltz have verified, for spin manifolds with finite fundamental group, a stable version of this conjecture.

4.7. Homotopy lens spaces

The classification of manifolds with the homotopy type of lens spaces with odd order fundamental group was one of the early great triumphs of surgery theory; see [BPW] and [Wal]. In an incredible, complex tour de force, these have been classified in every dimension. By direct comparison, one then gets the following:

DESUSPENSION THEOREM. *For k odd, suspension (equivariant join with rotation on S^1) provides a 1-1 equivalence between free $\mathbb{Z}_k$ actions on spheres of consecutive odd dimensions ≥ 5.*

Here I will sketch a more direct proof of the desuspension theorem, which can then be used to prove the classification theorem. Unfortunately, we will have to invoke certain algebraic facts regarding the L-groups.

CLASSIFICATION THEOREM. *A homotopy lens space with odd order fundamental group is determined up to homeomorphism by its Reidemeister torsion τ and the ρ invariant defined below.*

Recall the Reidemeister torsion was defined in 1.2.A. Before defining the ρ invariant, let us recall the definition of the **G-signature**:

DEFINITION. *If G acts orientation preservingly on a manifold M^{4k}, then we define $G-\mathrm{sign}(M) \in RO(G) = [H_+] - [H_-]$ where $H_\pm$ are maximal $\pm$-definite G-invariant pieces of the middle dimensional intersection pairing. If $g \in G$, we sometimes write $\mathrm{sign}(g, M) = Tr(g|_{[H+]}) - Tr(g|_{[H-]})$, i.e. for the character of this virtual representation.*

There is a similar definition [AS, pt. III, p. 579] for dimensions $4k+2$, except that in these dimensions one obtains a totally imaginary character.

REMARK. Atiyah and Singer prove as a corollary of the **G-signature theorem** (which we will return to in chapter 13) that for G acting smoothly (but it's true more generally) $\mathrm{sign}(g, M)$ only depends on the action of g on a neighborhood of the fixed set M^g. In particular, if g has no fixed points, the character of the representation $G-\mathrm{sign}(M)$ vanishes on g.

REMARK. There is another point of view on the G-signature. One can view it as assigning a number to every representation, namely, the signature of M with coefficients in the associated flat bundle. The reader should check that these different versions contain exactly the same information. (This is just the relation between characters and representations, i.e. how the traces of the matrices representing the group elements acting on the vector space determine a representation.) For instance, the signature of a manifold with coefficients in an arbitrary flat bundle is just the dimension of the bundle multiplied by the signature of the manifold, which is equivalent to the characters vanishing for all $g \neq e$, which is, in turn, equivalent to G-sign being a multiple of the regular representation for the free G-manifold obtained by taking a finite cover.

EXERCISE. Give another proof by bordism theoretic (or assembly theoretic) means that for a free closed G-manifold, G-sign is a multiple of the regular representation. (Hint: Reexamine the remarks about surgery obstructions on closed manifolds in 3.4.) Gilmer has given a proof of the G-signature theorem by purely bordism theoretic arguments in [Gi].

This invariant is an equivariant bordism invariant, by the usual argument for ordinary signature, and just depends on a G-invariant $\pm$-symmetric inner product, and therefore defines an invariant of $L_{2i}(\mathbb{R}G, +)$. This invariant was studied by Wall (in [Wal] by a rather more complicated, equivalent, algebraic definition). He calls it the **multisignature**. We will see important connections between this invariant and classification theory, largely because of the following:

THEOREM. *For finite G, the L-groups $L_*(\mathbb{Z}G)$ are finitely generated and, except for 2-torsion, are detected by the multisignature.*

Note this includes the statement that the odd L-groups are 2-torsion.

EXERCISE. Show that this is not true for infinite groups, e.g. free abelian groups. (A little harder: Give a group with odd torsion in its L-theory.)

The ρ invariant[12] is defined as follows. If L is a homotopy lens space, then a bordism argument shows that for some n, $k^n L$ bounds some manifold W with fundamental group $\mathbb{Z}_k$. Remembering that one can take the signature of a manifold with boundary, just as for a closed manifold (we just have to mod out by the torsion in the possibly singular bilinear form), we define

$$\rho(L) = \frac{1}{k^n}\big(G\,\text{-sign}(W)\big) \in \frac{RO(\mathbb{Z}_k)}{im RO(e)} \otimes \mathbb{Z}\left[\frac{1}{k}\right].$$

This is well defined because of Novikov additivity: the signature of two manifolds glued together along their mutual boundary is the sum (or difference, depending on orientation conventions) of the signatures of the manifolds.

The difference of the G-signatures of the cobounding manifolds will be the G-signature of a closed manifold, which by Atiyah-Singer (in the smooth case) is a multiple of the regular representation.

EXERCISE. Identify $RO(\mathbb{Z}_k)/im RO(e) \otimes l[1/k] \cong \mathbb{Z}[1/k][\xi]$ where ξ is a primitive k-th root of unity.

[12]It was essentially by a calculation of this invariant that [AB] showed that linear lens spaces are not h-cobordant (unless they're linearly diffeomorphic).

EXERCISE. Show that ρ can be defined for arbitrary odd dimensional manifolds with fundamental group G but that one might have to relax the ring to (in, say, dimension $4k-1$) $RO(G)/im RO(e) \times \mathbb{Q}$.

To prove Wall's theorem, we will rely on some algebraic facts, which we'll state as we go along, and some exercises.

FACTS. $Wh(\mathbb{Z}_k)$ is torsion free. The involution on it is trivial. L^s is torsion free except for the Kervaire element coming from $L_2(e)$.

EXERCISE. By modifying the argument given above for complex projective space, show that every quotient of the sphere by a free $\mathbb{Z}_k$-action is homotopy equivalent to a lens space. There are $\varphi(k)$ (the Euler φ-function) distinct homotopy types.

EXERCISE. Show that the Reidemeister torsions of homotopy lens spaces that are not homotopy equivalent cannot be equal. (Hint: Look at the formula for the linear case in 1.2 and recall that the torsion of a homotopy equivalence is the ratio of the Reidemeister torsions and is also in the image of $\mathbb{Z}[\xi]^*$.)

EXERCISE. Show that suspension induces a 1-1 correspondence between simple homotopy types of polyhedral homotopy lens spaces.

Now we can concentrate our attention on $S^{Top}(L)$. (In chapter 13 we will give another approach based on equivariant Bott periodicity [A1].) We can now see why desuspension is pretty surprising; just consider the surgery exact sequence

$$L_{2i}(\mathbb{Z}_k) \to S^{Top,s}(L) \to [L : F/Top] \to L_{2i-1}(\mathbb{Z}_k).$$

One quickly sees from the definition of ρ and the detection of L-groups by multisignature (aside from torsion, which we've been told vanishes here)[13] that aside from k torsion $S^{Top,s}(L)$ is detected by ρ.

But what about the k torsion? The normal invariant group which is k torsion is growing in size with dimension. (For k odd, the number of summands is fixed, but the exponent grows.)

What is happening is that the extension is not split and the ρ invariant is still detecting. Furthermore, the range of the ρ invariant is increasing (more denominators creep in) as normally cobordant homotopy lens spaces suspend into ones that are not normally cobordant!

Here is a way to reduce this phenomenon just to the algebra of L-groups. (This is fairly easy given the algebraic state of the art.) We

[13] Exercise: Deal with the Kervaire invariant (using a Kervaire manifold).

prove desuspension, and then we desuspend down to (virtual) dimension three,[14] where the normal invariant set vanishes, and we're all done!

EXERCISE. Use periodicity to eliminate the seeming violation of the dimension restrictions in surgery.

We have discussed the S^1 equivariant decomposition of odd dimensional spheres in 4.1. This gives rise to a decomposition of lens spaces

$$L^{2i+1} = E(\xi \downarrow L^{2i-1}) \cup S^1 \times D^{2i}.$$

Here ξ is the flat D^2 bundle associated to the representation of the fundamental group on the circle of rotation (that we're joining with). Now using the splitting exercise from 4.3, one can split along the $S^1 \times S^{2i-1}$. This leads to a fibration

$$S(S^1 \times D^{2i}, \text{rel } \partial) \to S(L^{2i+1}) \to S\big(E(\xi \downarrow L^{2i-1}) \text{ not rel } \partial\big).$$

We have seen in 4.6 that the fiber is contractible. Thus we have to deal with E. This falls nicely into a commutative diagram associated geometrically by pulling back the disk bundle:

$$\begin{array}{ccccccc} L_{2i}(\mathbb{Z}_k) & \longrightarrow & S^{Top,s}(L^{2i-1}) & \longrightarrow & [L : F/Top] & \longrightarrow & L_{2i-1}(\mathbb{Z}_k) \\ \downarrow & & \downarrow & & \downarrow \cong & & \downarrow \\ L_{2i+2}(\mathbb{Z}_k, \mathbb{Z}) & \longrightarrow & S^{Top,s}(E) & \longrightarrow & [E : F/Top] & \longrightarrow & L_{2i+1}(\mathbb{Z}_k, \mathbb{Z}) \end{array}$$

The arrow labeled an isomorphism is because $E \to L$ is a homotopy equivalence. We want to see the isomorphism of structure sets, but this follows from a purely algebraic fact:

FACT. The "L-theory transfers" above are isomorphisms for k odd.

EXPLANATION/EXERCISE. The L-theory for finite groups is a complicated business, but for the most part the ingredients of the calculations are assembled from the rational representation theory of the group. The real and quaternionic pieces behave differently in different dimensions. The unitary pieces behave the same in dimensions of the same parity. For a cyclic group of odd order, everything except the trivial representation is complex, so most of the periodicity is forced. The trivial piece is different in different dimensions ($\mathbb{Z}_2$ or $\mathbb{Z}$). Here the reader should check that the $\mathbb{Z}$ saves the day via Shaneson's formula from the previous section.

[14]That is, we do not have to consider the geometric meaning of the three dimensional object at all; we just consider it algebraically as part of an algebraic calculation of the structure set of the higher dimensional lens space. Alternatively, using Freedman's work, one can geometrically interpret the three-dimensional structure set as the topological homology s-cobordism classes of homology 3-spheres with free cyclic action. See [FrQ].

PROBLEM. For k even, there is exactly one other real representation. Show that desuspension fails. (You will rediscover the **Browder-Livesay invariant** in the process.) This leads to a rather different classification. See [Wal] or [LdM] for the case of $\mathbb{R}P^n$.

One can combine these calculations. One sees that the relative group $S(\mathbb{R}P^{2i-1}, L^{2i-1})$, where one maps $L^{2i-1} \to \mathbb{R}P^{2i-1}$ degree one, has periodicity. Thus there are the invariants of projective spaces (which are not periodic at all) and then the ρ and τ as for the odd order case.

4.7.A. Eta invariants

The ρ invariant studied in this section is closely related to Atiyah-Patodi-Singer's [APS, esp. pt. II] invariant for the signature operator. I will here briefly review this theory.

[APS] was an attempt to analyze the failure of Hirzebruch's signature formula for manifolds with boundary.[15] Recall that the (Chern's) Gauss-Bonnet formula

$$\chi(W) = \int_W \text{Euler form}$$

remains valid for manifolds with boundary if ∂W is Riemannian collared. This is not true of the Hirzebruch signature formula. Moreover, since signature is not multiplicative in coverings for manifolds with boundary, the error term cannot just be a local expression integrated over the ∂.

EXERCISE. Verify (using cobordisms between homotopy lens spaces!) that signature is not multiplicative in coverings for manifolds with boundary.

The remarkable theorem of [APS] is that the error is an explicit spectral invariant of the boundary.

EXERCISE. Use Novikov additivity to show that the error is an invariant of the boundary.

On the space of all exterior differential forms of even degree on an odd dimensional manifold V^{2l-1}, define the self-adjoint operator B by

$$B\varphi = i^l(-1)^{l+1}(*d - d*)\varphi$$

where $\deg \varphi = 2r$, $*$ is the Hodge operator, and d is exterior differentiation. The eigenvalues are then real, and they define

$$\eta(s) = \sum \text{sign}(\lambda)/\lambda^s$$

[15] They do, however, put their theory in the wider context of studying indices of operators with an appropriate global boundary condition.

where we sum excluding 0. We define $\eta(Y) = \eta(0)$ for this operator (after analytic continuation of what is shown to be a meromorphic function regular at 0!).

ATIYAH-PATODI-SINGER INDEX THEOREM. *Sign(W)* $= \int_W L + \eta(\partial W)$ *for L the differential form representative of the Hirzebruch class.*

We can do the same thing for signature coupled to a flat bundle. If we take the difference between this construction for a flat bundle α and that for a trivial bundle of the same dimension k, the characteristic classes wash out. Better yet, for an arbitrary Y we can define

$$\tilde{\eta}_\alpha(Y) = \eta_\alpha(Y) - k\eta(Y)$$

even if Y is not a boundary (or the flat bundle does not extend to a coboundary). This defines an invariant of the smooth manifold Y.

Atiyah-Patodi-Singer observe that by varying α one can use this invariant to distinguish lens spaces from one another. In particular, it is not a homotopy invariant.

The following exercises develop some of the role of η in the classification of manifolds.

EXERCISE (ATIYAH-PATODI-SINGER). Show that for (the holonomy of) α factoring through a finite group, $\tilde{\eta}_\alpha(Y)$ is rational. Show that for S^1 and the representation of $\mathbb{Z}$ by multiplication by θ one gets $1 - 2\theta$.

EXERCISE. Relate $\tilde{\eta}_\alpha(Y)$ as α varies to $\rho(Y)$ if the fundamental group is finite.

EXERCISE. Show that manifolds with finite fundamental group within a simple homotopy type are determined up to finite ambiguity by their rational L-classes and by their reduced etas.

Remember, though, that the simple type cannot in general be determined by intrinsic invariants (1.2.A).

I have recently shown that the deviation $\tilde{\eta}_\alpha(Y') - \tilde{\eta}_\alpha(Y)$ is rational for homotopy equivalent manifolds with arbitrary fundamental groups (and arbitrary representations). We will see later many examples of torsion-free groups for which $\tilde{\eta}_\alpha$ is actually a homotopy invariant (12.3.A). However, in the opposite direction, there is the following:

EXERCISE ([Wei6]). Show that if Γ is a residually finite or virtually torsion-free group with nontrivial torsion, then there are (simple) homotopy equivalent manifolds with fundamental group Γ and a representation α (flat bundle) of Γ for which $\tilde{\eta}_\alpha$ distinguishes these manifolds.

Eta will return in Part III when we look at signature operators on singular spaces. Indeed, we will study eta for manifolds via singular spaces.

4.8. The space form problem

I will now give a very rapid sketch (making use of the more functorial approach that we've developed to do some key steps) of one of the high points in surgery theory: the spherical space form problem. The problem is to classify those finite groups which can act freely[16] on a sphere. The classic criterion for doing this by orthogonal transformations is:

THEOREM. *A finite group π acts freely and linearly on some sphere iff all subgroups of order pq, where p and q are not necessarily distinct primes, are cyclic.*

See [Wo] for a proof. The first result in this direction topologically is due to P.A. Smith:

THEOREM. *If a finite group π acts freely on some sphere, then all subgroups of order p^2 are cyclic.*

Here's the idea. We have to eliminate $\mathbb{Z}_p \times \mathbb{Z}_p$. Without loss of generality, we can assume that we're on a rather high dimensional sphere. The quotient manifold is a $K(\mathbb{Z}_p \times \mathbb{Z}_p, 1)$ through a high dimension, so we can compute its cohomology, contradicting Poincaré duality.

EXERCISE. Show that the only finite abelian fundamental groups of 3-manifolds are cyclic.

Conversely, Swan [Sw] proved:

THEOREM. *If π is a finite group for which all subgroups of order p^2 are cyclic, then π acts freely on some finite complex homotopy equivalent to the sphere.*

The proof is by construction of an appropriate chain complex model for the quotient, and then using the ideas from 1.1 to geometrically realize it. In contrast to this Milnor [Mi4] gave a beautiful simple argument for:

THEOREM. *No dihedral group acts freely on the sphere.*

Ronnie Lee [Le] gave another, more algebraic proof of this (that also led to more precise results in other directions) and subsequently Jim Davis [Da1] found an elegant surgery theoretic argument, using calculations of $L(\mathbb{F}_2[D_{2p}])$ and the fact that if one had a manifold all surgery obstructions would be in the image of the assembly map (3.4). The contradiction is deduced by comparing the transfer in L-theory to that in homology.

It turns out that for our original question, Milnor's condition is the last word.

[16] A more refined version that I won't get to is to classify the manifolds whose universal cover is the sphere. See [DM] for a useful survey.

THEOREM (MADSEN-THOMAS-WALL [MTW]). *A group acts freely on some sphere iff all subgroups of order p^2 and $2p$ are cyclic.*

The proof is an application to surgery of the induction theory of A. Dress [Dr], which is itself a variation (or generalization) of the classical induction theory of representations of finite groups (see e.g. [Se2]).[17] We will outline an argument using a subsequent extension of this theory due to Nicas [Ni]. (His work, in turn, has been put into a useful general framework by [HmTW].)

Suppose we have a surjection $\Gamma \to \pi$ where π is a finite group. The theory relates the structures (and L-theory) of a manifold with fundamental group Γ to those associated to an appropriately thick set of finite covers corresponding to subgroups of π.

THEOREM. *An element of the structure set is trivial, localized at p, iff the transfer to the covers associated to p-hyperelementary subgroups of π is trivial at p. The element is rationally trivial iff its transfers to the covers corresponding to cyclic subgroups are rationally trivial.*

Recall that a group is p-hyperelementary iff it is a p-group extension of a cyclic group.

EXERCISE/PROBLEM. Use this, Bieberbach's theorem that a flat manifold has a finite cover which is a torus, and the fact that the Borel conjecture is known for Poly-$\mathbb{Z}$ groups (nilmanifolds; see 4.6.A) to deduce the rational L-theory Borel. (See [FH3].) Do the same (this is harder) using Brauer induction theory in representation theory to establish the C^*-algebra "rational Borel conjecture" for these groups.

At this point the method of proof can be described as follows. One finds a Poincaré complex (as Swan had) S/π whose 2-hyperelementary subgroups are homotopy equivalent to quotients of linear representations. (This uses the Milnor conditions as well.) The obstruction to being homotopy equivalent to a manifold is a component of delooped structure space[18](by the total surgery obstruction idea discussed at the end of 3.3). By assumption, at 2, the obstruction vanishes. Away from 2, thinking in terms of the surgery exact sequence, one sees the difficulty is just that of finding a normal invariant. However, this can be checked one prime at a time and therefore by transfer to an appropriate cyclic subgroup (see [Ad3] for why this transfer is injective), where the problem is trivial: that cover is a homotopy lens space (by an exercise in the previous section)!

[17]The relation between these is apparent on consideration of the multisignature, which links the theories.

[18]The reader can argue more classically with a surgery exact sequence and discussing induction on the structure set term.

PART II: THE GENERAL THEORY OF STRATIFIED SPACES

Having developed the classification theory of topological manifolds in Part I, we develop here the theory of stratified spaces. That is, we aim to solve the question, given two naturally occurring stratified spaces, when are they homeomorphic?

Our choice of central problem to some extent dictates the category of stratified spaces that we work in. The commonly used Whitney stratified spaces are not very suitable, because they include much additional structure that is not topologically invariant. (One could insist on homeomorphisms preserving that extra structure. With one interpretation this yields the Browder-Quinn theory, which we'll develop in chapter 7. In another interpretation, this project is well outside the capabilities of topology at this point in time.)

The idea of a Whitney stratification is that one has a filtration of the space X into pieces, called strata, such that differences between consecutive strata are open manifolds, and that each of these consecutive differences has a neighborhood given as a fiber bundle. Moreover, these neighborhoods are demanded to fit together nicely.

A great part of the difficulty involved in trying to classify the Whitney stratified spaces is that the structure groups of fiber bundles with manifold fibers are unknown for any fiber of dimension greater than three. In addition, this kind of structure is not very suitable for many applications. Often one would want the structure group to be restricted to a subgroup, with the reduction as part of the data: consider what one would need to view smooth embeddings as stratified spaces so that differentiably equivalent embeddings correspond to equivalent stratified spaces.

In the polyhedral category, it turns out that the Whitney idea is, in fact, more reasonable. With the conceptually slight difference of substituting block bundles for fiber bundles, "Whitney stratifications" (which will be called "PL stratified spaces" in the text) are PL invariant and lead to useful PL classification theories. The answer to the analogous question we asked about embeddings in the PL case would be that these sets are in a 1-1 correspondence. (The implications of this will be investigated in chapter 11.) The classification theory for the PL category is given in chapter 8 and is considerably simpler than the topological theory to establish.

In the topological setting, we shall use a (slight extension of a) class of spaces introduced by Quinn. What is most remarkable, perhaps, about these spaces is that one only assumes that the stratification is homogenous in some weak homotopical sense: i.e. that "normal links" at different points on a component of (pure) stratum are the same. Nonetheless, these are (most often) genuinely homogenous, have useful isotopy extension lemmas, and can be effectively classified. And, as in the PL case, this classification corresponds to interesting classifications when restricted to natural sources of stratified spaces.

The proofs in the topological category are based on a wonderful elaboration of topology to the bounded (and controlled) setting(s). What this means is that one imposes metrics on all topological spaces and tries to keep track of the "sizes" of various constructions, perhaps when measured in some auxiliary space. This idea is one of the most fruitful ones of topology, which is, unfortunately, nowhere adequately exposed.[1] (There are many fine research papers explaining one controlled problem or another or applying such theorems, but the big picture is yet to have its convenient exposition.) I have given a quick and dirty presentation in chapter 9, which is adequate for the purposes of this book, but the subject certainly deserves more attention than I could have given it. However, I do hope that the rough presentation I have given achieves two goals: that it encourages readers to try to work their way through the literature in this difficult and extremely rewarding area, and that it renders the following chapter, which completes the topological classification theory, readable.

The reader interested in only the applications presented in Part III can glance at chapter 5 for definitions and read chapter 6, which presents the classification theorems. Chapter 7 is devoted to the theory of Browder and Quinn, which is well adapted to (say) Whitney stratified spaces if one only tries to classify stratified spaces where the bundle data that glue the strata together are totally pinned down. It is a beautifully simple theory, and critical for what follows, but its defect is that too much of the stratification information is fixed in advance. The PL case is treated in the following chapter, and all but the most dedicated reader could stop there. Despite this, chapter 9 would be a mistake to miss, and indeed can be read immediately by the reader not particularly interested in stratified spaces! In this book, the material in chapter 9 is only used for the explanation of the proofs for the topological category in chapter 10 and in our discussion of rigidity.

---

[1]Steve Ferry is in the process of rectifying this with a forthcoming book and CBMS lecture notes.

5 Definitions and Examples

The definitions in this chapter are, of course, critical. The contents of the sections are self-explanatory. The point of 5.2 is that one needs some sort of transversality for many constructions that involve (self-) duality, intersection theory, and the like, but that one has some degree of choice in defining this critical notion. In some sense, transversality and its variants are at the heart of all of our work.

5.1. Stratified spaces

There are several different, but analogous, types of stratified spaces that we shall deal with. We shall define several classes:

$$\{PL \text{ stratified spaces}\}$$
$$\cap$$
$$\{PL \text{ weakly stratified spaces}\}$$
$$\cap$$
$$\{\text{manifold stratified spaces}\}$$
$$\cap$$
$$\{\text{stratified Poincaré spaces}\}$$
$$\cap$$
$$\{\text{stratified homotopy type}\}$$

Of the most intrinsic interest are the PL and manifold stratified categories, but the others have their uses.

At the very least, to begin we need the idea of a **filtered space**. X is said to be a filtered space (on a finite partially ordered indexing set S) if one has a closed subset X_s for each $s \in S$, so that $s \leq s'$ implies that $X_s \subset X_{s'}$. One also assumes that the inclusions $X_s \subset X_{s'}$ are cofibrations. The X_s are called the (closed) **strata** of X. A difference of the form $X_s - \cup_{t<s} X_t$ will be called a **pure stratum**. We will denote this pure stratum by X^s.

A **filtered map** between filtered spaces (filtered using the same indexing set) is a continuous function $f : X \to Y$ such that $f(X_s) \subset Y_s$.

Filtered spaces and maps form a category, and there are obvious notions of **filtered homotopy** in this category. For instance, if the indexing set $S = \{0, 1\}$ with $0 < 1$, then the category of S-filtered spaces is the usual notion of the category of pairs. Besides $S = \{0, 1, 2, \ldots, n\}$, other useful indexing sets are conjugacy classes of subgroups of a group G and the set of faces of the n-simplex.

A map $f : X \to Y$ will be said to be **stratified** if $f(X^s) \subset Y^s$. This also leads to a notion of homotopy, and therefore homotopy equivalence. In general it is much harder to deal with the homotopy theory of stratified maps than it is to deal with the filtered category. Nonetheless, all of the categories we will use demand that the morphisms be stratified; this seems more natural geometrically.

EXERCISE. Show that every object in the filtered category for $S = \{0, 1\}$ is filtered homotopy equivalent to one where X_0 has a neighborhood stratified homeomorphic to $X_0 \times [0, 1)$. Show that for such "boundary collared" X and Y, we have

$$[X, Y]_{\text{filtered}} = [X, Y]_{\text{stratified}},$$

where $[X, Y]$ means homotopy classes of maps.

From our point of view, this exercise explains why it was possible to obtain reasonable results on manifold pairs in Part I.

Another good example to keep in mind comes from embeddings. Given an embedding, one constructs a stratified space by viewing the submanifold as the bottom stratum and the whole ambient manifold as the next (top) stratum. Two embeddings of N in M give stratified spaces in the same filtered homotopy type[1] iff they are homotopic. They have the same stratified homotopy type iff they represent the same Poincaré embedding (see 4.4), and hence iff (assuming codimension at least three) they are isotopic. Concretely, it is quite simple to give embeddings of $S^a \cup S^b$ in S^c in which the components are linked, but all such are null-homotopic as maps.

REMARK. This example illustrates another important phenomenon that is not yet adequately understood. If c is larger than $2\max(a, b) + 1$, then all of the embeddings are isotopic, and the filtered theory coincides with the stratified homotopy theory. Presumably this demarcates the border between stable and unstable homotopy theory. It seems that "gap hypotheses" can be used to analyze stratified theory quite effectively in many situations. See the beginning of chapter 13 for some more discussion.

[1]Strictly speaking this is only correct if we assume that the filtered homotopy equivalence is homotopic to the identity as map $M \to M$ (forgetting stratifications).

In general it is very hard to deal geometrically with filtered spaces, and we must assume some form of homogeneity condition to make progress.

DEFINITION. *A filtered space X is a* **PL stratified space** *if all the X_s are polyhedra and for any two points x, y in a component of some pure stratum there is a PL isotopy $f_t : X \to X$, $f_0 = id$, and $f_1(x) = y$.*

It is automatic that one can choose the isotopy to be supported in the neighborhood of an arc connecting x and y within their pure stratum. A simple consequence of the definition is that the pure strata are open manifolds.

We leave to the reader the slight changes necessary to allow manifolds with boundary as "pure strata" without changing the indexing set.

Actually, for our purposes it is convenient to have a slightly weaker notion available.

DEFINITION. *A filtered space X is a* **PL weakly stratified space** *if all the X_s are polyhedra and for any two points x, y in a component of pure stratum there is a topological isotopy $f_t : X \to X$, $f_0 = id$, and $f_1(x) = y$.*

This is actually not very hard to check, as we shall soon see. The principal advantage is a technical one. While there is a reasonable classification theory for PL weakly stratified spaces, up to concordance (h-cobordism), within a stratified homotopy type, the theory for PL stratified spaces is more awkward. One requires an extra algebraic K-theoretic assumption on the local nature of the stratified homotopy equivalence implicit in setting up the classification problem.[2] Nonetheless, in many interesting cases, this condition on the map does hold, and one can then ignore the PL weakly stratified category.

The prototypical example of the phenomenon is the following. Let $(W; M, M')$ be a PL h-cobordism. Let X be obtained as the suspension of $W \cup cM \cup cM'$. If W were a product then one could PL stratify X with a circle (the suspension of the union of the cone points) as a stratum. In a PL-stratified space this would not be possible if the h-cobordism had nontrivial torsion, as the reader can check. However, we still could weakly stratify this example, as we shall see: the circle inside this polyhedron is topologically homogenous but not PL homogenous. (This can be seen by using the fact that for any h-cobordism $W - M \cong M' \times [0, 1)$, as shown in 1.3, and which also follows immediately from the proper h-cobordism theorem (1.5.A) or from an "Eilenberg swindle" (see 5.3 below).)

One could derive from the material of the next chapter an obstruction theory to concording a weakly stratified PL space to a stratified one.

[2] I.e. in surgery theory we classify manifolds simple homotopy equivalent to a given one and in the stratified theory one wants stratified simple homotopy equivalences, but one would be forced to add an additional local simplicity condition on the map as well.

In a slightly less general setting this was considered in [CW1,5], where "Rothenberg classes" were defined that describe the obstruction.

At this point, we should emphasize that our classification theory is not stratification independent; the stratification is part of the data. This will at times be useful, but at other times artificial (we cannot directly study when polyhedra are homeomorphic, for instance). For this reason, we introduce the idea of a **coarsening** of a stratification. A stratified space V with strata indexed by S is **coarsened** by a stratified space V' with strata indexed by T and maps $a : S \Rightarrow T$ and $f : V \to V'$ so that V_s is mapped into $V'_{a(s)}$, and the map $V_{a^{-1}(t)} \to V'_t$ is an equivalence. For instance, to consider a manifold with a distinguished submanifold as a stratified space with two strata, one can coarsen this stratified space so that it consists of just the ambient manifold, viewed as a space with just one stratum.

We now turn to the topological manifold analogue of these definitions, due to Quinn. Of course, in the topological case there is no need to make subtle distinctions between the types of isotopies allowed. Actually, Quinn discovered that one can assume some local homotopical conditions that prima facie seem much weaker than homogeneity but actually imply all the isotopy extension principles one could want. Consequently, a PL weakly stratified space is simply a topological homotopically stratified space, where all strata have PL structures. We will not re-prove all of Quinn's results, but they serve to enable one to think of these spaces as being essentially not so different from PL stratified spaces. The main difference, as we will see, is the lack of existence and uniqueness of regular neighborhoods.

DEFINITION. *A subspace $A \subset X$ is called* **tame** *if there is a neighborhood N of A and a one-parameter family of functions $f_t : N \to X$ that are the identity on A and on all of X for $t = 0$, and that for $t \neq 1$ are a stratified map of the pair $(N, A) \to (X, A)$, but $f_1(N) \subset A$.*

PROPOSITION ([Q3, §2.5]). *A is tame iff A is locally tame.*

Examples of tame embeddings are given by PL embeddings. Observe that the definition is topological in nature, and according to the proposition, local. A good example to keep in mind is the one point compactification of a noncompact manifold. The inclusion of ∞ is tame iff the end is tame in the sense of Siebenmann (see 1.4). One can verify tameness by local fundamental group conditions (i.e. conditions about the ends of small deleted neighborhoods of points in A) together with local homology conditions.

REMARK. We also sometimes need "reverse tameness" for certain constructions. That means that $X - A$ can be pulled outside neighborhoods of A. If all local fundamental groups are finitely generated, then tameness is equivalent to reverse tameness. See [Hu3]. We assume this in all of our examples and theorems.

The following definition is a homotopical analogue of the deleted regular neighborhood of $A \subset X$.

DEFINITION. *The* **homotopy link**, *abbreviated* **holink**, *is given as the space of stratified maps* $(I, 0) \to (X, A)$, $I = [0, 1]$.

This is really semilocal (i.e. local around A) since one can shrink the size of any path to lie within a prescribed neighborhood. For us, it provides a homotopical shadow of the geometry of a neighborhood of a stratum. For instance, for the cone cP, the holink of the cone point is homotopy equivalent to P.

DEFINITION. *A filtered space X is* **homotopically stratified** *if whenever $s \le t$, $X^s \subset (X^s \cup X^t)$ is tame and the natural map holink $(X^s \subset (X^s \cup X^t)) \to X^s$ is a fibration.*

This should be thought of as saying that there is a local type of homogeneity that holds for X. Quinn shows that homotopy versions of all of the compatibility conditions one often assumes in other types of stratified spaces are a consequence of this compatibility just for pairs of strata.

It is sometimes more useful to think about the homotopical stratification condition like this. Let S be a finite index set. Let $o \in S$ be a minimal element. An **S-stratified homotopy type** is inductively a pair of T-homotopically stratified homotopy types, where $T = S - o$, (A, B) with a T fibration (i.e. the homotopical structure group involves only T-stratified maps) $p : B \to X_o$. (Notice that this map does not lie in the category!) The stratified space corresponds to $A \cup_B Cyl(p)$ (the mapping cylinder of p). It is obviously an S filtered space and it satisfies Quinn's conditions. Conversely, a filtered space is homotopically stratified iff it is stratified homotopy equivalent to a stratified homotopy type. We will sometimes refer to B (as a T-stratified homotopy type) as the **(stratified) holink** of X_0, and the (T-stratified) fiber as the **local (stratified) holink**. (If we are not on a minimal stratum, just remove all lower strata.) It is the analogue of the PL notion of the link of the highest codimensional face of a pure stratum.

A **manifold stratified space** is a homotopically stratified space for which all pure strata are manifolds. A **manifold stratified space with boundary** is the obvious thing, with the additional condition that if one glues on an open collar to the boundary, the resulting object is still a manifold stratified space. (This again has a local homological interpretation.) This condition implies that the boundary is collared.

REMARK. A PL filtered space is a PL weakly stratified space if all pure strata are manifolds and it is stratified homotopy equivalent to a stratified homotopy type. This can be proven by a type of engulfing or by the isotopy theorems in [Q3], such as:

THEOREM. *If p and q are points in a single component of a pure stratum of a manifold stratified space, X, then there is a stratified isotopy $X \times I \to X$ from the identity to a homeomorphism moving p to q. (The isotopy can be taken to be supported in any neighborhood of an arc connecting p and q in the given pure stratum.)*

The inductive approach also enables one to formulate the homotopical ideas of Poincaré duality. Of an **S-stratified Poincaré space**, one demands that the minimal strata are Poincaré complexes in the conventional sense (2.1), that (A, B) is a T-Poincaré pair, and that the local holinks are T-Poincaré. (The last actually implies that B is T-Poincaré.) A standard example for $S = \{0, 1\}$ would be a Poincaré embedding or a Poincaré pair.

We leave to the dutiful reader the notion of stratified Poincaré pair.

In the next chapter we will study which stratified Poincaré spaces are stratified homotopy equivalent to manifold (or PL weakly) stratified spaces, and when a stratified homotopy equivalence between two manifold (or PL stratified) stratified spaces is stratified homotopic to an equivalence.

REMARK. We shall sometimes be interested in a variant of manifold stratified spaces wherein all pure strata are ANR homology manifolds (see 9.4.D). These should be called homology manifold stratified spaces (of course).

5.2. Transversalities

Corresponding to the different types of spaces considered in the previous section there are different types of transversality conditions that one can impose on a map.

Suppose first that one has a PL stratified space. Then around each stratum one has the structure of a stratified block bundle with fiber the (geometric) stratified link (stratified homotopy equivalent to the stratified local holink). The proof is similar to that for the existence of block bundle structures for submanifolds of a PL manifold; see [RS2] or [Sto] for a complete proof.

Knowing this, define a map $f : X \to Y$ to be **stratified transverse** (sometimes the adjectives are used in the reverse order) if f is a PL map, stratified, and f defines a PL homeomorphism between the boundary of a regular neighborhood of each pure stratum in X and the pullback of the corresponding stratified block bundle over the target stratum in Y.

This is the notion that Browder and Quinn use in [BQ] to set up a stratified surgery theory. We will review their theory in chapter 7 because it is very elegant and is foundational for our main results. This same notion is used in [FM] and [GM1] and is called "normal smoothness".

REMARK/EXERCISE. This condition is valuable for defining various types of functorialities. For instance, show that such a map induces wrong-way maps in homology and cohomology. (Remember that PL stratified spaces, as defined in the previous section, are automatically manifold stratified.)

The notion that is of more interest to us is **homotopy transverse**. It is best defined for stratified homotopy types. Observe that a stratified map $f : X \to Y$ induces a map on all (stratified) holinks of all (pure) strata. One can now insist that over each pure stratum the map from the holink to the transverse homotopy pullback of the holink fibration over the target is a stratified homotopy equivalence.

The significance of this notion is underscored in the following simple result of supreme ideological importance:

THEOREM. *A stratified homotopy equivalence is automatically homotopy transverse.*

PROOF. This is true almost by definition,[3] because stratified maps take the system of deleted neighborhoods in any stratum touching a point to the corresponding neighborhood system in the target space. By using a map and its homotopy inverse, one sets up an equivalence between the holink fibrations in domain and range.

For instance, in a stratified map between cones of manifolds, the link of one vertex is mapped into a deleted neighborhood of the other vertex, which deform retracts to *its* link. If the map were a stratified homotopy equivalence, this process would quickly give a homotopy equivalence of links and hence a homotopy transverse map between the cones.

REMARK. Similar discussions appear in [CW1,2, Q10].

Unfortunately, in the PL case, this will not suffice if we are interested in notions that enable an isomorphism classification; it is necessary, for the reasons discussed in 2.4, to have simple homotopy equivalences. Happily, it is possible to make sense of a stronger notion of homotopy equivalence that gives the type of **simple stratified transversality** we need. Moreover, a PL stratified map homotopic to a PL homeomorphism is automatically one of these.

---

[3]Indeed, a stratified homotopy equivalence sets up an equivalence in the category of stratified homotopy types discussed in the previous section, and part of the data of this homotopy category is the collection of holink fibrations.

First we must define in the PL category the notion of a **closed pure stratum**. We define $\overline{X}^s$ as the complement in X_s of the interior of a regular neighborhood of the union of the lower strata. It is well defined up to PL homeomorphism. Now the condition of simple homotopy transversality on a map already known to be homotopy transverse is that the homotopy equivalence from each closed pure stratum of each local link to the corresponding closed pure stratum of the target local link is a simple homotopy equivalence.

EXERCISE. Show that simple homotopy transversality is invariant under stratified PL homotopies of maps.

5.3. Examples

The main difference between the PL and topological categories is that in the former there are links of simplices and there are regular neighborhoods, while in the latter these do not exist. In an appendix to chapter 10, we shall describe the local structure of the neighborhoods of a minimal stratum in a manifold stratified space. In this section I would like to give some examples of stratified spaces and some phenomena that occur in their study.

Examples of intrinsic mathematical interest were mentioned in the Introduction and will be the subject of Part III. The examples to follow are of pedagogical interest as prototypes of certain behavior.

EXAMPLE 1. A one-point compactification of a noncompact manifold is a manifold stratified space iff the noncompact manifold is tame. It has the structure of a PL stratified space iff the complement has a PL structure and can be compactified as a manifold with boundary; i.e. the Siebenmann obstruction of the end vanishes (1.4).

If this exists, the PL structure need not be unique. One can modify the boundary structure by gluing in an h-cobordism (1.5). (This is, incidentally, a variant of Milnor's original examples of counterexamples to the hauptvermutung [Mi5].)

EXAMPLE 2. There are nontrivial PL h-cobordisms that are not topologically products. This will show a difference in simple homotopy theory in different categories. Let (W, M) be a manifold with boundary with $\pi_1 M \to \pi_1 W$ an isomorphism. Erect an h-cobordism on W relative to M. Glue onto $M \times I$ a copy of $cM \times I$. One easily sees that this is not a product in the piecewise linear sense.

However, it is topologically a product. Consider figure 14.

The product structure between two copies of M near the I (which is the rightmost vertical line in fig. 14) can be made to contain the negative of the torsion of the original h-cobordism. Consequently the regions bounded by the broken lines are products! These successive copies

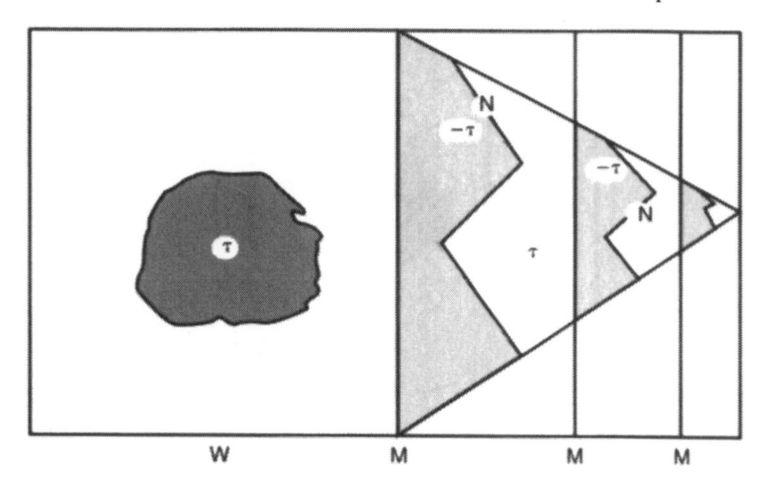

Figure 14. Eilenberg swindle.

of τ and $-\tau$ are constructed to be smaller and smaller. The sizes of the homeomorphisms produced by the h-cobordism theorem are consequently getting smaller themselves as we approach I, so the process converges.

This is our first example of how we can do geometric constructions (here a quite simple one) with control on the sizes of the various modifications to get information about the topological category. We will see many more. Chapters 9 and 10 study this systematically.

EXAMPLE 3. A more interesting example of how invariants of closed pure strata can be killed on gluing in lower strata is this. We will look at the analogue of a coning example, but involving a circle.

Consider an h-cobordism on the manifold with boundary $M \times I \times S^1$. Now one glues onto the bottom boundary component $cM \times I \times S^1$. Are these all homeomorphic? Certainly all of these have homeomorphic pure strata. (Why?)

The answer (due to Steinberger [St] and Quinn [Q3], and which also follows from Anderson and Hsiang [AH2]) involves the Bass, Heller, and Swan formula [BHS] (mentioned in 1.6):

$$Wh(\mathbb{Z} \times \pi) \cong Wh(\pi) \times \tilde{K}_0(\pi) \times \text{ Nils.}$$

In terms of h-cobordisms, there is a transfer map induced by taking k-fold cyclic covers. The k-transfer is multiplication by k on the $Wh(\pi)$ piece (h-cobordisms which are supported near a copy of M), the identity on the $\tilde{K}_0$ piece, and, for each element of Nil, there is a k which kills it. We know how to kill the $Wh(\pi)$ piece using an infinite process (which converges at a point on the circle). There is a more subtle infinite process, pictured in

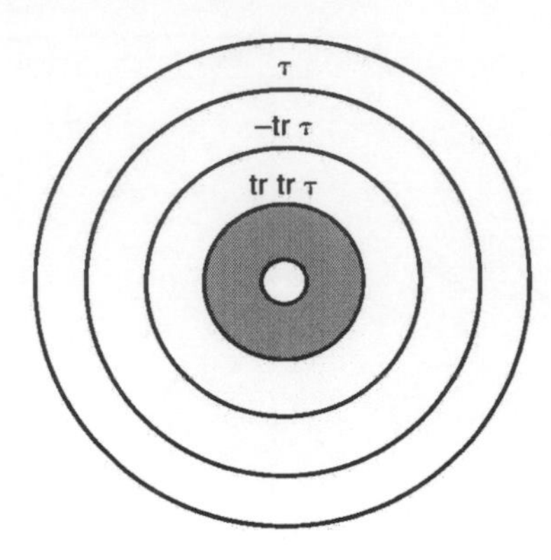

Figure 15. Transfer invariant Eilenberg swindle.

Figure 15, using transfer invariance that kills the $\tilde{K}_0$ piece. (This figure is from [RtW], where it is called the transfer invariant Alexander lemma.)

There are an infinite number of layers in the shaded region. Transfer invariance makes neighboring regions cancel. With a little care, the resulting homeomorphism converges on the innermost circle.

The reason for the convergence is that each transfer of an h-cobordism makes the twisting of the handle structure closer and closer to preserving the projection to S^1.

The issue remains: Are the Nil pieces topological invariants? The answer is yes. The most conceptually direct way to see this is via the "controlled end theory" discussed in chapter 9.

For now, we shall be content in having seen that obstructions involving the K-theory of the interior can "leak off" the noncompact part over the lower strata, and that there are infinite processes that make use of points, circles, tori, etc. (and, by some toral geometry, intervals, 2-disks, gluing) that dig deeper and deeper into the K-theory of fiber germ (= local holink). (There are similar $\mathbb{Z} \times \pi$ formulae for the negative K-groups; in fact, Bass defines them to make the Bass-Heller-Swan formula true.) Analogous phenomena arise in L-theory. The main results regarding "leaking" consist geometrically in the assertion that these infinite constructions generate all phenomena, and in some sort of organization of all of these possibilities. Of course, it will be stated in a more systematic algebraic fashion.

EXAMPLE 4. As mentioned, because of the $L(\mathbb{Z} \times G)$ formula, the same type of leaking displayed in the previous examples also occurs in L-theory (although not in a particularly more striking fashion topologically than in PL). The first naturally occurring examples of this phenomenon were examples by Cappell and Shaneson of linear representations of cyclic groups which are topologically conjugate, although linearly (and PL) distinct [CS3].

Another interesting example that follows from the method of example 2 is the following: Schultz has produced differentiable $\mathbb{Z}_p$ actions on the sphere with the 2-sphere as fixed set, and for which the normal bundle has (with a certain natural almost complex structure) nontrivial Chern classes [Scz2]. These actions are all topologically equivalent to the linear action that has trivial Chern classes.

5.4. Notes

There are many different notions of stratified spaces in the literature. It seems that most of the examples outside the topological literature are at least as refined as PL stratified, and the examples studied within the topological literature seem to be included within the class of manifold stratified spaces (or, as I feel most convenient, the homology manifold stratified spaces). Siebenmann [Si4] introduced and studied another interesting category of stratified topological spaces, which have local cone-like structure. These are included in the manifold stratified category.

We shall discuss the role of transversality again in chapters 7 and 12.

Examples like the first two were first discovered by Milnor. Milnor actually used Mazur's thesis (see the exercises in 1.5 and 1.7) to produce homeomorphisms between the one point compactifications of different lens spaces × Euclidean space. Stallings formalized the Eilenberg swindle in a very pretty paper [Sta1]. One can equally well use engulfing.

The transfer invariant Eilenberg swindle (Alexander lemma) was motivated by many uses in the literature of using transfers to gain control. A notable use is [FH1], where certain rigidity theorems are proven by this technique. In [RtW] this was done to have a more explicit hold on how PL nonisomorphic polyhedra become homeomorphic, so that we could make the homeomorphisms Lipschitz and apply analytic techniques to the topological category (see [RsW1]).

The nonlinear similarities discussed in example 4 were a major impetus in the development of much of what has been learned about topological group actions in the last decade. (For instance, the papers [HsP] [MR] [Q3] [RtW] [St], as well as the theory presented here, provide developments of the general theory that were designed to shed light on the existence and nonexistence of such examples.) The classification of nonlinear similarities still remains a valuable test case of machinery developed in this subject.

6 Classification of Stratified Spaces

This chapter will state the main results of the general theory. Because of the lack of homogeneity in the spaces we are classifying, it is necessary to make use of (generalized) homology theories whose nature changes from point to point. This is formalized in the notion of a cosheaf of spectra. (Spectra give rise to generalized (co)homology theories. With sheaves, we take cohomology; cosheaves, the dual notion, permit the construction of homology. We have already seen that structure sets are naturally covariant functors, so that homology is the correct notion.)

While this is a fairly abstruse language, the objects produced are not actually that exotic. We shall discuss this idea briefly in an appendix. Very often, the types of homology that enter can be fairly explicitly computed, as we will see in Part III.

The PL and topological classifications, while quite similar, have different peculiarities. The PL classification has some additional codimensional conditions that the topological case does not have because of low dimensional considerations. (In both cases one must avoid low dimensional strata, but in the PL case, low codimensional ones cause trouble as well because of the existence of links.[1] In the topological case, the impossibility of such constructions and the consequent nonuniqueness of links – even when they exist – allow the possibility of complete classification!) Also, there is a bump in the theory because of Rochlin's theorem (or equivalently, the nontrivial k-invariant at 2 for F/PL); see 2.5.B.

The topological case is a bit unusual in that the structure sets are not fibers of assembly maps. This is perturbed at the prime 2 by a requirement of first stabilizing and then destabilizing. This is related to the obvious notions of controlled surgery (in the sense of chapter 9) not being calculable as homology theories.

Both theories are plagued by a formal ugliness: the failure of Siebenmann periodicity (3.4) to be an isomorphism. Consequently, there

[1]More precisely, since there is a PL construction of links, they are well defined as PL objects.

are some additional $\mathbb{Z}$'s, one for each component of pure stratum, that change the theory. In short, the usual structure sets inject into "periodicitized structure sets" which have a nice form, and they correspond to elements where certain $\mathbb{Z}$ obstructions vanish.

Rather than correct for these difficulties, we will classify *homology manifold* stratified spaces up to s-cobordism. The difference between these and manifold stratified spaces is determined by a collection of "resolution obstructions" (see 9.4). Thus, there is a classification theorem for a more general class of objects, and the classification of more conventional objects in terms of them must be obtained by examining where some particular invariants vanish. Those who are just interested in the manifold case will almost never have much difficulty pulling this out of the general classification.

6.1. *PL* Classification

We start with the Whitehead theory. If X is a PL stratified space, we define $Wh^{BQ}(X) = \bigoplus Wh(\overline{X}^s)$. Since PL stratified homotopy equivalences preserve closed pure strata, one can define for a PL stratified homotopy equivalence $f : Y \to X$ a torsion $\tau(f) \in Wh(X)$. An induction on strata enables one to prove the following:

h-COBORDISM THEOREM. *Let X be a PL stratified space. Then*

$$\begin{aligned}\tau : \{&h\text{-cobordisms with one boundary component } \cong X \text{ that}\\ &\text{are products on strata of dimension } \leq 4\}\\ &\leftrightarrow Wh^{BQ}(X, \text{ rel strata of dimension } \leq 4\}.\end{aligned}$$

There are similar extensions of all the theorems in chapter 1. It is interesting, though, to observe that the involution on $Wh^{BQ}(X)$ obtained by turning h-cobordisms upside down does not necessarily preserve the product decomposition. It is not difficult to see why. If one has a stratified h-cobordism, then one can map two obvious subsets into the closed pure stratum of the largest space. One can map in the corresponding closed pure stratum of the bottom space, or alternatively one can map in the union of this set with the boundary of a regular neighborhood of the h-cobordism on the lower strata (see fig. 16).

These torsions differ by the "transfer" (to be made a bit more clear later) of the torsion of the h-cobordism on the lower strata. The Milnor duality formula holds verbatim (with respect to a naive $\oplus$ involution) if on one side the decomposition of the Whitehead group is given in terms of one of these decompositions, and on the other it is the other.

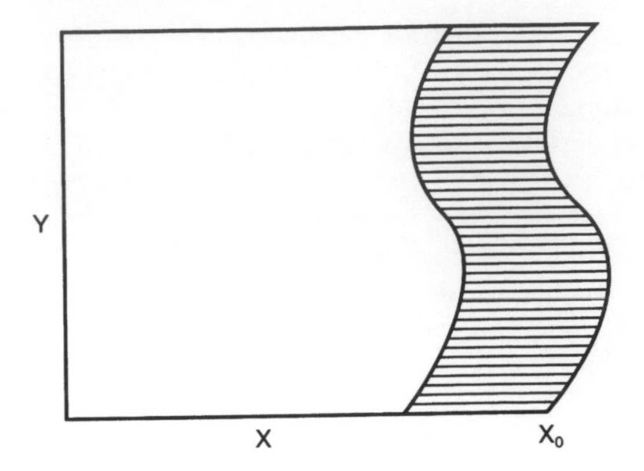

Figure 16. Two decompositions of WH^{BQ}. One can include either just x^{-i} or x^{-i} with the boundary of the striped region into y^{-i}.

EXERCISE. Show that if we view a manifold with boundary as a two-stratum stratified space, and one is in the $\pi - \pi$ case, then the Tate cohomology vanishes, $H^*(\mathbb{Z}_2; Wh^{BQ}(X)) = 0$.

Transfer is a very important construction. (A special case of this appeared in 4.7.) If one has a certain type of geometric problem (say trivializing an h-cobordism) and a fiber bundle or a block bundle or even a stratified system of fibrations (the reader can supply the definition) and if the problem would be automatically solved if solved downstairs, then one can often define a "transfer" from the obstruction group of the base to that of the total space. All one needs formally is a realization theorem for the obstruction group: each element must arise for some problem associated with that given base. One realizes, lifts to the total space, and computes there.

We warn the reader that this is only loosely related to what algebraic topologists call transfer, which has to do with "wrong way maps." For the geometrician, the question of whether or not a map is "wrong-way" is often a matter of taste. In a homological version of surgery the "transfer" on normal invariants is indeed "wrong way," but in the more old-fashioned cohomological form (which is the only form one knows for the smooth category, for instance), it is simply the induced map on cohomology.

There is a large literature on giving purely algebraic definitions of different transfers. It is quite important to isolate to what extent the detailed geometry of the bundle is (ir)relevant. A classic instance of this is the issue of computing the signature of a fiber bundle [A2]; in that case, arbitrary bundles over a simply connected space have as signature the product of the signatures of base and fiber, while there are "monodromy corrections" for nonsimply connected bases.

For our purposes, we just shall assume that we have a transfer. For the spectra that arise here that have geometric interpretations, one obtains an infinite loop space map realizing the transfer when needed.

For the L-theory, we define **stratified L-groups**, sometimes called **Browder-Quinn L-groups**, inductively as follows. They will be homotopy groups of certain spectra. If X_0 is a minimal stratum, and we form the closed complement $cl(X - X_0)$, its boundary $\partial\, cl(X - X_0)$ is a stratified block bundle over X. Now we let

$$\begin{aligned}&\mathbf{L}^{\mathbf{BQ}}(X) := \text{ homotopy fiber of the composite,}\\ &\mathbf{L}(X) \to \mathbf{L}^{\mathbf{BQ}}(\partial\, cl(X - X_0)) \to B\mathbf{L}^{\mathbf{BQ}}(cl(X - X_0) \operatorname{rel} \partial),\end{aligned}$$

where the first map is induced by transfer, and the second by inclusion.[2] In other words, the L-groups fit into a long exact sequence and are built up out of the L-groups of the various closed pure strata. While this is similar to the situation for the Wh theory, it is not the case that $\mathbf{L}^{\mathbf{BQ}}(X)$ breaks up into a product of conventional L-spectra.

EXERCISE. Give an example of $\mathbf{L}^{\mathbf{BQ}}(X)$ not breaking up into L-spectra of the strata. (Hint: Consider manifolds with boundary.)

THEOREM. *If $f : X \to Y$ is*

(a) *a simple homotopy transverse map, then there is an induced map $\mathbf{L}^{\mathbf{BQ}}(X) \to \mathbf{L}^{\mathbf{BQ}}(Y)$, or*

(b) *the inclusion of a closed (impure) stratum, then there is an induced restriction map $\mathbf{L}^{\mathbf{BQ}}(Y) \to \mathbf{L}^{\mathbf{BQ}}(X)$, or*

(c) *the projection map of a stratified system of fibrations, then there is an induced map $\mathbf{L}^{\mathbf{BQ}}(Y) \to \mathbf{L}^{\mathbf{BQ}}(X)$.*

These functorialities are critical for conceptual analyses of these functors.

PROPOSITION. *In case (b) above, the fiber is $\mathbf{L}^{\mathbf{BQ}}(cl(Y - X) \operatorname{rel} \partial)$.*

This is immediate from the definition for a minimal stratum. All of these results will be somewhat clearer after they recur in the next chapter in a more geometric guise. Only case (a) of the theorem demands any comment. It is proven by observing that the $\mathbf{L}^{\mathbf{BQ}}$ definition makes sense in the stratified Poincaré category, by induction.

Note also that we have been working with simple L-groups. One can also work with other decorations and change the type of transversality

[2] Note that the L-spectra are indexed here by dimensions of spaces. The map $\mathbf{L}^{\mathbf{BQ}}(\partial\, cl(X - X_0)) \to B\mathbf{L}^{\mathbf{BQ}}(cl(X - X_0) \operatorname{rel} \partial)$ is then the delooping of the map induced by the codimension zero inclusion $\partial\, cl(X - X^0) \times I \subset cl(X - X^0)$. Multiplication by I has the effect of looping the L-spectrum. Recall, too, that all L-spectra are their own fourth-loopspaces, so there is no difficulty in delooping any maps.

accordingly. For instance one can use the L^h version to classify PL weakly stratified spaces within a stratified homotopy type up to concordance.

Let X be a stratified space. For each open subset U one has $\mathbf{L}^{\mathbf{BQ}}(U \operatorname{rel} \infty)$.[3] If $U \subset V$, then there is an induced corestriction map $\mathbf{L}^{\mathbf{BQ}}(U \operatorname{rel} \infty) \to \mathbf{L}^{\mathbf{BQ}}(V \operatorname{rel} \infty)$. This leads one to see an L-cosheaf on X (see the appendix). We will denote this cosheaf as $\underline{\mathbf{L}}^{\mathbf{BQ}}$. As mentioned in the introduction to this chapter (again, see the appendix), one can then take homology of such a gadget.

CLASSIFICATION THEOREM. *Let X be a PL stratified space with no four dimensional strata and no neighboring strata whose dimension differs by less than five. Then there is a fibration for computing $S^{PL}(X) =$ {simple homotopy transverse simple homotopy equivalences $Y \to X\}/PL$ homeomorphism:*

$$\mathbf{S}^{\mathbf{PL}}(X) \to \mathbf{H}_0(X, \underline{\mathbf{L}}^{BQ}) \to \mathbf{L}^{\mathbf{BQ}}(X) \times \bigoplus [\mathbf{H}_{i-4}(X_i; \mathbb{Z}_2) \times \mathbb{Z}].$$

(For simplicity, we have labeled $\dim X_i$ *by i.)*

There is a similar existence theorem, where obstructions to existence, given a stratified simple Poincaré complex, lie in the component set of the delooping of $\mathbf{S}^{PL}(X)$.

The codimension five restrictions are a nuisance and can often be removed at the cost of complicating the sequence some more. They mask our ignorance of low dimensional topology and the failure of low dimensional h-cobordism theorems. An example of where this is possible is the case of manifolds with boundary, which certainly violates the codimension condition! Also, one can certainly deal with situations where these bad situations occur, if one works relative to them.

The $\bigoplus \mathbf{H}_{i-4}(X_i; \mathbb{Z}_2)$ is basically a bunch of Kirby-Siebenmann (triangulation of manifold) obstructions of pure strata. A homology calculation shows that these classes can actually be made to lie in the homology group of the closed stratum, due to the codimension assumption.

The $\mathbb{Z}$'s are a reflection of the difference between G/Top and $\mathbf{L}(e)$. The map from $\mathbf{H}_0(X, \underline{\mathbf{L}}^{BQ})$ into the i-th $\mathbb{Z}$ can be computed as the result of applying a restriction of the cosheaf to the i-th pure stratum, and there restricting further to any small copy of $\mathbb{R}^i$. In the topological case, these $\mathbb{Z}$'s can be removed at the cost of allowing homology manifold stratified spaces, as mentioned above.

We will refer to the homology term as the stratified normal invariant set. Certain simple homotopy transverse maps with bundle data give

[3]For noncompact spaces we usually work with surgery groups and spaces that are rel ∞. This means that we consider the inverse limit of $\mathbf{L}^{\mathbf{BQ}}(U \operatorname{rel} U - K)$ where K ranges through compact sets. If the end is tame, then this reduces to a conventional L-space.

rise to elements of this group.[4] This then makes it possible to define characteristic classes for arbitrary stratified spaces. It is necessary first to introduce coefficients into the theory.

DEFINITION. $\mathbf{L}^{\mathbf{BQ}}(X; \mathbb{Z}[1/2])$ *is obtained using the above definitions using* $\mathbb{Z}[1/2]\pi$ *instead of* $\mathbb{Z}\pi$ *throughout. One can extend this to produce a similar cosheaf of spectra.*

DEFINITION. $\Delta(X) \in H_0(X, \underline{\mathbf{L}}^{BQ}(\mathbb{Z}[1/2]))$, *called the* **signature class** *of* X, *is the class associated to the degree two surgery problem* $X \cup X \to X$.

If X is an oriented manifold and one inverts 2 and identifies the cosheaf homology with $KO[1/2]$-homology via the characteristic variety theorem, one obtains the Sullivan orientation or, if X is smooth (or topological), the class of the (Teleman) signature operator. Other cases where one can interpret this class will appear in Part III.

REMARKS. 1) Using Weiss's visible L-theory [Ws] one can define this class in a related homology group that is quite useful at the prime 2. One can also rephrase the whole classification theory in these terms. The main thing lost is that L-theory will no longer be a functor of fundamental groups, so we won't consider this extension here.

2) Using the signature class, one can describe the normal invariant for a map as a difference of intrinsic invariants, away from the prime 2.

3) Finally, we remark that the PL simple structure set is functorial with respect to simple homotopy transverse stratified maps.

6.2. Topological classification

In this section we would like to explain how to calculate the structure set of a manifold stratified set, i.e. the (homology) manifold stratified spaces simple homotopy equivalent to a given X, up to homeomorphism (or s-cobordism).

To begin with, we need to understand the simple homotopy theory of such spaces. To motivate the h-cobordism theorem, realize that there is one nontrivial stratified space for which we have already described the h-cobordism theorem, namely one point compactifications of noncompact manifolds with tame ends. In that case, $Wh^{Top}W^+ = Wh^pW$. Actually there is a nontrivial point here. The map $Wh^{Top}W^+ \to Wh^pW$ is fairly clear. To build the map the other way, one point compactify a proper h-cobordism and glue collars on the boundaries (see fig. 17).

This produces a map from the proper Whitehead group to the topological Whitehead group of the one point compactification. One needs

[4] As usual we will interchange spaces and groups; the group is usually a homotopy group of the space (which is almost always a spectrum).

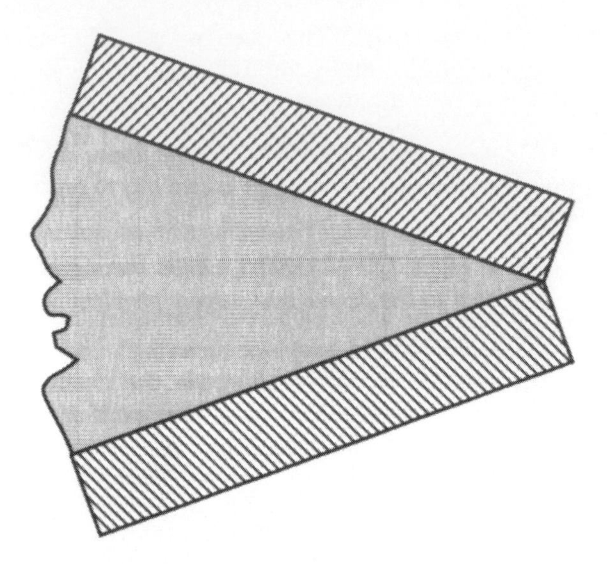

Figure 17. Singular h-cobordism from proper ones. One point compactify an h-cobordism and glue collars on the boundaries.

an engulfing argument to show that this map is onto. Trusting this, we have

$$\begin{gathered} Wh(\mathbb{Z}\pi_1(\epsilon)) \to Wh(\mathbb{Z}\pi_1(W)) \to \\ Wh^{Top}W^+ \to \tilde{K}_0(\mathbb{Z}\pi_1(\epsilon)) \to \tilde{K}_0(\mathbb{Z}\pi_1(W)) \end{gathered}$$

We have seen in chapter 5 that the topologically invariant part of torsion involves modding out by terms involving the lower K theories of holinks that can be coupled with cells of appropriate codimensions. This is very suggestive of a spectral sequence for computing a generalized homology theory.[5] In addition, we are used to, by now, viewing long exact sequences as homotopy groups of a fibration. Consequently, we should build a spectrum whose negative homotopy groups are the negative K-groups. We will do this in chapter 9 geometrically. Recall from the previous section the notation $Wh^{BQ}(X) = \bigoplus Wh(\bar{X}^s)$, which we extend to all K-groups and spectra. As in note 3, we mean the Whitehead group

[5]In orbifold cases one could also get the essential ingredients of a topological s-cobordism theorem by combining the fact that (1) triangulable things have the Browder-Quinn type of h-cobordism theorem, (2) there is an equivariant triangulation theory [LR2] (parallel to the theory sketched for manifolds in an exercise in 2.4), and (3) there is a stratified triangulation theory [AH2] that involves bundle lifting (on a bottom, closed manifold stratum), which is cohomological, and is expressible in terms of lower algebraic K-theory (and, therefore, by Poincaré duality, capable of being described homologically).

of the noncompact manifold X^s rel ∞. Supposing this, one is led to

$$\mathbf{Wh}^{Top} X \to \mathbf{H}(X; \tilde{\mathbf{K}}_0^{BQ}(\text{neighborhood})) \to \tilde{\mathbf{K}}_0^{BQ}(X).$$

The spectrum $\mathbf{K}_0^{BQ}$ is a nonconnective delooping of a sum of algebraic K-spectra,[6] so that we can get all of the negative K-groups of links arising. In this sequence, the coefficient spectrum changes from point to point. This is only to be expected, because the nature of the space also so changes. It is remarkable that this wasn't necessary in the PL case! (That was due, ultimately, to the existence of closed regular neighborhoods, which form a barrier to "leaking" of obstructions from one stratum onto another.) Notice that the middle homology theory is entirely supported on the singular set (at least as far as homotopy groups in dimension ≤ 0, which is all that is relevant for us).

THEOREM ([Q2, St]). *$\pi_0\mathbf{Wh}^{Top} X$ is in a 1-1 correspondence with h-cobordisms on X provided that all strata of X are of dimension ≥ 5.*

Note that there is now a beautiful parallelism between surgery and Whitehead theory that was not apparent in the nonsingular case: $\mathbf{Wh}^{Top} X$ is now also the fiber of an assembly map. Actually, the same is true in an interesting way in the nonsingular case when one pays appropriate attention to higher homotopy groups: this is Waldhausen's parametrized version of the s-cobordism theorem [Wald1]. Also, we begin to realize that the parallel of the Whitehead group is not the surgery group but the structure set!

Unfortunately, the surgery theory is not quite so simple. We need two fibrations, one for a "stable calculation" and one for destabilizing. This is related to the difficulties described in 2.4.B in trying to describe proper surgery in terms of absolute groups.

First, some formal setup (closely paralleling discussion in the previous section):

For the L-theory, it is necessary to describe stabilization first and then stratification. We define $L^{-\infty}(\mathbb{Z}\pi)$ as the direct limit of the transfer invariant part of $L(\mathbb{Z}[\mathbb{Z}^i \times \pi])$. We now define **stratified L-groups**, or **Browder-Quinn L-groups**, inductively as follows for a stratified Poincaré space. They will be homotopy groups of certain spectra. If X_0 is a minimal stratum, we form the closed complement[7] $cl(X - X_0)$ (this is explicitly part of the definition) and $\partial cl(X - X_0)$, a stratified fibration over X. Now

[6] For group rings, this is the delooping presented in [PW1].

[7] Note that after crossing with tori, as is implicit in forming $L^{-\infty}$, it is possible to form closed complements in the topological case as a consequence of Siebenmann's end theorem (1.4 and 1.6).

we let

$$\mathbf{L}^{\mathrm{BQ},-\infty}(X) := \text{ homotopy fiber of the composite,}$$

$$\mathbf{L}^{-\infty}(X) \to \mathbf{L}^{\mathrm{BQ},-\infty}(\partial cl(X - X_0)) \to B\mathbf{L}^{\mathrm{BQ},-\infty}(cl(X - X_0) \operatorname{rel} \partial)$$

where we are using stable L-spectra throughout. In other words, the L-groups fit into a long exact sequence and are built up out of the L-groups of the various closed pure strata.

THEOREM. *If $f : X \to Y$ is*

(a) *a homotopy transverse map, then there is an induced map $\mathbf{L}^{\mathrm{BQ},-\infty}(X) \to \mathbf{L}^{\mathrm{BQ},-\infty}(Y)$, or*

(b) *the inclusion of a closed (impure) stratum, then there is an induced restriction map $\mathbf{L}^{\mathrm{BQ},-\infty}(Y) \to \mathbf{L}^{\mathrm{BQ},-\infty}(X)$, and the fiber is $\mathbf{L}^{\mathrm{BQ},-\infty}(Y - X \operatorname{rel} \infty)$, or*

(c) *the projection map of a stratified system of fibrations, then there is an induced map $\mathbf{L}^{\mathrm{BQ},-\infty}(Y) \to \mathbf{L}^{\mathrm{BQ},-\infty}(X)$.*

The only difference between this and the analogous theorem in the previous section is that there are no closed strata here, and one can work only with the stable spectra, because the unstable ones simply do not exist!

Now we can describe our analysis of $\mathbf{S}^{\mathrm{Top}}(X)$. The $-\infty$ in the structure part of the next theorem is parallel to its meaning in the L-group. It is the limit of various transfer invariant structure sets.

STABLE CLASSIFICATION THEOREM. *Let X be a manifold stratified space with no four dimensional strata. Then there is a fibration for computing $S^{Top}(X) = \{$Topologically simple homotopy equivalences $Y \to X\}/$homeomorphism:*

$$\mathbf{S}^{\mathrm{Top},-\infty}(X) \to \mathbf{H}_0(X, \underline{\mathbf{L}}^{\mathrm{BQ},-\infty}) \to \mathbf{L}^{\mathrm{BQ},-\infty}(X) \times \bigoplus \mathbb{Z}.$$

If one works with homology manifold stratified spaces up to s-cobordism, the sequence is a fibration without the $\bigoplus \mathbb{Z}$.

We already described in §6.1 how to compute the map to $\bigoplus \mathbb{Z}$.

We will refer to the homology term as the stable stratified normal invariant. Sometimes we will neglect to emphasize the stability, but it will always be tacit, unless otherwise stated. Homotopy transverse maps with appropriate "bundle data" give rise to elements of this homology group. However, we remind the reader that for the stratified spaces considered here there isn't quite a conventional normal bundle of pure stratum in the ambient space.

As before, one can introduce a characteristic class for stratified spaces using the theory with coefficients (or visible theory). An important consequence of defining it in the topological theory is the topological invariance of this class.

DEFINITION. $\Delta(X) \in H_0(X, \underline{\mathbf{L}}^{\mathbf{BQ},-\infty}(\mathbb{Z}[1/2]))$, *called the* **signature class** *of X, is the class associated to the degree two surgery problem* $X \cup X \to X$.

This does indeed generalize the classes discussed in Part I for the manifold case and, in particular, tacitly includes the topological invariance of rational Pontrjagin classes.

To destabilize, we truncate the $\mathbf{Wh}^{\mathbf{Top}}$ spectrum above dimension zero. This new spectrum still has an involution. Given a spectrum with involution one can take its Tate cohomology [GrM, WW, ACD]. With such a notion, we have

DESTABILIZATION THEOREM. *For X a stratified space all of whose strata are of dimension at least five*[8] *there is a fibration*

$$\mathbf{S}^s(X) \to \mathbf{S}^{-\infty}(X) \to \mathbf{H}^*(\mathbb{Z}_2; \mathbf{Wh}^{Top}(X)^{\leq 0}),$$

where $\mathbf{Wh}^{Top}(X)^{\leq 0}$ *is the result of killing the homotopy groups of* $\mathbf{Wh}^{Top}(X)^{\leq 0}$ *in dimensions greater than 0.*

There is a spectral sequence starting from the Tate cohomologies of the various negative K-groups that abuts the homotopy groups of $\mathbf{H}^*(\mathbb{Z}_2; \mathbf{Wh}^{Top}(X)^{\leq 0})$. (See [GrM] for this spectral sequence and a discussion of when it converges, and [HsM] for examples where it does not.)

I should also remark that one can extend the whole theory to "stratified spaces" where some of the strata are not manifolds but are only ANRs,[9] provided we work relative to those strata. This can be of great use in applications (when the stratified space occurs in the middle of some construction, rather than as an object of primary interest).

REMARK. We will usually skip the $-\infty$ decoration in our notations if it is clear from the context that we are dealing with an application of the stable topological surgery sequence.

6.2.A. Homology with coefficients in a cosheaf of spectra

This appendix is just about terminology. Additional references are [BG, Tho, Q2, pt. II].

Let us first give definitions for cohomology with coefficients in a sheaf of spectra. (Since we will only be taking cohomology and homology, there is no need for us to distinguish between presheaves and sheaves.)

DEFINITION. *A presheaf of spectra $\mathcal{S}$ is a contravariant functor from the category of open subsets of X to spectra.*

[8] One can work relative to strata of lower dimension; some four dimensional strata can be allowed using [FrQ].

[9] Even the ANR condition can sometimes be weakened, but then the meaning of the terms involved requires more care.

The idea is now this. For each open set $\mathbb{O}$ we can form $H^*(\mathbb{O}; \mathcal{S}(\mathbb{O}))$, if one likes, concretely, by taking mapping spaces. (That way cohomology groups are homotopy groups of the cohomology space.) Now, if we have two open sets, $\mathbb{O}$ and $\mathbb{O}'$, we can form the pullback

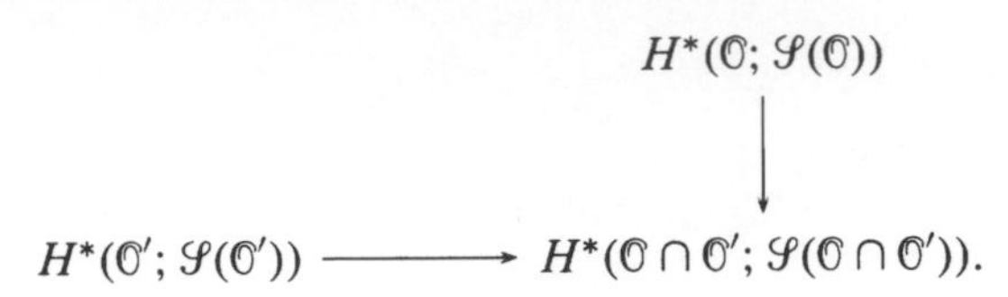

$$\begin{array}{ccc} & & H^*(\mathbb{O}; \mathcal{S}(\mathbb{O})) \\ & & \downarrow \\ H^*(\mathbb{O}'; \mathcal{S}(\mathbb{O}')) & \longrightarrow & H^*(\mathbb{O} \cap \mathbb{O}'; \mathcal{S}(\mathbb{O} \cap \mathbb{O}')). \end{array}$$

The maps are given by the composition $H^*(\mathbb{O}; \mathcal{S}(\mathbb{O})) \to H^*(\mathbb{O}; \mathcal{S}(\mathbb{O} \cap \mathbb{O}')) \to H^*(\mathbb{O}\cap\mathbb{O}'; \mathcal{S}(\mathbb{O}\cap\mathbb{O}'))$ given by the presheaf data and the restriction on cohomology. Of course, one can do this for an arbitrary cover $\mathcal{U}$ of X and get $H^*(\mathcal{U}(X); \mathcal{S})$, taking an appropriate pullback (holim). If $\mathcal{U}' < \mathcal{S}$ is a refinement then there is a canonical map $H^*(\mathcal{U}(X); \mathcal{S}) \to H^*(\mathcal{U}'(X); \mathcal{S})$. We then define $H^*(X; \mathcal{S})$ as the limit of these cohomology spaces under refinement.

There are many natural examples. Constant sheaves of spectra are just spectra. Many other examples come from functorial constructions on spaces: $\mathcal{S}(U) = L(\mathbb{Z}\pi_1(U))$. Or if one has a map $f : Y \to X$, one can push forward sheaves on Y to X. (It is actually often best to use pushforward rather than the derived functor.) An important example comes from a fiber bundle over X with an infinite loop space as fiber. Then the "twisted cohomology" makes sense, defined as sections of this bundle, and is the cohomology with coefficients in the sheaf.

The above example is basically what one would call a locally trivial sheaf of spectra. From this, it is quite simple to go further and describe what one means by a constructible sheaf of spectra. (We'd say that $\mathcal{S}$ is constructible with respect to a stratification of X if whenever $U \subset V$ is a stratified homotopy equivalence of open subsets, the induced map on spectra is a homotopy equivalence.)

Now, cosheaves are exactly the dual notion, with all arrows reversed. For them, one has corestriction maps induced by inclusions. The maps induced by refinement go in the opposite direction, so that one has to take inverse limits, which makes the general theory a little less pretty.

More precisely, one forms, for an open cover $\mathcal{S}$, hocolim $(\mathbb{O} \wedge \mathcal{S}(\mathbb{O}))$. The homotopy groups of this are the Cech homology groups associated with the given cover. For a refinement $\mathcal{U}' < \mathcal{U}$ there is an induced map $H_*(\mathcal{U}(X); \mathcal{S}) \to H_*(\mathcal{U}'(X); \mathcal{S})$, so we take the holim, to define the spectrum $H_*(X; \mathcal{S})$, whose homotopy groups are the homology groups of the cosheaf $\mathcal{S}$.

With constructibility assumptions the homology theories are quite computable using Mayer-Vietoris exact sequences or Leray covers and Atiyah-Hirzebruch type spectral sequences. This also includes "simplicial" homology if the cosheaf is described simplicially. One also can define morphisms associated to cosheaf preserving maps between cosheaved spaces, and anything else that seems reasonable.

The cosheaves that arise for us are (controlled, sometimes) h-cobordism or surgery obstruction spectra for the open sets rel infinity. Note that the corestriction maps follow from the relativity condition when we consider open manifolds. Constructibility follows from assertions regarding how these types of functors just depend on homotopy type and the stratification conditions on our spaces.

EXERCISE (TO GAIN FAMILIARITY WITH THE RELEVANT NOTIONS AND HOW THEY APPLY TO US). Reformulate the theorem in 3.2. Associated to a fibration $E \to B$ construct a twisted homology and an assembly map $H_*(B; \mathbf{L}(F_b)) \to \mathbf{L}(E)$. Show that the obstruction to block fibering a manifold simple homotopy to E over B is an element of the component group of the homotopy fiber of this map.

As another example, consider a codimension one submanifold of W, and map W to the interval (or circle) with the complementary regions going to endpoints. Show that there is an assembly map $H_*(I; \mathbf{L}(W_i)) \to \mathbf{L}(W)$. The obstruction to codimension one splitting lies in this fiber. Relate this to Mayer-Vietoris sequences in L-theory as in exercise 4.6.A.2.

REMARK. In chapter 9 we will find it useful to use certain restricted classes of open covers. In particular, we will want to study properties of spaces that hold "in the large" and will thus be interested in studying the effects of coarsening covers, rather than refining them. In that case homology will be a direct limit and cohomology an inverse limit!

6.3. Notes

The BQs throughout this chapter denote Browder and Quinn, whose early work is explained in the next chapter. As the reader can see, their definition was sufficient for the algebra of describing surgery groups.

The PL Whitehead theory was discovered by many people [BQ, Rot; and see Luck1]. The fact that the involution does not preserve the pieces enters in many people's work and leads to a number of subtle phenomena [CL, DoR1, Luck2, Wei7]. At first this was described by saying that even smooth G-manifolds do not satisfy simple duality, but that there is a correction formula in terms of fixed sets of various subgroups. I prefer the point of view that they have the simple duality that is appropriate to themselves. Substantively, there is no difference, of course.

Transfers appear in many places. The fact that they lead to infinite loop maps probably appeared first in Quinn's thesis [Q1]. There has been a large literature, both for calculations and for purely algebraic descriptions. See [LuR, Luck1, O1], the section on transfers in volume 2 of the conference series of the Canadian Mathematics Society, and the references in these places. [LM] applies these ideas to giving an algebraic description of equivariant surgery obstruction groups, which are a special case of Browder-Quinn groups (but are described in terms of more convenient equivariant data).

The topological simple homotopy theory was developed in the orbifold case by Steinberger and West and led to an h-cobordism theorem in [St]. The general case is due to [Q3]. The beautiful paper [Q3] develops manifold stratified spaces and proves a large number of their important properties. Their homogeneity properties were foreshadowed in an example in [Q2, pt. II] and the generality of the method was applied to solve some natural problems in group actions in [Wei2]. In [Q3] Quinn does not prove the realization part of the h-cobordism theorem but promises that it will appear in a future installment of [Q2]. One can prove realization by other techniques. Steinberger did succeed in giving a proof of realization for the case of orbifolds by using a special trick. It seems to me that his trick is general enough for the application of the simple homotopy theory to destabilization, at least if all of the algebraic K-theory of all of the holinks vanishes below dimension -1. In any case, we will sketch a different proof (without extra assumptions) of the realization theorem, following [HTWW], in 10.3.A.

The surgery theory is new. A number of special cases were already known, although they were written in a very different language; see Part III. I would like to mention my work with Cappell, especially [CW2], as having been critical in guessing and verifying these results.

The PL perspective follows a basic idea of [Sto]. The first version of that work was actually called "block bundle sheaves". The topological case follows the PL case quite closely, despite the absence of most of the usual PL constructions, as the reader will discern. Essentially, "controlled topology" fashions alternatives to these constructions.

7 Transverse Stratified Classification

7.1. Browder-Quinn theory

The theory presented here, due to Browder and Quinn, is very elegant, uses the "right" surgery groups (and spectra), and, unlike the other material in Part II, has application to the smooth category. Also, the whole theory only takes a couple of pages to set up. What a bargain!

Consider a category of manifold stratified spaces where neighborhoods are given structures which can be pulled back. (In the smooth case, for instance, this might be genuine fiber bundle structure.) We will call a space stratified in this way a **strongly stratified space**. Given the notion of pullback, one can discuss transversality. A map is transverse if there is an isomorphism between a neighborhood of the stratum and the pullback such that the composite of the identification with the maps of total spaces agrees with the restriction of the original map to the neighborhood.

We consider maps that are stratified and transverse to each pure stratum. These maps are called "transverse" by Browder and "normally smooth" by [FM, GM1]. We shall denote by $S^{ns}(X)$ the structure set in this category. It consists of (Cat) manifold strongly stratified spaces with a map of this sort and a simple homotopy equivalence (in this category, i.e., the homotopy inverse must also lie in this category[1]) up to transverse stratified homotopy.

Interestingly, the classification theorem is independent of the type of local structure with pullback.

The h-cobordism theorem in this category is identical to that in 6.1. This is because in the PL case inclusions of boundary components are automatically transverse, and PL homeomorphisms, by definition, preserve all block structures.

The main result is the following:

THEOREM. *There are groups $L^{BQ}(X)$ for a strongly stratified space X such that the following is a long exact sequence:*

$$\ldots \to [\Sigma X; F/Cat] \to L^{BQ}(X\times I) \to S^{ns}(X) \to [X; F/Cat] \to L^{BQ}(X).$$

[1] **Exercise**. This is equivalent to the map restricted to all pure strata being a (not necessarily proper!) homotopy equivalence. Can you find circumstances where pure strata can be replaced by strata?

The groups fit into an exact sequence

$$\ldots \to L(X_0 \times I) \to L^{BQ}(X - X_0 \operatorname{rel} \infty) \to L^{BQ}(X) \to L(X_0)$$

if X_0 is the bottom stratum.

See 6.1 for more exact sequences for strata besides the bottom. Such generalizations can be proven in exactly the same way. Also, this whole theory spacifies nicely (3.1), so that all sequences are really exact sequences on homotopy groups of a fibration of spaces denoted by the same character in bold.

To prove the above theorem one needs to do three things. First, identify normal invariants with $[X; F/Cat]$. This is very neat in the spacified version. Second, one has to prove a $\pi - \pi$ theorem (2.4). This is an induction on strata and the usual $\pi - \pi$ theorem. Having this, one can use the spacification techniques and define **L(X)** as in chapter 3 (or, more elementarily, using [Wa1, chap. 9] for the groups). The proof of this does the third thing as well: proving the exact sequence asserted in the theorem.

Normal invariants are defined exactly as they were in 2.3.

PROPOSITION. $\mathbf{NI}^{ns}(\mathbf{X}) \cong \mathbf{Map}[X : F/Cat]$.

This can be proven by induction over strata or all at once by redoing the constructions of [RSu] (2.3). Simply observe that a normal invariant of X is exactly equivalent to a normal invariant of X_0 together with an extension of the "transfer" of the normal invariant of X_0 to the boundary of the closed complement of X_0 to the closed complement. This "transfer" is just the map on $[: F/Cat]$ induced by projection. By induction on the number of strata, this is simply (homotopy equivalent to) $[X : F/Cat]$.

PROPOSITION. *(**Transverse isovariant $\pi - \pi$ theorem**). Suppose (Y, X) is a strongly stratified pair, $X = \partial Y$, and each pure stratum of Y touches exactly one stratum of X for which the inclusion is a 1-equivalence (**Stratified $\pi - \pi$ condition**). If all strata of X are of dimension ≥ 5, then any normal invariant of $(W, V) \to (Y, X)$ can be surgered to a simple homotopy equivalence.*

We prove this by induction on the number of strata on which the normal invariant is not assumed to be homotopy equivalence. By hypothesis, one can do surgery on $(W_0, V_0) \to (Y_0, X_0)$. Glue onto $W \times I$ the pullback over the normal cobordism of the bundle over Y_0. This is possible because of the strong stratification of Y and the transversality of f. See figure 12. This produces a normal cobordism to another $\pi - \pi$ situation where there are fewer strata that must be fixed.

EXERCISE. Complete the proof of the theorem. Also, show that the definition given in this chapter for L-groups is equivalent to the definition used in chapter 6 as the fiber of the composition of a transfer with (the map induced by) inclusion.

EXERCISE. Re-prove Browder's splitting theorem in terms of $L^{BQ}(X)$.

EXERCISE. Show that with the good H-space structure on *F/Top*, for any X the surgery obstruction map is a homomorphism.

PROBLEM. How can one compute this homomorphism? That is, what replaces aspherical spaces as the terminal object to which one should map X, in order to use the resulting map to factor the assembly map. The main difficulty already occurs in the general two stratum situation where the holink fibration over the bottom stratum will not pull back from any fibration over an aspherical space. (The ray of hope is that quite often, although not for, say, the Hopf fibration, the cosheaf of spectra comes from an aspherical space.)

EXERCISE ([Wei8]). Show that if X is a smoothly strongly stratified space, the map $S^{ns,\,\mathrm{Diff}}(X) \to S^{ns,\,\mathrm{Top}}(X)$ has finite kernel, and that the image contains a subgroup of finite index.

These exercises pave the way to using BQ theory to prove nontrivial results in the smooth category. See [Wei8] for an application to smooth transformation groups.

EXERCISE. By varying bundle data on a Poincaré object in the BQ category, classify smooth knots of codimension ≥ 3 as the relative homotopy of $(F/O, F_{c-1}/O_c)$. (This is due to Haefliger and Levine; see [Ha, Lv2].) As a corollary, show that the group (under connected sum) of embedded $S^3 \subset S^6 \cong \mathbb{Z}$. This is in striking contrast to Zeeman's unknotting theorem (1.7).

7.2. Notes

This chapter is just an exposition of [BQ]. Some of the exercises are my own creation.

The main difficulty with applying the Browder-Quinn machinery is that it presumes a great deal of local structure on both the spaces (often not so serious) and the maps (usually very serious). In one sense, the theory is just the most straightforward extension of surgery to the nonmanifold setting. However, the real phenomena that one needs to understand often lie in the neighborhood theory.

Consequently, most further work concentrated on the case of group actions (orbifolds) and avoided the transversality (and stratification) hypothesis. This theory is very complicated. I will discuss some of its aspects in chapter 13.

As far as I know, the only paper in transverse isovariant theory after [BQ] by Katz [Ka], where he shows that the surgery obstruction map can have a fairly large image for problems coming from closed G-manifolds, unlike the phenomenon we saw in 3.4 for surgery on closed manifolds with finite fundamental group. Oddly enough, this was viewed by many as evidence that one would not get a good calculational hold on the obstruction map, since it could not be factored rationally through something of the form $B\pi/G$ (for then the arguments of 3.4 that detect certain equivariant surgery obstructions for normal maps between closed free G-manifolds in terms of the unequivariant obstruction would apply and limit the possibilities enormously). What happens is that the functoriality of the Browder-Quinn theory is with respect to transverse maps, and there is no equivariant $B\pi$ that serves as a terminal object for all spaces with fundamental group π. This is the motivation for the one problem in this chapter.

Browder and Quinn envisioned that their theory would be applied in ways similar to the final exercise. Our approach in the next chapter is somewhat different. In addition to the surgery, the "bundle theory" itself will also be inductively handled by BQ-surgery spectra. The "normal invariants" themselves will not be of any direct geometric importance to us: they can be viewed, for all practical purposes, as just the not rel boundary structures on $X \times D^3$. (Of course, it would be nice to see some explicit nontransverse normal invariants arising naturally and having computable surgery obstructions...)

8 *PT* Category

This chapter is devoted to explaining how the main theorems are proven for the PL category. However, we will not want to deal with the nuisance of the exotic k-invariant in F/PL (2.5.B) so we will modify the category slightly to allow a bit of nontriangulability in. The reader can deduce the PL theory from the PT theory presented here.

The PT category is defined inductively, as follows. An object X (with boundary) of type n has a decomposition into two pieces $A \cup B$, where A is a mapping cylinder of a (type $n-1$) block bundle over a topological manifold (with boundary) glued via a type $n-1$ homeomorphism to (part of) the boundary of B, which is a type $n-1$ space. Morphisms are stratified maps preserving the block structure. Notice that as a consequence of a theorem of Stone [Sto] asserting general existence and uniqueness results for block structures in the PL case, the PT category really differs from the category of polyhedra and stratified maps only by way of manifolds with nonzero Kirby-Siebenmann obstruction.

For instance, a PT space with two strata is the same thing as a manifold whose boundary is given as block fibered over another one. If the strata were triangulated, the PT space would boil down to a PL stratified space with two strata.

The h-cobordism theorem in this category is identical to the Browder-Quinn theorem, as in the PL case, because the inclusion of a boundary component in a PT stratified space is automatically transverse, and PT homeomorphisms, by definition, will preserve the block bundle structure. Of course, this does not settle the problem of when a PT h-cobordism is topologically trivial; that is the point of the topological s-cobordism theorem in chapter 10 (or 6.2).

Before dealing with the proof in section 3, we need two sections of preliminaries.

8.1. Surgery obstructions for homotopy transverse maps

Our program depends critically on two simple facts: that stratified homotopy equivalences are automatically homotopy transverse and that homotopy transversality suffices for defining Browder-Quinn obstructions.

Furthermore, these obstructions are bordism invariant and vanish for homotopy equivalences. In this section we consider the surgical steps.

The key example to keep in mind occurs for maps between spaces with one point singularity, and the point there is that the surgery theory for working relative to a homotopy equivalence is identical to the one for working relative to an isomorphism.

In fact, since polyhedra are built up inductively from manifolds by coning, taking products, and gluing, one could deduce the result from this.

The most natural way to proceed uses ideas from Poincaré surgery. We have discussed in 3.3 the fact that one can define surgery obstructions for maps between Poincaré complexes. According to [J1, Q8, HV] there is even a $\pi - \pi$ theorem for Poincaré surgery, and the obstructions that are defined actually measure the obstruction to normal cobordism to a homotopy equivalence. If one took this approach to its logical next step, one would define a Poincaré-Browder-Quinn category, which would be equivalent to the stratified Poincaré complexes in chapter 5.1. In this category, transversality (in the strongest possible sense) is simply homotopy transversality, and the methods of the previous chapter apply.

A natural way to get around the reliance on geometric Poincaré surgery is to use the algebraic theory of surgery and use an algebraic analogue of the Browder-Quinn groups, defined as fibers of (the composite of inclusion with) the transfer from the bottom stratum to the boundary of its regular neighborhood.

A final way to do this is to set up surgery theory and use the [Wa1, chapter 9] approach, as we did in the previous chapter for Browder-Quinn theory.

DEFINITION. *Let X be a PL stratified space. An isovariant normal invariant for X is a 4-tuple consisting of the following:*

(a) *a normal invariant for X_0, the bottom stratum,*

(b) *a normal invariant for the closed complement $cl(X - X_0)$,*

(c) *a block bundle over X_0 fiber simple homotopy equivalent to the fibration of the boundary of $cl(X - X_0)$ over X_0, and*

(d) *a normal cobordism between the pullback of the block bundle in (c) and the boundary restriction of the normal invariant in (b).*

Note that this is a 4-tuple in name only! It encodes a much more elaborate set of data by induction, which we suppress. Note, too, that an isovariant homotopy equivalence gives a normal invariant. And, also, note that to make sense of the definition, one needs to verify that the pullback of a block bundle over a normal invariant is a normal invariant with fewer strata (using the definition for fewer strata).

We leave the proof of the following to the reader; it is a modification using the bundle data in (c) of the Browder-Quinn argument given in the previous chapter:

THEOREM (PT $\pi - \pi$ THEOREM). *Suppose (Y, X) is a strongly stratified pair satisfying the stratified $\pi - \pi$ condition. If all strata of X are of dimension ≥ 5, then any isovariant normal invariant $(W, V) \to (Y, X)$ can be surgered to a simple homotopy equivalence.*

Now, one can define Browder-Quinn groups as before. Another induction shows that the inclusion of the transverse isovariant category into the isovariant category induces an isomorphism between the two sets of surgery obstruction groups.

8.2. Homology as cohomology with vanishing conditions

This section is more a philosophy – it is so general (and trivial) – than mathematics. Since I like the concept so much and it is widely useful,[1] I will isolate it. The idea is that to define homology classes on a singular object one often succeeds by having cohomology classes on the nonsingular part, with some type of control on their nature at ∞. As the simplest example, cohomology with compact supports is the reduced homology of the one point compactification.

A much more sophisticated variant of this point is that often growth (or vanishing) conditions on cohomology classes suggest compactifications of spaces where they give rise to homology classes. We will not discuss this here, but refer the reader to [Che1, Lo, SS].

To be concrete, here is an example:

PROPOSITION. *If X is a stratified space with two strata, $\mathbf{H}_0(X, \underline{\mathbf{L}}^{BQ})$ is canonically given as the fiber of*

$$[X - X_0 : F/Top] \to \mathbf{H}_{-1}\big(X_0, \underline{\mathbf{L}}^{BQ}(germs)\big).$$

The cosheaf occurring in the right-hand side is the restriction of the cosheaf $\underline{\mathbf{L}}^{BQ}$ to X_0, which is quite different from the $\underline{\mathbf{L}}^{BQ}$ sheaf of X_0 (viewed as a singular space in its own right).

One Poincaré dualizes the right-hand side and recognizes this as the exact sequence of a pair in sheaf theory. In general,

$$\ldots \to H_i(X) \to H_i(Y) \to H_i(Y - X) \to H_{i-1}(X) \to \ldots$$

[1] So much so that I think it is probably hopeless to try to trace its early history.

Note that as is usual in sheaf theory, we are following the convention that homology of a noncompact object is defined using the locally finite chains[2] (Borel-Moore homology).

Geometrically this proposition and variants of this are important in that $[X - X_0 : F/Top]$ represents unrestricted normal invariants of the top pure stratum and the $\mathbb{H}_1$ $(X_0, \underline{\mathbf{L}}^{\mathbf{BQ}}$ (germs)) term can be identified with block fibering obstructions through some inductive scheme (although we will argue somewhat differently). Thus, we will have a process for defining global classes inductively.

8.3. The inductive proof

Let us now turn to the proof of the theorem of 6.1. Recall that we have seen in 3.1-2 that knowing relative surgery for spaces formally implies the theory of blocked surgery. This means we are free to invoke blocked surgery for less complicated stratified spaces. Equivalently, the inductive method of proof works just as well for the spacified version.

First we need to define a map from $\mathbf{S}^{\mathbf{PT}}(X)$ to $\mathbf{H}_0(X, \mathbf{L}^{\mathbf{BQ}})$. (Actually, this cosheaf homology group is isomorphic to the normal invariants introduced in 8.1, but this is not necessary.) We produce a **characteristic class** in this homology more generally for any homotopy transverse map that has normal data on the pure strata, henceforth called a **geometric stratified normal invariant**. This is done, characteristically enough, inductively. Consider the sequence

$$\mathbf{H}_0(X, \underline{\mathbf{L}}^{\mathbf{BQ}}) \to \mathbf{H}_0(X - X_0, \underline{\mathbf{L}}^{\mathbf{BQ}}) \to \mathbb{H}_1(X_0, \underline{\mathbf{L}}^{\mathbf{BQ}}(\text{germ})).$$

To produce an element of the fiber of the second map it suffices to have a normal invariant of $X - X_0$ which restricts on its boundary (i.e. has a neighborhood of ∞ with boundary) to a surgery problem over the holink of X_0 that extends to a problem blocked over X_0 with prescribed singularities. (The reader is recommended to think about this for a while.)

Now suppose the local structure near X_0 looks like $\mathbb{R}^i \times cL$. Then $\mathbb{H}_1(X_0, \underline{\mathbf{L}}^{\mathbf{BQ}}(\text{germ})) \cong \mathbf{H}_i(X_0, \underline{\mathbf{L}}^{\mathbf{BQ}}(cL)) \cong [X_0 : \mathbf{L}^{\mathbf{BQ}}(cL)]$ (or actually sections of a bundle over X_0 with fiber $\mathbf{L}^{\mathbf{BQ}}(cL)$). However, this nullcobordism in homotopy is precisely what is given by the homotopy transversality of the map as one moves near the singularity. But we have seen in 8.1 that homotopy is good enough for surgery purposes.

From the above description it follows that the map from $\mathbf{S}^{\mathbf{PT}}(X)$ to $\mathbf{H}_0(X, \underline{\mathbf{L}}^{\mathbf{BQ}})$ factors through the fiber of $\mathbf{H}_0(X, \underline{\mathbf{L}}^{\mathbf{BQ}}) \to \mathbf{L}^{\mathbf{BQ}}(X)$. Let us show that this is an equivalence.

---

[2]A definition of locally finite homology of a noncompact space is the reduced homology of the one point compactification. A little better is $\lim h_*(U, U - K)$ as K runs through compact subsets of U. For nice U, these coincide, and nice U suffice for all of our applications.

Consider the sequence of fibrations (written horizontally)

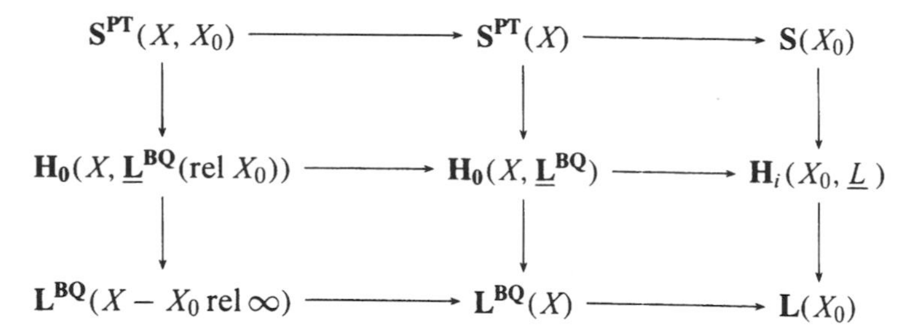

The rightmost vertical composite is a fibration by ordinary surgery. To prove the middle one is a fibration it suffices to deal with the leftmost. (Remember the relative form of stratified surgery still deals with a cosheaf for all of X; it just reflects the relative nature of the problem in its stalks.) Consider the diagram

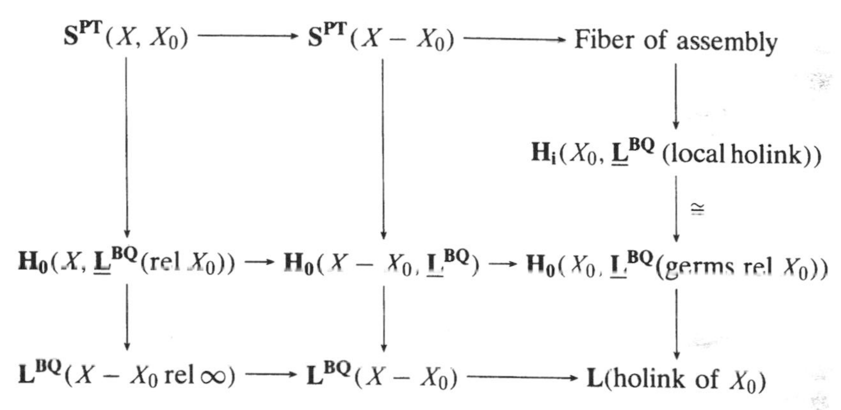

The space denoted **Fiber of assembly** is just the fiber of the vertical assembly map. By blocked stratified surgery with fewer strata it measures the obstruction to homotoping to stratified block fibering a map stratified homotopy equivalent to a fibration with fiber the local holink of X_0. The definition of the *PT* category then shows that the top line is a fibration and therefore the leftmost vertical sequence is, since all other rows and columns are now seen to be fibrations. This leftmost sequence is precisely the leftmost sequence of the previous diagram, so that diagram, too, consists of fibrations, and the proof is complete.

9 Controlled Topology

This chapter is a frankly revisionist approach[1] to a subject that has been flowering over the past decade. Basically, the idea is to redo the constructions of the first four chapters while keeping track of sizes of things.

I cannot overemphasize the importance of these ideas. In addition to the role they play in the theory of stratified spaces, they also have had many applications both within and without[2] topology. Several appendices will develop a small part of this.

There are many different types of control used in the literature. Just to give an idea of what's involved, let's reconsider the theory of noncompact manifolds. (See 1.5.A and 2.4.B.) Rather than study homotopy equivalences between these, we found it preferable to study *proper* homotopy equivalences. Now, every noncompact manifold has a (unique up to proper homotopy) proper surjection to a ray. Using this map we can compactify all of our spaces (this is simply the one point compactification). The condition of propriety is equivalent to the continuity of the map obtained between compactifications defined by extending our original maps via the identity at ∞. We have seen that as far as Whitehead theory is concerned, the point at ∞ contributes via $\tilde{K}_0$, and that in surgery theory it behaves (modulo decorations) just as if it were a boundary. In both cases, it is the "functor shifted one degree."

This suggests using maps to other "control spaces" and using other compactifications in the course of our investigations. If one does this, not all proper maps will continuously extend, but if the sizes of point inverses get smaller (from the point of view of the control space) as one moves out to ∞, this continuity can be achieved. We will see that the reduced homology (with coefficients in a certain spectrum) of the "space at ∞" computes the contribution of the compactification points.

A good example to keep in mind for understanding the way controlled topology enters the theory of stratified spaces is the topologically trivial PL nontrivial h-cobordism, with the singularity set a circle, constructed in 5.3. Before invoking transfer invariant Eilenberg swindles (or Alexander

[1] I suppose the earlier ones were as well, but I am more aware of my tendencies in this direction in discussing more recent work.

[2] In the sense of the Beatles.

trick), we already had seen that an h-cobordism is trivialized on removing a boundary component, a fact that suffices to trivialize the one point compactification (i.e. the mapping cylinder of the map from the boundary to a point). When we inflated that point to be a circle, we had to be more careful in constructing this homeomorphism of the complement of the boundary component, because not all homeomorphisms of the complement will automatically extend continuously to the circle. Indeed, we were only successful for h-cobordisms that had no Nil component in the Bass-Heller-Swan decomposition of the Whitehead group. The lesson we learn from there is the importance of measuring the size of the "handle slides" (used in proving the s-cobordism theorem) as one approaches the circle and demanding that they become smaller and smaller. (Note too that Nil is what is left over when one tries to approximate $\mathbf{Wh}(\mathbb{Z} \times \pi)$ by $H_*(S^1; \mathbf{Wh}(\pi))$.)

As I said, the subject of controlled topology reexamines all of the problems studied in Part I from the point of view of measuring the sizes of their solutions. There are many ways to do this. The idea sketched above is called "continuously controlled at ∞." We will also have use for "controlled" and "bounded categories." Of course, the subject and methodology suggest new problems as well.

The idea that such a subject might exist was first broached in a paper of Connell and Hollingsworth [CoHo]. They made a number of conjectures that they showed would imply many of the then unknown properties of topological manifolds, but these conjectures seemed quite difficult. Unfortunately this work was not followed up on for quite some time. Instead, Kirby and Siebenmann directly dealt with the foundational problems of topology, without, at least explicitly, invoking controlled theory. For example, Edwards and Kirby and Cernavski proved the local contractibility of homeomorphism spaces, which is essentially a controlled theorem: it gives canonical isotopies to the identity of homeomorphisms close to the identity. Chapman and Ferry proved very obviously controlled theorems[3] (such as the α-approximation theorem [ChF]) but, again, did not place them in the general context of geometric groups. In some ways, one could view their approach to their theorems as akin to proving the GPC without first getting the h-cobordism theorem (1.6). The results were very impressive and useful, but one did not realize that they were "universal problems."[4] Quinn realized this in his well-known "end papers" [Q2] and set up controlled algebraic K-theory and the analogues of the results of chapter 1 in this setting. Furthermore, he provided a number of signif-

[3]In some, but not all, cases, modeled on the ideas of Kirby and Siebenmann.

[4]By a "universal problem" I mean one whose solution, after the development of an appropriate general framework, solves all problems of that sort. For instance, once one knows the h-cobordism theorem, one realizes that the GPC is no simpler or harder than trivializing arbitrary simply connected h-cobordisms.

icant applications in these papers and others that led to the extensive development of both theory and applications by many others,[5] some of which will be discussed below.

Similar ideas were pursued by operator theorists with rather different applications in mind (see appendix 9.4.A). The analogue of the boundedness in the topological bounded category is the finiteness of the propagation speed for kernels of certain geometric elliptic operators; see [Roe 4].

I should point out that during this period of explicit development of controlled topology there were many competing redevelopments of the foundations, other settings (one of which, the bounded category, seems quite useful and will be central to our presentation), and, what I believe to be most important, working out of examples. Many topologists, especially those working in transformation groups, worked out examples that were either "universal" or important motivation for what the general theory should look like. Because of this, it sometimes becomes difficult to assign the credit for an idea to any one specific worker. After all, is it the first one to use the idea, the one to explicitly recognize its importance, or the one who puts it into its right general contextual framework (and what if there are several such frameworks)? Who is the hero?

Be that as it may, the goal of this chapter is to explain the basic theorems of bounded and controlled K-theory and surgery and describe how they get used. I will not give much in the way of proofs for some things but will instead try to be the reader's guide through some of the literature.

9.1. The bounded and controlled categories

Let X be a metric space. We shall use X as a place to measure sizes of geometric constructions on (other) spaces. There are (at least) two different ways to use X for measurements.

DEFINITION. *Y is a space over X if we endow Y with a map $p : Y \to X$.*

So far, nothing interesting. We need to decide what are the morphisms we allow and, especially, what are the equivalent spaces over X.

DEFINITION. *The bounded category of spaces over X consists of spaces over X and morphisms maps $f : Y' \to Y$ such that there is a uniform bound on the deviation* $\mathrm{dist}(p', pf)$.

---

[5] I should also mention the work of Anderson and Hsiang [AH1] and its immediate pre-Quinn applications by Lashof and Rothenberg [LR2] to equivariant smoothing theory. These, too, formed universal problems in K-theory and also are part of a cohomological tradition in the subject which is often quite useful in thinking about problems. However, as in surgery (3.4 superseding 2.4), for many purposes the homological viewpoint is the more natural.

This category was originally introduced to study the controlled category which follows:

DEFINITION. *The* **controlled category of spaces over** *X consists of spaces over X and one parameter families of maps f_t, $0 \leq t < \infty$, such that* $\text{dist}(p', pf_t) \to 0$ *(uniformly) as t gets large.*

(The category mentioned in the introduction is called "continuously controlled at infinity" and was introduced in [ACFP]. In many ways it is more elementary and more suitable for the applications of controlled topology to stratified spaces, but we will, in any case, concentrate on the theories already described.)

There are two relations between these categories: The first is the inclusion of the controlled category into the bounded one. Surprisingly, there are situations where this inclusion is quite interesting to study. However, if X is compact, this inclusion is in some sense almost devoid of interest.[6]

Usually it is more interesting to relate the controlled category of a metric space X to the bounded category on cX, the open cone on X. The open cone is usually metrized by the same formula that cones of subsets in one Euclidean space are given when embedding in Euclidean space one dimension higher. The functor inflates a space $Y \to X$ to $Y \times [1, \infty) \to cX$ in the obvious way. Then a bounded map between the inflations corresponds quite directly with a controlled map. (Actually, weighted cones can be used quite effectively as well.)

IMPORTANT REMARK. In 9.3, we will discuss an important phenomenon that ultimately justifies the controlled theory: for many problems there is a critical threshold, ϵ, for which problems with data below ϵ can be made controlled. This principle is behind many, but not all, of the applications of the bounded theory to ordinary topology of manifolds. See e.g. 9.4.B.

Note the bounded category is interesting only when X is a space of infinite diameter; finite diameter spaces can be replaced by a point. On the other hand, the controlled category is very sensitive to the nature of X as a space. Indeed, we do not really need to use continuous maps (as control maps) when studying problems in bounded topology, since the equivalence relation on control maps allows a fixed "modulus of discontinuity." Thus, for instance, there is no essential difference, in bounded topology, between using $\mathbb{Z}^n$ and $\mathbb{R}^n$ as control spaces. Of course, without continuity, controlled topology would be chaos.

Now the general program should be more comprehensible. For reasonable spaces over (reasonable) X redo the theories of Part I. Which

[6]That is, it can often be interpreted as the assembly map in some algebraic theory, a map we're already familiar with.

spaces are finite complexes over X (i.e. homotopy equivalent, in the category, to spaces that have only a finite number of cells over any bounded region in X)? What is the difference between homotopy equivalences over X and simple homotopy equivalences over X? When are homotopy equivalent manifolds over X homeomorphic over X? Etc.

Why should we care about all this? Is this all just some meaningless generalization? Or, are there some interesting examples?

EXAMPLE 1. If M is a compact manifold, then there is a natural map $\tilde{M} \to \Gamma$ where $\tilde{M}$ is the cover of M associated to Γ. Compact problems with target M give rise to bounded problems over Γ. We will find this quite useful when Γ is the fundamental group of M.

REMARK. This example can be generalized to leaves of foliations.

EXAMPLE 2. For many Riemannian manifolds, e.g. Hadamard manifolds (complete simply connected manifolds of nonpositive curvature), or groups Γ given the word metric, there are natural compactifications that are useful for bounded problems. (There might, for instance, be a good M with that fundamental group for which $\tilde{M}$ has a natural compactification and one is then in both situations.) Then the bounded problem often gives rise to a controlled problem over the space at infinity. (See [ACFP] for a systematic study of this relationship.)

REMARK. These two examples are implicit or explicit in many approaches to the Novikov conjecture as we'll see in chapter 14.

REMARK. The analogue for an elliptic operator of being in the bounded category of the Riemannian manifold itself is having bounded propagation speed. In that setting, this relation to compactifications is essentially Higson's corona idea. (See 9.4.A.)

EXAMPLE 3. Collapsing Riemannian manifolds [ChGr1, Fu] give rise to manifolds with interesting control spaces, namely the Gromov-Hausdorff limit of the collapsing spaces. We will discuss a related interesting differential geometric example in 9.4.B.

EXAMPLE 4 (THE REASON THIS CHAPTER IS IN THIS BOOK). We will see many examples related to stratified spaces. The fact that each stratum is an ANR implies that one has a control map for small neighborhoods of strata in X over the the strata. A variant of this map controls any stratum over the cone on the next lower one. (The bulk of the higher stratum is mapped to the cone point.) If one solves problems on the pure stratum controlled with respect to this map, then the solutions will extend continuously to the lower strata.

We have not quite given examples of specific problems of the sort discussed in Part I that occur naturally, but this too will come.

REMARK. It might seem more natural to the reader to deal with a category where the maps to X genuinely commute. We call this the parametrized category, and there is a natural map to the controlled category. The parametrized category (1) is much harder to analyze in practice (e.g. *higher* algebraic K-theory enters, and one can compute only in a stable range; cf. [Wald1]) and (2) seems to arise less frequently in nature. However, this category is the one relevant to the *smooth* classification of stratified spaces.

9.2. Geometric algebra

Whereas the solutions of the problems in Part I involve algebra of group rings, the problems of controlled and bounded topology involve a more complex type of algebra, "geometric algebra."

Parallel to the two types of control categories, there are also two types of geometric algebra, bounded and controlled. This section will be devoted to the reduction of controlled and bounded geometric problems to the geometric groups; the next section will deal with the calculation of certain geometric K- and L-groups. In view of some of the examples of 9.1, it is of quite great interest to technically refine the theorems described below to more general situations (such as nonlocally constant fundamental group), but I am only trying to give the general shape of the theory, which will suffice for the applications to stratified spaces in the following chapter.

For convenience, let us concentrate on K-theory in the bounded category. (We refer the reader to [Q9] for the controlled situation.)

To establish a setting for this, we "recall" that there is a general formal procedure for dealing with simple homotopy theory, a procedure that generalizes to many other settings and can be used to define a formal solution to the simple homotopy problem over X. One defines the obstruction group as being made up out of problems, with the 0 element representing problems that can be solved![7] In other words, the obstruction group is built up out of pairs over X (with a given fundamental group over X). The pairs are to be bounded deformation retractions. We view two pairs as being equivalent if there is a series of expansions and collapses using cells of uniformly bounded size that takes one to another. The trivial elements are those equivalent to an identity pair. Cohen, in his beautiful book, verifies that this forms a group, and that it solves the problem of simple homotopy theory, i.e. if something is equivalent to a trivial element, then it's trivial itself.

With this approach, the Wall finiteness theory can also be set up purely geometrically along the lines of Ferry's [Fe2] method (described in 1.6).

[7] Of course, in L-theory, this is essentially the Wall, chapter 9, approach (see 3.1).

If one defines the obstruction groups this way, then the entire burden of theory is calculation! In other words, what is the algebraicization of the Whitehead group or the finiteness group? If we set up the algebra correctly, the proofs that the algebra and geometry appropriately reflect each other should be close to the classic case.

Let's think a little before proceeding. A very first requirement is to be able to recognize homotopy equivalences. Of course, in such a task the fundamental group plays a special role (which is why the group ring $\mathbb{Z}\pi$ is so important). Suppose that we start with a manifold $M = \coprod M_i$ (with the M_i connected), which we map to $\mathbb{N}$, the natural numbers. On reflection, we discover that anything bounded homotopy equivalent to this is of the form $\coprod M_i'$ where each M' is homotopy equivalent to some M according to a reindexing that takes no integer more than a fixed number D away. A bounded map is a bounded homotopy equivalence iff it is a homotopy equivalence on each component.

However, checking this (say, by trying to compute with twisted homology, as we do throughout chapter 1) requires using a different fundamental group at different places.

Furthermore, this is not just something artificial due to disconnectedness. Imagine taking the connected sum of all of these manifolds (with tubes attaching manifolds with adjacent subscripts).

For machinery handling these complications see [AM1].

We shall make as an assumption (which we will, in fact, sometimes have to violate) that the map $p : M \to X$ has **boundedly constant fundamental group π.**[8] What does this mean? We want a map $\varphi : M \to B\pi$ which in some sense is an isomorphism on the fundamental groups of the fibers of p. More precisely, we assume that there is a positive real valued function $f : \mathbb{R}^+ \to \mathbb{R}^+$ such that for every x and r

$$\varphi* : Im[\pi_1(p^{-1}(B(r, x))) \to \pi_1(p^{-1}(B(r + f(r), x)))] \cong \pi.$$

EXERCISE. This condition just depends on the bounded homotopy type of M.

To understand what happens under a bounded homotopy equivalence, we must consider the sequence of balls and follow them around. One cannot find any particular scale where M' and M look the same. However, whatever deviations there are between the (inverse images of the) r-balls of the two spaces die by the time we get to balls of specified radius, say, $r + f(r)$. That is, the homology of the map restricted to the r-ball maps trivially into that of the $r + f(r)$ ball. This condition on the pro-system of the homologies of maps on the various balls, at all points of X, can

[8] The formulation I am giving here also avoids twisting. I recommend that the reader on a second or third reading try to think through how to deal with more complicated situations using sheaf theoretic ideas.

be used to describe a Whitehead and Hurewicz theorem to recognize homotopy equivalences.

Let us return to our bounded homotopy equivalence and try to algebraicize on the chain level, based on our experience with the classic case. What type of object is the cellular chain complex of such a guy?

Obviously, a good first step is to insist that all of the cells are relatively small, say of radius 1 when projected into X. (After all, our description of a simple expansion is gluing on locally finite collections of cells along faces, where the cells have a uniform diameter.) Then we obtain a free $\mathbb{Z}\pi$ module, based by the cells, with a point in X given for each of the generators of this complex. (The precise location of the basis element does not matter; it can be placed anywhere in the image of the ball.) The boundary maps send generators to linear combinations of generators, of which none are more than some uniform distance D away.

We obtain "finitely generated" modules if we impose propriety on $M \to X$. Finite generation follows from the fact that there are only finitely many cells that map into any particular ball of finite radius. (Much more would be unreasonable;[9] any less would flirt with killing the group by an Eilenberg swindle.)

The set of such chain complexes (with D varying) form a category with morphisms that are chain maps of bounded diameter (i.e. no generator gets mapped to an element whose components, in terms of the range basis, are more than a specific distance away). We also now inherit a notion of bounded chain homotopy equivalence.

A map (between spaces with constant bounded fundamental group π) is a bounded homotopy equivalence iff it induces a bounded chain homotopy equivalence of the cellular chain complexes.

DEFINITION. *$K_1^{bdd}(\pi \times X \to X)$ is the Grothendieck group of bounded chain homotopy contractible based $\mathbb{Z}\pi$ chain complexes over X.*

Now one can formulate and prove a bounded simple homotopy theory and h-cobordism theorem involving $Wh^{bdd}(\pi \times X \to X)$ directly along the lines of 1.5. For concreteness:

BOUNDED h-COBORDISM THEOREM. *Suppose X is a boundedly simply connected space, and that $M \to X$ is a Cat manifold of dimension at least five, with uniformly bounded fundamental group π over X. Then bounded h-cobordisms with one boundary component M up to Cat equivalence are in a 1-1 correspondence with $Wh^{bdd}(\pi \times X \to X)$.*

REMARK. The controlled theory takes on a quite similar form. L-theory can be described by using the above technique to vary the usual defini-

[9] Actually, it is quite reasonable to insist on uniform boundedness of the number of generators of the modules as a function of the radius of the balls. See [At, ABW, BW1-3, Ge2].

tions. In particular, the bounded $\pi - \pi$ theorem holds (see e.g. [FP1]) by a more careful version of the usual proof. The calculation of these theories is the topic of our next section.

9.3. Recognition as homology

Again for simplicity, I will *assume that we are in a situation of constant bounded (or local, for the controlled case) fundamental group*.

The main result is that, in appropriate situations, the geometric K-groups are homology theories.

In this game the theories end up being quite similar in the bounded and controlled cases, but the issues involved in setting things up seem to be a bit different. First let me state the main theorem:

MAIN THEOREM OF CONTROLLED TOPOLOGY ([Q2,6, PW2, Ya, FP1, ACFP; cf. HTW1, RtW]). *The controlled K-spectrum of a finite dimensional ANR X is just* $\mathbf{H}_0(X; \mathbf{K}(\pi))$.[10] *The bounded K-spectrum over the open cone cX is also isomorphic to this (with a shift).*

The same holds true for L-theory by the time we go to $L^{-\infty}$, but is not true unstably.

Before continuing, I should point out that the precise metric used in metrizing the open cone is a bit of a red herring. One can reparametrize the open cones and use various weightings and the proof can be completed in extremely similar terms. This is done in [HTW2] to allow one to deal with hyperbolic space as a cone of the sphere and more generally in [FP1 and FeW2] (see also [ACFP]). However, it is useful and interesting to study the bounded topology of much more general spaces than open cones, as we shall see later. One cannot expect the theorem to be correct as stands for general X: there is too much topology in the compact sets that we ignore in bounded topology. (After all, $\mathbb{R}^n = \mathbb{Z}^n$ in this setting.) The following definition describes a class of spaces that do not suffer from this defect.

DEFINITION. *A space X is* **uniformly contractible** *if there is a function f such that for each x, the inclusions of the ball around x of radius r is nullhomotopic in the ball of radius f(r).*

It has been conjectured that for such a space, the bounded K-theory defined as in the previous section $\cong H^{lf}(X; \mathbf{K})$ (and similarly for L). However, Dranishnikov, Ferry, and I have recently disproved this [DFW]. Nonetheless, for most common examples, one can use this as a guide of what should be true.

Note that for a space satisfying the conjecture, one knows that the inclusion of controlled into bounded is an equivalence! This means that

[10] When X is noncompact we intend locally finite homology.

any bounded problem is boundedly equivalent to a controlled one, and bounded solutions suffice for controlled problems, a very useful principle indeed!

Notice that open cones are uniformly contractible, and the conjecture is affirmed by the main theorem. This conjecture is indeed quite strong; in [FeW2] and [CP] it is shown to imply the integral Novikov conjecture for a group Γ for which $B\Gamma$ is a finite complex.

There is a slightly different conjecture that applies to metric spaces that are not uniformly contractible, discussed in 9.4.A. (It also accounts for the examples of [DFW] and suffices for the applications to the Novikov conjecture.)

There are two main approaches to the proof of the main theorem. The first is via splitting theorems. It is easiest to describe this method for the case of L-theory.

Consider a manifold M mapping to a control space of the form cP, P a polyhedron. For simplicity, let us suppose that M is simply connected over cP. Let $Q \subset P$ be a codimension one subpolyhedron. Take $N \subset M$ a codimension one submanifold lying over cP. (This can always be arranged.) We consider an element of $S^{bdd}(M)$, e.g. a bounded structure on M, $M' \to M$. The techniques of splitting theory (see e.g. [Ca1] and 4.4, 4.6.A) adapt nicely to this situation (with some extra work) and the obstruction lies in a bounded K-group (see [FP2]). We will cross with a circle to kill this obstruction (the source of stability in the bounded L-theory version). Doing this allows us to assume solvable all codimension one splitting problems associated to codimension one subpolyhedra of the end.

The idea of the proof is now this: to show that $\mathbf{L}^{bdd}(cP)$ is a homology theory; the only difficult part is to prove excision or, equivalently, Mayer-Vietoris. As discussed in 4.6.A, Mayer-Vietoris sequences for L-groups are equivalent to codimension one splitting theorems. (Thus, one could organize the proof by induction on the cells of P.) Following [Ca5] (in another context) one would start with cases of $\emptyset$ and $*$ (a point) and show that the assembly map is an isomorphism. The case of $\emptyset$ is ordinary surgery theory. The case of $*$ is handled by an Eilenberg swindle (the $0 = 1 - 1 + 1 - 1 + - \ldots = 1$ trick; see e.g. 5.3).

The second approach to the main theorem is via systematic gain of control,[11] which gives much more information. This phenomenon is called "sucking" in the vernacular of many of the workers in the field and has been referred to as such in print in the paper [HTW3]. I prefer to

[11] Squeezing seems to be stronger than the main theorem, as far as I can see. Applying controlled techniques to "approximate" problems involves squeezing. Having squeezed, we are fortunate that we can then use squeezing again to identify the obstruction group as homology.

follow Ferry's advice and call this "squeezing." We state this informally as the following metaprinciple (requiring separate proofs in different situations):

SQUEEZING PHENOMENON (= STABILITY; cf. [Ch2,3, Q2, pt. II, Hu2] AND OTHERS). *For any type of problem over X and $\epsilon > 0$ there is a $\delta > 0$ such that any δ controlled problem over X can be ϵ-approximated by some controlled problem. The δ controlled problems are ϵ-solvable iff the controlled problem is controlled solvable.*

Rather than explain how this is done in many different cases, I would rather discuss an example of squeezing, the original, and then explain why squeezing is sufficient for the main theorem.[12]

EXAMPLE (KIRBY'S ORIGINAL TORUS TRICK [Ki]). We will prove that any orientation preserving homeomorphism of $\mathbb{R}^n$ homotopic to the identity is isotopic to the identity ($n \geq 5$). We presume known the same fact for PL homeomorphisms.

KEY LEMMA. *If f is a homeomorphism of $(\mathbb{R}^n, 0)$, then there is a homeomorphism of $(T^n, 0)$ whose germ at 0 is identical to the germ of f.*

The theorem follows from this and two simple lemmas.

LEMMA 1. *Two homeomorphisms of $(\mathbb{R}^n, 0)$ that have the same germ at 0 are isotopic.*

Apply an Alexander trick to the composition of one homeomorphism with the inverse of the other. Radial expansion expands the region where the homeomorphisms agree.

The next lemma explains why we want to gain control.

LEMMA 2. *Any homeomorphism of $(\mathbb{R}^n, 0)$ that deviates a bounded amount from the identity is isotopic to the identity.*

One can compactify $\mathbb{R}^n$ by adding on a sphere at infinity that corresponds to directions of affine linear rays. A homeomorphism that boundedly deviates from the identity continuously extends to the ball as the identity on the boundary. Now one can apply the usual Alexander trick.

Granting the key lemma, one arranges for the homeomorphism of the torus to be homotopic to the identity (here is where orientation preserving is used – one composes with the inverse of another homeomorphism that is the identity near 0 and induces the same map on homology). The lift of the toral automorphism is then boundedly far from the identity (the bound is in terms of how far the fundamental domain is moved, which is finite, since the fundamental domain of the $\mathbb{Z}^n$ action is compact, and the

[12] Ferry and Pederson do not actually use squeezing in verifying that controlled theory is homology.

action is proper) and agrees with the germ of the original homeomorphism at 0, so the easy lemmas complete the proof. As squeezing type statements, we made an arbitrary homeomorphism related to bounded ones and also took arbitrary bounded homeomorphisms and moved them boundedly to ones with smaller bound (until they were ultimately the identity).

Kirby reduces the key lemma to a fact that can be proven using PL surgery theory (see 4.6): that any homotopy equivalence[13] between PL homotopy tori is homotopic to a homeomorphism after taking a finite sheeted cover (actually a 2^n-fold cover is all that's needed).

Immerse, using immersion theory, a punctured torus in a small ball around 0. (Note that a punctured torus is parallelizable, so this is possible.) Use f to pull back a new PL structure on the punctured torus. According to Siebenmann's thesis (1.4) one can put a boundary on this manifold; it is a homotopy sphere, so that by GPC (1.7) it is a sphere. Kirby cones this off to form a manifold τ. We can extend our map to a homotopy equivalence between τ and T by extending the identification near 0. If we had a PL homeomorphism between τ and T we'd be done by the relatively easy PL version of the key lemma. Well, we don't know this for τ and T, but we do know the corresponding fact for certain finite sheeted covers, so we win.

This example is prototypical of the way in which one analyzes many problems in controlled topology. A problem over $\mathbb{R}^n$ is furled over a torus (filling in the hole is sometimes difficult: here we used Siebenmann's thesis and lucked out using the strong information surgery theory gives). One gains control by taking finite covers and rescaling.

If your goal is to get invariants of the controlled situation over $\mathbb{R}^n$, you'd examine the nonsimply connected invariants over the torus. (We will see more examples of this in the next section.) Now, if the local fundamental group were (the constant group) π, the relevant functor would be associated to $\mathbb{R}^n \times \pi$. By repeated application of the Bass-Heller-Swan formula in K-theory (1.6) or Shaneson's thesis in L-theory[14] (4.6) one sees a piece that is isomorphic to shifted K and L theory. The only piece that remains on taking covers of the torus is this piece, which is isomorphic to $H^{lf}(\mathbb{R}^n, \mathbf{K} \text{ or } \mathbf{L}(\pi))$.

Therefore, we understand what the spectrum involved must be, provided that we can see that the controlled theory is a homology theory.[15] In order to see that it is a cosheaf homology theory, what one really needs is to be able to restrict to open sets, and to see that a solution of

[13] Kirby uses the Schoenflies theorem of M. Brown [Brnm] to extend this to be a homeomorphism, but we do not need this.

[14] Modulo K-theoretic difficulties that necessitate a change of decoration.

[15] Rigorously, this only shows that the spectrum has the expected series of homotopy groups; more argument is necessary to prove the spectrum is exactly what one expects.

the original controlled problem is equivalent to a solution for each element of an open cover that agrees (in compatible ways) on the overlaps. (In the bounded case, one cannot use arbitrary decompositions of the control space, but the idea is still to decompose with respect to appropriate covers.) The trouble is that one cannot simply restrict to an open set since a controlled solution for an open set uses ϵ-functions that decay to 0 at the frontier of an open set. However, squeezing achieves this: i.e. it enables one to restrict controlled objects to open sets. Then the result is quite formal. The spacification (cf. chapter 3) of the problem for the whole space is simply the homotopy pullback of pieces corresponding to an open cover. (This is the spectral sequence associated to a cover. It gives the usual calculation for a Leray cover, i.e. the Atiyah-Hirzebruch spectral sequence (when the cosheaf is a constant spectrum).)

REMARK. For a verification of a squeezing principle and its use in proving a calculation of a controlled object, I recommend the highly readable [HTW3]. They work, however, with a twisted cohomology (twisted by the structure group of the tangent bundle) rather than homology. Their main result is discussed in the next section.

That it is a genuine homology theory in the untwisted case we're discussing in this section, and not just a cosheaf homology theory, is a consequence of the covariant naturality of our functors with respect to arbitrary continuous maps, rather than the contravariant functoriality with respect to open inclusions. (In the bounded case, one needs Lipschitz maps or one loses the bounds!) If one maps one control space to another, then one can use the target as the place to measure smallness. Of course, by doing this, the local fundamental group will change, if one works geometrically, so one has to use some process that forgets some extra piece of a groupoid (i.e. corresponding to taking just the simply connected signature of a perhaps disconnected manifold with nonsimply connected components). Then one verifies directly the Eilenberg-Steenrod axioms.

The extreme case is mapping the control space to a point, which corresponds to the assembly map. Then one is going from homology of X with coefficients in (say) the L-spectrum to $L(X)$. We've spent much of chapter 3 looking at this case. From the controlled point of view, the normal invariant in homology associated to a degree one normal map is simply its controlled surgery obstruction. The main result of surgery theory is that structures on a manifold are given just by all possible controlled surgery obstructions subject to the restraint that they assemble to (i.e. become, on forgetting control) the trivial obstruction. The same holds, at least stably, for stratified spaces according to the classification theorem (6.2).

9.4. Selected applications

For those readers who, like me, need to see some concrete geometric examples to understand the meaning of all the formalism of the previous sections and to appreciate the difficulties involved in proving such theorems, I am including a number of applications in this section and some of the appendices. None of these are absolutely necessary for the reader who's only interested in classifying stratified spaces. At the least, the reader should observe the recurring themes of squeezing and applied controlled algebraic K- and L-theories, in conjunction with the techniques developed in Part II for the classic theory of manifolds.

9.4.1. The α-approximation theorem

This theorem, due to Chapman and Ferry [ChF], is one of the earliest examples of a theorem in controlled topology and is in many ways prototypical.

THEOREM. *Let M^n, $n \geq 5$,*[16] *be a closed manifold, and let $\epsilon > 0$ be given; then there is a $\delta > 0$ such that any δ-homotopy equivalence $f : M \to N$ is ϵ-homotopic to a homeomorphism.*

A δ-homotopy equivalence has a homotopy inverse $g : N \to M$ such that gf is homotopic to the identity in such a way that the homotopy does not take any point outside a δ-ball around itself, and similarly for the image of the homotopy of fg to the identity. When one writes this theorem for noncompact manifolds it is necessary to use open covers α and β replacing ϵ and δ, which is where the theorem gets its name. One can view it as answering the question of when a map is close to a homeomorphism. The following theorem of Ferry [Fe1], proving a conjecture of Kirby and Siebenmann, is often easier to use:

THEOREM. *Let M^n, $n \geq 5$,*[17] *be a closed manifold, and let $\epsilon > 0$ be given, then there is a $\delta > 0$ such that any map $f : M \to N$ to a manifold of no higher dimension with the diameters of all inverse images $f^{-1}(x)\, x \in N$ no larger than δ is ϵ-homotopic to a homeomorphism.*

To see the difficulty, try to show that such an f is necessarily onto. If it were, then it would be pretty easy to see how to build g and be in a situation where one could try to use the α-approximation theorem.

Nowadays, there are at least two rather different ways to prove the theorem geometrically. The first is the original method, which is closely modeled on Siebenmann's CE approximation theorem [Si3], which is based on a variant of Kirby's torus trick (and its extension as the main diagram

[16]This is also true for $n = 4$; see [Au] and [FeW1].

[17]This is also true for $n = 4$; see [FeW1].

in [KS]), with surgery theory used as an essential component at the last step. The second proof, due to Chapman [Ch2], uses controlled engulfing (also established via torus geometry) to squeeze an α-approximation and make it an ϵ-homotopy equivalence for all ϵ. This means that the inverse g is a CE map, so Siebenmann's theorem or, better, Edwards's disjoint disk theorem [Dvr, Ed] (see also 9.4.D for the statement), implies that g is approximable by homeomorphisms. Edwards's theorem is proven by pure geometry[18] and this variant is a little more convenient for parametrized purposes.

In terms of the machinery, the proof is almost mindless. For small enough δ, one has an approximate structure set that is isomorphic to the controlled one (squeezing). Now the controlled surgery exact sequence is

$$[\Sigma N; F/Top] \to L^c(N \times I \to N) \to$$

$$S^c(N \to N) \to [N; F/Top] \to L^c(N \to N).$$

However, by the main theorem, and the equivalence $\mathbf{L}(e) \cong \mathbb{Z} \times F/Top$, the map $[\Sigma N; F/Top] \to L^c(N \times I \to N)$ is an isomorphism and $[N; F/Top] \to L^c(N \to N)$ is an injection. These immediately yield the vanishing of the controlled structure set, and hence the theorem.

9.4.2. *Approximate fibrations*

This subsection is both a substantial generalization of the previous one and useful in one approach to the analysis of controlled surgery and germ neighborhoods in a stratified space. (See 10.3.A.)

A fibration is, of course, a map $P : E \to B$ which has the homotopy lifting property. An α-approximate fibration is a map which has the homotopy lifting property within α. (In other words, given a map $X \to E$ and a homotopy of the composite map to B, one can lift the homotopy back up to E, at least up to a small error.) An approximate fibration is one which is an α-approximate fibration for all α. These were first introduced and studied by Coram and Duvall [CD1,2] and then were studied in [Ch2, Q2, pt. I, Hu1,2, HTW1]. An α-homotopy equivalence as in the previous subsection is an example of an α-approximate fibration. If the total space and base are manifolds, we call this a manifold approximate fibration (MAF).

The first theorem is Chapman's squeezing theorem:

THEOREM. *Given B and $\epsilon > 0$ there is a δ for which every manifold δ-approximate fibration over B is within ϵ of an approximate fibration.*

---

[18] And amazing geometry it is! He makes good use of the following object: take a simplex and barycentrically subdivide infinitely often and then take the 2-skeleton of this.

Chapman applied this to do some special MAF classification problems that are useful for various purposes (such as homotopical characterization of local flatness; see [Ch2, Q2, pt. II] and 11.3 below). These problems determine the "coefficients" for controlled surgery:

THEOREM. *Any manifold bounded homotopy equivalent to $\mathbb{R}^n \times M$ over $\mathbb{R}^n$ is boundedly homeomorphic to the $\mathbb{Z}^n$-fold cover of a manifold homotopy equivalent to $T^n \times M$. This manifold can be taken homeomorphic to its own finite covers (i.e.* **transfer invariant***) and is well defined up to homeomorphism.*

The reader should compare our remarks from the last section. Note that we're relating a bounded surgical classification question, via squeezing, to one about MAFs. This is the general point of view in [HTW2].

The classification of MAFs with a general base B given in [HTW1], which extends earlier work of the first author (e.g. [Hu1,2]), is "cohomological" in nature and takes some preparation to state.

Suppose that $p : E \to B$ is an MAF. Then we can restrict the MAF to any open set in B (this is, of course, not obvious). Thus, picking a small ball, we get a local type MAF ($N \to \mathbb{R}^b$). It is not obvious, but is nonetheless true, that the controlled homeomorphism type of this local germ is independent of the open ball chosen. Note that the homeomorphisms of $\mathbb{R}^b$ act on MAF ($N \to \mathbb{R}^b$), so that one can take the MAF ($N \to \mathbb{R}^b$) bundle over B associated to the tangent bundle of B.

THEOREM. *There is a homotopy equivalence between the space of MAFs over B with $N \to \mathbb{R}^b$ the local germ and sections of the induced MAF ($N \to \mathbb{R}^b$) bundle over B.*

The relevant homotopy groups are given by a generalization of Chapman's calculation. For instance, π_i MAF ($F \times \mathbb{R}^b \to \mathbb{R}^b$) $\cong$ MAF ($F \times D^i \times \mathbb{R}^{b-i} \to \mathbb{R}^{b-i}$) if $i \leq b$. It is slightly more difficult to phrase the result in general since one cannot peel off some directions in the fiber germ.

Again, these results can be approached via the general methodology of the previous sections. δ fibrations are "δ structures" on the total space of a fibration over B. As such, δ surgery theory is appropriate. However, for small δ δ surgery is independent of δ (L-theory squeezing), so that one obtains Chapman's first theorem. The next theorems don't quite follow from the material of the previous section, because we've only analyzed stable controlled surgery, but weaker stable versions of them do follow from this material.[19]

[19]Also, those theorems describe spaces of parametrized MAFs, while our methods do not give information about the higher homotopy groups of these spaces.

How does one reconcile the cohomological view implicit in this classification of MAFs with the homological view we've been describing 'til this point? If one already believes in controlled surgery (which is an issue of proving an appropriate $\pi - \pi$ theorem as in [Wal, chapter 9] or 3.1) then the connection is quite simple. If X is an arbitrary control space, we can thicken it to be a manifold with boundary. (Imagine then a "universal" problem which is boiled down to be the construction of an MAF structure.) Then, cohomologically, one would get the *rel boundary* cohomology with coefficients in a shifted L-theory (because the dimension of the putative fiber goes down), which is Poincaré dual to homology.

EXERCISE. Using the results of this section, show that if all the K-groups of π vanish below dimension two (i.e. $Wh(\pi \times \mathbb{Z}^n) = 0$ for all n) then every MAF with homotopy fiber having fundamental group π can be approximated by a block bundle. This result was first proven by Quinn [Q2, pt. I] as an application of his end theorem.

9.4.3. *End theorems*

I will here review some of the work of Quinn [Q2, pts. I,II] on the controlled version of the problem considered in 1.4. Indeed, this was the problem chosen to motivate the geometric algebra! The reader would do very well to read §3 of [Q2, pt. I] for a very lucid account of how to use end theorems effectively. The historical remarks in that paper are also very interesting. They show how well he understood the shape of the theory even in its embryonic form (although it developed, in detail, somewhat differently).

THEOREM [Q2, pt. II]. *Let $f : W \to X$ be a map with a tame end with a constant local fundamental group at ∞ of π and X a finite dimensional ANR; then the obstruction to embedding W in V as the interior of a manifold with boundary and extending f as a proper map to X lies in $H_0(X; \mathbf{K}(\mathbb{Z}\pi))$.*

The boundary is unique up to a controlled h-cobordism which is classified by an element of $H_0(X; \mathbf{Wh}(\mathbb{Z}\pi))$, whose loopspace is $H_0(X; \mathbf{K}(\mathbb{Z}\pi))$.

REMARK. Quinn also allows the possibility of the end having a stratified system of fundamental groups, but we will not consider that difficulty here. Also, there are approximate end theorems and h-cobordism theorems, but I will leave the formulation of these to the reader.

The proof is a combination of the results of the previous sections. (ANRs are treated by viewing them as subsets of a decreasing sequence of finite dimensional polyhedra.)

Here's an example of how this gets applied:

COROLLARY. *An ANR subspace X of a manifold W which is $1 - LC$ embedded (i.e. the local fundamental group of the complement at each point of X is trivial) has a mapping cylinder structure. (It is unique in an appropriate sense.)*

A **mapping cylinder structure** is a manifold V with a map to X whose mapping cylinder is homeomorphic to a neighborhood of X. PL embeddings always have mapping cylinder structures because of regular neighborhood theory. (See 1.2.)

The proof is that we can complete the retraction that's defined on a small deleted neighborhood of X. Since we are $1 - LC$, the coefficient spectrum is contractible through dimension one, so there is no obstruction to existence or uniqueness of the completion. The mapping cylinder of this map can be identified with a neighborhood of X using the collar directions and the control.

COROLLARY (WEST'S THEOREM [We]. *Any finite dimensional ANR is homotopy equivalent to a finite complex.*

The mapping cylinder neighborhood is a nice compact manifold with boundary. (By starting with an embedding into a PL manifold, one can even avoid some of the difficulty of the theory of topological manifolds.)

EXERCISE. Use the uniqueness of the mapping cylinder structure to show that the Whitehead torsion (1.2) of a homeomorphism between polyhedra is trivial. This is originally due to Chapman [Ch1]. This proof is more in the spirit of his second proof [Ch4]. More direct is to view the torsion of a homeomorphism between polyhedra as a controlled homotopy equivalence whose controlled torsion lies in a group that on inspection is 0! A fortiori, the uncontrolled torsion vanishes as well.

REMARK. The nontrivial end obstructions occur if one is interested in equivariant mapping cylinder structures around the fixed sets of group actions and equivariant finiteness of ANRs. In [Q2, pt. II] Quinn gave the first example of a locally linear group action for which equivariant finiteness failed. This led to a very vigorous development by many authors and to other different examples that demonstrated how central these sorts of actions are to an understanding of the equivariant topological category (see [DoR2] and [Wei2]). All of these phenomena can now be said to be understood as a result of the stratified h-cobordism theorem (see 6.2 and 13.0.)

9.4.4. *Triangulation*

After triangulation of manifolds was settled through the work of Kirby and Siebenmann [KS] it became natural to analyze the question of when *locally triangulable* spaces are triangulable. We have already seen in 5.3

nonuniqueness of triangulations due to K-theory phenomena (Milnor's examples, for instance). The first result I know of showing nonexistence is due to Siebenmann: an unpublished example referred to in his ICM talk on topological manifolds, reprinted in [KS].

The problem was systematically analyzed (cohomologically, presaging[20] some of the work on MAFs already described) in papers of D. Anderson and W. C. Hsiang [AH1,2]. Quinn also analyzed the problem, but, somewhat differently, from the end point of view. I will not give his analysis or that of Anderson and Hsiang, although the reader would do well to read these papers; instead I'll just point out one result that occurs in all of their approaches.

THEOREM. *Suppose all the pure strata of a stratified space have all holinks (see 5.1) with vanishing lower K-theory, and suppose there are no low codimensional situations; then the space has a PT (see 8.1) structure, which is unique if the Whitehead groups vanish as well.*

REMARKS 1. It is straightforward to put in Kirby-Siebenmann obstructions (2.5.B) to get a PL theorem.

2. In light of this theorem, one can now obtain the topological classification of such stratified spaces using chapter 8. It should not be surprising that we almost have enough technology to do the general case. (The ingredients will be assembled in the next chapter.)

In deference to laziness, and not wanting to go through complicated inductions as in chapter 8, let's just look at the case of two strata. How shall we triangulate? By the end theorem, we will have no trouble putting a boundary (uniquely) on the complement. So let's examine the mapping cylinder structure. First consider the case where the bottom stratum is a circle. We subdivide the circle into many arcs, intersecting in points. Over the interior of these arcs, we have exactly the inverse images we would desire for a block bundle projection. Now, to get a good inverse image for a point, we put an end on the manifold inverse image of an arc this point is incident to. (We must combine the different choices of "ends" of the different arcs this point is incident to, but this is really not that hard.) A controlled s-cobordism argument shows that the original mapping cylinder neighborhood is homeomorphic to one produced from the cylinder of the block bundle projection just constructed.

To deal with the more general bottom stratum, imagine the base put together rather like a brick wall, so that triangulating is essentially the same as finding the "blocks" over codimension one submanifolds of codimension one submanifolds of.... Each of these is inductively solving an

[20]The precise connection between MAFs and triangulation is discussed in [HTWW] and the appendix to the following chapter.

end problem by taking the inverse image of a corresponding open set bordering on the relevant face. As before, this is always possible, uniquely, under the hypotheses.

9.4.A. Index theory on noncompact manifolds

I would like to discuss an interaction between bounded topology and some aspects of analysis on noncompact manifolds[21] that I find very interesting. Some aspects of the analogy in the compact case have been dealt with in the appendices to 2.5 and 4.6. Further, equivariant extensions of these parallel developments will be discussed in 14.2.

This appendix demands more familiarity with index theory than any other section of this book. I recommend [Bla, Hig3] as useful references for aspects of the K-theory of C^*-algebras.

I believe that the general theory of stratified spaces developed in this book will also have an interesting analytic counterpart.

Let me reiterate some of the basic ideas of the analogy between surgery and index theory. The collection of surgery problems and elliptic operators are both combined in homology theories. Miraculously these agree (using Real operators) away from the prime 2. The theories are given by the L-spectrum (F/Top) and by K-theory respectively. (See 2.5.A.) The obstructions (indices) associated to these objects lie in an algebraically defined group associated to (a completed) group ring. Furthermore, consideration of the index of the signature operator leads to a map from the topology to the analytic theory.

We'll summarize the situation for closed manifolds in the chart on the following page, much of which we have already seen. The reader might want to consult [Wei5] for more information.

If we study the question of which manifolds have positive scalar curvature, then one can expand the table slightly, with Gromov-Lawson's spin cobordism invariance [GL2] (see also [ScY1])[22] being a more fitting companion to the $\pi - \pi$ theorem. This is a lovely test problem for index theory.

Now our goal is to extend this diagram to noncompact manifolds somewhat. Many of the problems that have been studied thus far have reasonable extensions to general operators and connect nicely to work

---

[21] Oliver Attie is doing interesting work on the analogies between Roe's work [Roe1] and topology regarding the complexity of diffeomorphisms on noncompact manifolds. Block and I have refined some of this index theory partially exploiting the analogies described in this appendix. Whether there are geometric topological cognates of the index theorems of Cheeger-Gromov [ChGr2] or Stern [Ste] I do not know.

[22] They have also done the analogue of the Cappell, Waldhausen, and Pimsner line for this problem.

Index Theory	Topology	Algebraic K-Theory
Operator	Surgery problem	Chain complex
C^*-algebra	Ring	Ring
$C^*\pi$	$\mathbb{Z}\pi$	$\mathbb{Z}\pi$
K-theory	L-theory	K-theory
Commutative algebra	Space	
K-homology	$\mathbf{L}(e)$-homology	$\mathbf{K}(\mathbb{Z})$-homology
$D^*D > 0$	Homotopy equivalence	Acyclicity
Cobordism invariance	$\pi - \pi$ theorem	
Pimsner-Voiculesciou [PV]	Shaneson's formula (4.6)[1]	Bass-Heller-Swan formula (1.6)[2]
Pimsner [Pi]	Cappell [Ca4] (see exercise in 4.6)	Waldhausen [Wald3]
Strong Novikov conjecture	Novikov conjecture	K-theory Novikov (4.6) conjecture
Special case of Baum-Connes	Borel conjecture	Assembly isomorphism conjecture[3]
BC injectivity[4]	Equivariant Novikov (see 14.2, 14.4)	See 14.4
BC conjecture	Equivariant Borel[5]	See 14.4
?	Stratified Borel	See 14.4
η-invariant	ρ-invariant	"Semitorsion," e.g. [DR]
[APS]	(4.7)	
[Lo]	Higher ρ (14.4)	

[1]This has a twisted generalization for $\mathbb{Z} \times_\alpha G$.
[2]This has a twisted generalization for $\mathbb{Z} \times_\alpha G$.
[3]If one writes the assembly map for arbitrary rings, this cannot be true because of Nils.
[4]Formally, this is another problem, but the connection was realized in work of [RsW2] and, subsequently, [BDO].
[5]This is false; see [Wei5]

done on complete Riemannian manifolds.[23] Information travels in both directions.

We'll begin with the use of almost flat bundles. These made their debut in papers of Gromov and Lawson for the positive scalar curvature problem [GL1,3] and were recently adapted by Connes-Gromov-Moscovici for use on the Novikov conjecture on the homotopy invariance of the higher signatures [CoGM].

To explain the idea, I will be more or less historical.

[23]It is fairly routine to develop the analogue in index theory of the proper theory as developed in §§1.5.A and 2.4.B.

THEOREM [GL1].[24] *Hyperbolic manifolds do not have positive scalar curvature.*

SKETCH PROOF. We begin by recalling (again, see the exercises in 4.6.A) the Lichnerowicz method. If D is the Dirac operator on a spin manifold[25] $D^*D = \Delta + \kappa$ where Δ is the usual Laplacian and κ is the scalar curvature. Thus, positive scalar curvature implies that $\ker D^* = \ker D = 0$, so the Atiyah-Singer index theorem implies vanishing of the $\hat{A}$-genus.

Suppose we were to pair with a flat bundle. We'd repeat the argument to get vanishing but we'd learn nothing new because the higher Chern classes of a flat bundle vanish. If we take a bundle which isn't flat, though, we'd lose because the curvature of the bundle could defeat the positive scalar curvature of the manifold, and the coupled Dirac operator might gain index.

However, if the curvature of the bundle were small enough compared to the curvature of the manifold, we'd be able to repeat the argument and get a new obstruction to positive scalar curvature.

If M is hyperbolic, then one can find large finite sheeted covers with large injectivity radius (using the residual finiteness of the fundamental group of M). Collapsing the complement of a large ball onto the sphere and using the Bott bundle on the sphere, we have a bundle on the cover with interesting topology and small curvature. The scalar curvature of the cover is as large as it was downstairs, so we win.

REMARK/EXERCISE. This proof only makes sense as written for dimensions $\equiv 0 \bmod 4$. Modify it to apply in all dimensions.

In [GL3] this idea was put into a noncompact setting, where it gained much additional power.[26] In that paper they prove a **relative index theorem** for manifolds and operators (with a positivity condition at infinity) that are identified outside some compact set. (The index so defined is independent of which compact set the manifolds are identified in the complement of). The formula for the index is the same as the Atiyah-Singer formula. One subtracts the integrand on one manifold from the corresponding integrand on the other (using the identifications outside the compact sets to avoid $\infty - \infty$ problems). This is the index of a certain Fredholm operator which I will not define. Here is an application.

[25] Exercise: Show that hyperbolic manifolds are spin.

[26] Especially when combined with the families version (see §13 of [GL3]). The use of families to do descent arguments from "noncompact Novikov conjectures" on a universal cover to ordinary Novikovs on a manifold is explained (in differing degrees of coherence) in [FRW, Wei5, FeW1,2] and 14.3.

THEOREM [GL3]. *Nonpositively curved manifolds do not have positive scalar curvature.*

The difficulty in applying the previous argument stems from the fact that we do not know whether such manifolds are spin or whether they have any finite covers at all. However, they certainly have a universal cover, which is (diffeomorphic to) Euclidean space, and hence spin.

The inverse of the exponential map is a Lipschitz diffeomorphism to Euclidean space, so we can pull back any model of the Bott element without increasing its curvature too much. On Euclidean space, by rescaling we can make the Bott element have as small a curvature as we like (at the cost of increasing its support) so the relative index of Dirac coupled to the pullback of the (rescaled) Bott bundle (and the trivial bundle) is 0, but it is the integral of the Atiyah-Singer integrand, with the Chern class of the Bott element tossed in, and is therefore not 0. This contradiction proves the theorem.

The topological use of Lipschitz diffeomorphisms that do not have Lipschitz inverses is mentioned in [FRW]. We recognize this as being a feature of the bounded category. (There is a morphism induced by a Lipschitz map of the bounded categories over differing control spaces.) See [Gr2] for more discussion of such maps in differential geometry.

It seems possible to adapt the almost flat bundle idea to the purely topological situation.

The idea following [CoGM] is this. Again let's start from flat bundles. Let's try to get an invariant of quadratic forms over $\mathbb{Q}\pi$ (which is more or less equivalent to the higher signature problem, because L-groups are made up out of such forms and they measure the entire obstruction to constructing homotopy equivalences; see 4.6.A). We take the holonomy representation and transform the matrix of our quadratic form into one over $\mathbb{R}$ (of larger size) and then take the signature of that.

Now if the bundle has a little curvature, we still can use an expression of group elements as words in fixed generators and use holonomy to get a matrix over $\mathbb{R}$. Fixing our initial matrix, one can estimate how small a curvature is necessary to guarantee that the image will be nonsingular. If one has a Witt equivalence between two matrices, a yet smaller threshold of curvature might be necessary to get the same signatures for both. Thus, if we take limits with curvature going to 0, we obtain a priori homotopy invariants of manifolds which a generalized Hirzebruch signature formula (proven by an extension of the methods developed in [CM]) identifies with a higher signature. In [CoGM] they assert that for many interesting groups the Chern classes of almost flat bundles in fact generate $H^*(B\Gamma; \mathbb{Q})$, and they deduce the Novikov conjecture.

I would like to point out that one can do this rescaling trick on any open cone cX, to obtain trivialized at ∞ almost flat bundles corresponding to shifted K-theory of X. Pairing this with bounded algebraic Poincaré complexes over cX one obtains[27] (assuming a combination generalized Hirzebruch formula/relative index theorem) at least rationally, although probably $\otimes\mathbb{Z}[1/2]$ if one uses care, the injectivity of the map

$$H_{*-1}(X; \mathbf{L}(\mathbb{Z})) \to \text{ Bounded } \mathbf{L}(\mathbb{Z})\text{-theory over } cX.$$

We of course know, by the main theorem, that for all rings this is an isomorphism (if we decorate with $-\infty$). Nonetheless, even this injection is a very useful result.

EXERCISE. Show that, as a corollary of this injection, if M is a manifold with a proper map to $\mathbb{R}^n$, then the signature of the transverse inverse image of a point is a bounded homotopy invariant. As a corollary, deduce the topological invariance of rational Pontrjagin classes.[28] (See also [Wei10] for more details of this step and another, variant, analytic proof.)

It seems to me that almost flat bundles usually measure the bundle theory of a space at infinity (shifted). The same sort of thing arises in "exotic" theory, which I will now turn to. (I do not know the connection between these two theories, although one can certainly make conjectures, especially over uniformly contractible manifolds and the like.)

REMARK. This method does not seem to suffice to prove the main theorem of bounded algebraic K-theory, or even an injectivity version of it. However, I believe, because of the work of [BHM, Ti] that the overall cyclic homology techniques that are used in the details of [CoGM] can be used to verify this (and the conjecture made in 9.3) for special rings (and metric spaces, with only very slight additional hypotheses).

PROBLEM. Prove the Bass-Heller-Swan injection using cyclic techniques.

Let me now describe some work of Roe, which is based on [CM] and [Hig1].[29] Let's talk topology before discussing indices.

If X is a space, Roe forms the coarse cohomology as follows.[30] Take Cech covers of X. Then one can take the cohomology associated to the given cover. By taking coarsenings, one gets an inverse limit system. Without restricting the covers, this limit is uninteresting; one simply has a terminal object corresponding to the cover of X by itself, and the cohomology vanishes. However, if we give X a metric, one can insist that

[27]Recall that $\mathbf{L}(\mathbb{Z})\otimes\mathbb{Z}[1/2] \cong BO\otimes\mathbb{Z}[1/2]$.

[28]Recall the Thom-Milnor construction of L-classes. This will be reviewed in the appendix to 12.4.

[29]Higson and Roe have succeeded in proving for operator theory the analogue of the calculations of bounded L-theory of weighted cones.

[30]He defines it otherwise, to make the definition of an index simpler, and then shows that it can be computed by the prescription I'm giving.

all covers have some uniform bound on the diameter of their elements (the open sets). Then one obtains something interesting. This inverse limit is what Roe calls the coarse cohomology.

The following proposition shows that coarse cohomology isn't exotic.

PROPOSITION [Roe3]. *If X is uniformly contractible, then the exotic cohomology of X is the cohomology with compact supports.*

We will denote exotic cohomology by HR since it is the result of a kind of reduction of all bounded topology to contractibility and also because it is related to Rips's construction of a certain cell complex for the study of hyperbolic groups. In his paper Roe describes how to pair exotic cohomology classes with operators of bounded propagation speed (like the lift of an elliptic operator from a compact manifold to its universal cover) to define an index and then provides an index theorem (following from [CM]) giving a topological expression for this index.

Unfortunately, this index is not of much use unless the operator extends to a certain completion of the algebra of bounded propagation speed operators. (This algebra is then the closure of the algebra generated by kernels supported within a finite distance of the diagonal.) The reason for this is that one knows how to define homotopy invariant signatures for Poincaré complexes only over a C^*-algebra or, similarly, that positivity of D^*D only gives a vanishing of the index when one is dealing with a C^*-algebra. (See [KaM1] and [Roe2,3].)

It is therefore important to understand the K-theory of the completion of these operators. Higson [Hig2] has shown that if X is an open cone cP, this is the shifted K-theory of P.

More generally one considers compactifications of X by gluing on ideal points V (called a **corona** for X) with the property that functions of bounded variation on X extend to the compactification. Then there is a map $HR^{i-1}(V) \to HR^i(X)$. These classes give rise to indices that extend over the C^* completion of the bounded propagation speed operators.

COROLLARY [ROE3]. *If M is a manifold with positive scalar curvature, then the evaluation of the A-class of M on any class coming from a corona vanishes.*

Similarly, the value of $L(M)$ on such a class is a homotopy invariant in the category of bounded homotopy equivalences over M. (We remark that one can put the same corona onto quasi-isometric spaces.)

This discussion certainly should remind readers of the main theorem of bounded topology.[31]

[31] In fact, Higson's proof for open cones seems to me to be reminiscent of the Pederson-Weibel proof of the corresponding result in algebraic K-theory. He and Roe have given another proof based on some nice homotopy invariance principles for the K-theory of these algebras.

Let me shift perspective slightly. Rather than trying to pair operators with cohomology classes, one wants to assign to operators homology classes. (This is in keeping with the ideas of [A4, Kas1, BDF, and BD] described in 2.5.A.) This just involves dualizing, and one in the happy circumstance of having direct limits rather than inverse ones in the construction of our "exotic" objects.

The analogue of the transgression map is a boundary map associated to a corona $HR_i(X) \to H_{i-1}(V)$, under whose image one finds obstructions to putting on metrics of positive scalar curvature, and whose image should be an appropriate homotopy invariant.

Also, as usual in operator theory, one should really be working with K-theory rather than homology. There is no real difference in the construction. To produce KR (not to be confused with Atiyah's object denoted by the same letters) one takes the K-homologies of the geometric realizations of the nerves of the various covers and then takes direct limits under coarsenings.

(Incidentally, there is a type of surgery theory where $\mathbf{L}R$ naturally arises. Suppose that one is in a category of spaces over X, and let's assume for simplicity that the bounded local fundamental group is trivial. If, instead of trying to produce a bounded homotopy equivalence of manifolds over X, one tries to do surgery to achieve a **decomposed homotopy equivalence** over X, i.e. a map that preserves some particular bounded decomposition of X, and on each of the pieces of the decomposition, the map is actually a homotopy equivalence. I will leave the verification to the reader. The basic theorem of controlled topology describes situations where bounded is equivalent to decomposed.)

Finally, it is clearly advantageous to allow our operators to keep track of their "semilocal" fundamental groups. After all, in the compact case, we've seen how to use $K_*(C^*\pi)$ to great effect. That is, one should be using a cosheaf of spectra. (See 6.2.A.) For each simplex in the nerve of an open cover one assigns the spectrum which gives the K-theory of the C^*-algebra of the fundamental groupoid of the relevant open set. If one believes in the standard (Novikov, Baum-Connes, Borel...) conjectures, one might replace this cosheaf of spectra by one that is more homologically defined in terms of the groupoid. I will call this the enriched exotic K-homology (or algebraically enriched if we use the C^*-algebra) or enriched exotic homology. Let's denote these by $EKR(X)$, $EHR(X)$, $EKR^{\mathrm{alg}}(X)$, $EHR^{\mathrm{alg}}(X)$, etc. I warn you though that these functors are not bounded homotopy invariants any more and are somewhat more sensitive to the actual geometry of the spaces. (They are homotopy invariants in the bounded category of spaces over X.)

EXERCISE. What is the geometric significance of $E\mathbf{L}R^{\mathrm{alg}}(X)$?

PROBLEM. Using Stoltz's theorem [Stoltz] and the Gromov-Lawson surgery theorem [GL2], show that if M is quasi-isometric to a Hadamard manifold of sufficiently smaller dimension (say, five dimensions lower) and M has free abelian local fundamental group of not too large a rank relative to its dimension (e.g. it's simply connected), then M has positive scalar curvature with a Lipschitzly larger metric iff the class of the Dirac operator in $EKOR_*(M)$ vanishes.

PROBLEM. Suppose we look at a one-ended noncompact manifold quasi-isometric to a ray. Then $EKOR^{\mathrm{alg}}(M) \cong KO(C^*\pi, C^*\pi')$ where π is the fundamental group of the interior and π' that of the end. Use this theory to produce obstructions in this group to metrics of positive scalar curvature on noncompact manifolds. The boundary map to $KO(C^*\pi')$ describes "bad ends," i.e. ends that cannot be ends of positive scalar curvature manifolds. (See [GL3].) Show that the punctured torus is a manifold with a good end but no metric of positive scalar curvature.

EXERCISE. Prove that if $\pi' \to \pi$ is an isomorphism in the last exercise, and one is in a high dimension, one can always produce a metric of positive scalar curvature.

Thus one should continue the chart as follows:

Index Theory	Topology	Algebraic K-Theory
Algebra of bounded propagation speed operators	Bounded algebraic Poincaré complexes	Bounded projective chain complexes
$EKR(X)$	$E\mathbf{L}R(X)$	$E\mathbf{K}^{\mathrm{alg}}R(X)$
K (completion of the enriched Higson algebra)	Bounded algebraic L-theory over X	Bounded K-theory over X
[Hig2]	[FP1]	[PW2] [Q2, pt. II]

In the constant fundamental group situation it is reasonable to make an "exotic Novikov conjecture" that if X is quasi-isometric to a uniformly contractible space, then the map from the exotic object to the bounded algebraic object is an isomorphism. We will see (following [FeW2, CP]) situations where this implies usual Novikov conjectures in chapter 14. In any case, here it just seems like a useful analogy.

Presumably for a more general case one wants a condition that makes X quasi-isometric to a uniformly contractible stratified space, where the stratification is constructible and relates to the semilocal fundamental group of the corresponding pieces of X.

In the topological case (i.e. for algebraic K-theory and L-theory) certain cases of this package of conjectures are available to the perspicacious reader of these notes (for instance, when there is a Lipschitz map to a stratified open cone with the system of fundamental groups constructible with respect to the stratification).[32]

REMARK. The use of KR, LR, etc. essentially allows one to avoid the assumption of being coarse quasi-isometric to a uniformly contractible space. Such metric spaces do, in fact, arise quite naturally in the study of S-arithmetic groups. However, while I certainly believe in these cases of the isomorphism conjectures, I am not aware of any proofs.

9.4.B. The rigidity package for infranilmanifolds (after Farrell-Hsiang)

Farrell and Hsiang gave a beautiful argument for the following:

THEOREM [FH1]. *A closed manifold of dimension $n \neq 3, 4$ has a flat structure iff it is aspherical and its fundamental group contains an abelian subgroup of finite index. It has an almost flat structure iff it is aspherical and its fundamental group contains a nilpotent subgroup of finite index.*

Almost flat means that one can find a uniformly bounded series of metrics on the manifold whose curvatures are going to zero. This theorem is equivalent, in light of the Bieberbach theorem (characterizing flat manifolds as manifolds with finite cover a torus) and Gromov's work [Gr3] on almost flat manifolds, to the Borel conjecture for flat manifolds and infranilmanifolds.

Their argument in full detail and generality is a little long and complicated, but the idea is simple and easily explained. I'll focus on the structure set result; the K-theory goes rather similarly.

The key case is really the understanding of crystallographic groups, and the general infranilmanifold case follows from this by an analysis of block bundles over such bases. (One really shows inductively that the assembly map for blocked surgery theory problems over a Bieberbach base is an isomorphism if the K-groups of the fiber vanish. Infranilmanifolds tend to fiber over Bieberbach manifolds with smaller fibers. In particular, if the fibers have contractible structure spaces, then the total spaces will as well.)

So let's concentrate on understanding Bieberbach (= flat) manifolds; and for simplicity, I'll only describe the case of odd order holonomy. (The general case is a bit more involved and can be elucidated from a stratified point of view; see the sketch of a related result in [Q6].) These have fundamental group $1 \to \mathbb{Z}^n \to \Gamma \to G \to 1$ where G is a finite group called the holonomy of the manifold.

[32]This theorem, in K-theory, is closely related to the result on controlled K-theory of a stratified system of fibrations in [Q2, pt. II].

If G is cyclic, for instance, then Γ is poly-$\mathbb{Z}$ and the arguments in 4.6.A suffice. Now, using Dress induction on structure sets, more or less as in 4.8 (see [Ni]), one can see that for such a manifold $S(M)\otimes\mathbb{Z}_{(2)} = 0$ (since the only 2-hyperelementary subgroups of the odd order group G are, by definition, cyclic).

However, we will not do this immediately. Instead, we will do double induction on the rank of Γ and the order of the holonomy, making use of the following algebraic fact:

FACT. If Γ has odd order holonomy, then either there is a nontrivial homomorphism of Γ to $\mathbb{Z}$ with a lower rank Bieberbach kernel, or there is an infinite sequence of quotients G_k of Γ which project onto G so that any hyperelementary subgroup of G_k maps surjectively to G.

Now the proof can be completed. In the first case, one uses 4.6.A again. In the other case, one wants to use Dress induction, which means that we have to worry about the hyperelementary subgroups of these quotients. All of these manifolds are homeomorphic to M (they are rigged by the fact to have the same fundamental group, and a little geometric analysis of the situation shows that the diffeomorphism statement is correct). On going to these larger and larger covers and rescaling (think about the case of a torus) one arranges for the induced homotopy equivalence to have smaller and smaller diameter, (on the same manifold!), so Ferry's theorem (see 9.4.1) can be applied to give the homeomorphism on these covers and, hence, by induction for the original homotopy equivalence.

The careful reader will note that here I've cheated: the induction requires that one handles bundles instead. This refinement follows from squeezing for approximate fibrations (9.4.2) and use of the assumption that all fibers involved have vanishing structure sets to prove the suitable generalization of Ferry's theorem. (One could just as well use end theorems to do this: see [Q2, pt. I].)

EXERCISE. Prove a variant of Shaneson's formula (4.6) for extensions of infranilgroups $1 \to \pi \to E(\Gamma) \to \Gamma \to 1$ whenever there is no K-theory in the fiber. (In other words, generalize the discussion of block bundles to the case where the fibers might not be rigid.)

9.4.C. The Grove-Peterson-Wu finiteness theorem

Another remarkable application of controlled topology in recent years is the following theorem:

FINITENESS THEOREM ([GPW]). *The number of homeomorphism (diffeomorphism) types of closed n-manifolds, $n \neq 3$ (and $n \neq 4$), with given upper bound on diameter and lower bounds on volume and curvature is finite.*

For more complete expositions of the fact that there are a finite number of homotopy types, let me recommend [GP] and [Che5]. The result on finiteness of homeomorphism types is a slightly delicate further step where the controlled topology is used.

The idea is this. We take an infinite sequence of such manifolds; they will converge in the Gromov-Hausdorff sense [Gr4] to a metric space X. (This simply means that there are a metric space (X, d) and metrics on the disjoint union $\cup M_k \cup X$ for which each point of M and X is within ϵ of the other space, at least for k large enough.)

Then one in effect (see 9.4.D for the relevant concepts) shows that (1) X is a finite dimensional ANR homology manifold, and (2) the natural homotopy equivalences among these manifolds are all the result of a natural map to X with an "approximate resolution." (This means that, for each ϵ, by going sufficiently far out in the sequence, we can arrange that the inverse images of tiny balls become contractible in the inverse images of ϵ-balls.) One could prove a squeezing result and apply uniqueness of resolutions (see the following appendix), but instead they make do with varying Quinn's argument and proving uniqueness of approximate resolutions.

The result on diffeomorphism is a consequence of the fact [KM] that except in dimension four, every compact topological manifold has at most finitely many differentiable structures.

REMARK. The result on diffeomorphism is certainly more natural than the one on homeomorphism. The counterexamples to the hauptvermutung in 5.3 and the examples of infinitely many smooth embeddings that are topologically the same (1.7, Haefliger knots) show how little we understand the mechanism of finiteness. Is there differentiable finiteness for embeddings or isometric group actions under hypotheses similar to the Grove-Peterson-Wu theorem? (The topological part should not be too difficult to get using the material presented here.)

REMARK. T. Engel-Moore has given examples in her thesis of sequences of metric manifolds that have a local contractibility function (a major consequence of the hypothesis of the [GPW] theorem) but that converge to an infinite dimensional homology manifold. Subsequently, Dranishnìkov and Ferry [DF] analyzed the situation and proved that in this case (1) one can have different (i.e. nonhomeomorphic) sequences of manifolds converging to the same space, and (2) this reflects a nonuniqueness of resolutions of infinite dimensional homology manifolds. Ferry has also shown that only finitely many different manifolds can be converging to this space.

The nonrigidity displayed in [DF] is responsible for the "flexible $\mathbb{R}^n$'s" constructed in [DFW] that disproved the conjecture discussed in 9.3.

There, rescaled metric spheres form a sequence converging to an infinite dimensional homology manifold.

9.4.D. Homology manifolds

Let X be an ANR homology manifold; i.e. X is an ANR and its local homology $H(X, X-x)$ agrees with that of Euclidean space at every point. We will assume that X has dimension at least five and is connected. A fundamental question is whether X can be resolved. That is, is there a manifold $M \to X$ which is an approximate homotopy equivalence (equivalently, with cell-like inverse images)?

If the answer to this were yes, then homology manifolds would indeed be quite close to manifolds. As things develop, I believe that they will be viewed as being a natural, geometric extension of manifolds, i.e. basically spaces that should be regarded as being on an equal footing with manifolds.

In any case, to return to a more or less historical approach: although they were early viewed with great interest through the work of Bing and his school, resolutions gain much practical value from the following theorem of Edwards (see [Dvr, Ed]).

DISJOINT DISK THEOREM.[33] *A resolution of an ANR homology manifold X of dimension ≥ 5 can be approximated by homeomorphisms iff X has the disjoint 2-disk property, i.e. that any two maps from the 2-disk into X can be separated by small homotopies.*

EXERCISE. Deduce from this the double suspension theorem ([Can]) that the double suspension of any manifold homology sphere is a (manifold and therefore a) sphere.

Indeed, during the course of our constructions of group actions and embeddings, it will be clear that our constructions produce ANR homology manifolds. We are, of course, more interested in manifolds. One way to deal with this problem is to produce resolutions of the constructed homology manifolds and then it will usually be a simple matter to verify the disjoint disk condition and Edwards saves the day. (For our applications, there is another approach, obtained by comparing Top and PT classification theory; see 11.3.)

THEOREM [Q5].[34] *X has a resolution iff some open subset of X does. In fact, there is an integer valued obstruction to resolution that can be computed by restricting to any open set.*

---

[33]See note 18.

[34]Suppose that X is connected.

A proof can be constructed as follows.[35] The total surgery obstruction (see 3.3, 3.4) of X in $B\mathbf{S}(X)$ vanishes, because that can be looked at as the obstruction to taking all of the local deviations from self-duality that are given assembled to 0 (since X satisfies Poincaré duality) and cobording them to something with trivial self-duality. Since X satisfies duality for local reasons, this obstruction vanishes. Again, since the vanishing is completely local, we can make each open set homotopy equivalent to a manifold and do all of these compatibly, so that one gets a(n approximate or, by squeezing, genuine) resolution. Now, the difference between the genuine obstruction to being homotopy equivalent to a manifold and the algebraic version is a $\mathbb{Z}$ that comes from the difference between G/Top and $L(e)$.[36] Thus we have a $\mathbb{Z}$ for each component, and we can measure it locally around any point.

The same argument actually shows:

THEOREM [Q2, pt. I] (UNIQUENESS OF RESOLUTIONS). *Let $f_i : M_i \to X_i = 1, 2$ be two resolutions of X; then for each $\epsilon > 0$, there is a homeomorphism $g : M_1 \to M_2$ such that $d(gf_2, f_1) < \epsilon$.*

In [Q2, pt. I] there is a more elementary proof of this using Edwards's theorem and no controlled surgery. By Edwards, $T^2 \times f_i$ are approximable by homeomorphisms (this is not quite obvious). We can now unwrap a circle and apply the end theorem (local fundamental groups are trivial) complete and see a controlled h-cobordism, which is a controlled product for fundamental group reasons, giving us the result that the products $T^1 \times M_i$ are homeomorphic in a controlled way. We can then unwrap again and remove the second circle as well.

However, now let us return to the question of whether nonresolvable homology manifolds exist.

THEOREM [BFMW]. *For every integer there is a homotopy n-sphere, $n \geq 6$, with arbitrary resolution obstruction.*

In fact, the following stronger result is correct:

THEOREM [BFMW]. *S-cobordism classes of homology manifolds simple homotopy equivalent to X are in a 1-1 correspondence with the fiber of the assembly map, i.e. exactly as predicted by surgery theory.*

COROLLARY. *Siebenmann periodicity (3.4) holds for homology manifolds.*

In other words, we've seen that periodicity fails for manifolds because F/Top differs from $\mathbf{L}(e)$ by a factor of $\mathbb{Z}$. Homology manifolds fill in

[35] Quinn uses the end theorems more extensively and makes use of Edwards's theorem in the proof of his own, but this is not actually necessary.

[36] The differences coming from negative homotopy groups in periodic L-theory are relevant only in infinite dimensional situations. See [DF, DFW].

this lacuna. Indeed with homology manifold stratified spaces, the classification theorem of (6.2) takes a more beautiful form, but only up to s-cobordism at this date (August 1993).[37]

The proofs of these theorems use controlled surgery on certain types of Poincaré spaces to produce better and better approximate homology manifolds. The homology manifolds are produced by an appropriate limit of these approximations. Any sufficiently good constructive method given a homotopy type, when combined with controlled refinements, allows one to produce s-cobordisms.

More precisely, take a reduction of the Spivak fibration of Poincaré spaces X. If the controlled surgery obstruction of this were 0 (which is equivalent to the vanishing of the resolution obstruction) we'd normally cobord to a resolution. The idea is to now take a very fine 2-skeleton of the domain, Wall-realize the negative of the controlled surgery obstruction on the boundary of a regular neighborhood of this skeleton, and glue the "other side" of the cobordism to the complement of this neighborhood of the 2-skeleton. This gives a Poincaré complex, with small duality from X's point of view (although not from the complex's point of view), which, with effort, can be surgered to a (different scale intended!) homotopy equivalence to X.

Now, we take better and better approximations and begin measuring the control over previous stages in the constructions since these are, after all, now somewhat controlled Poincaré complexes. The limit, as I said, is our space. The result is a homology manifold, essentially because the i-th approximation has small duality over the $(i-1)$st stage, giving the limit self-duality over itself, which is the same thing as being a homology manifold.

There's one amusing technical detail I'd like to point out. We're using the interpretation of controlled surgery as homology = normal invariants to make this argument. This requires the control map to have locally constant trivial fundamental group of the fiber. We achieve this using the following beautiful theorem of Ferry (which is partially responsible for the unfortunate dimension assumption in the construction of homology manifolds):

THEOREM [Fe3]. *Let $M^m \to P$ be a map from a compact manifold to a compact polyhedron whose homotopy fiber has a finite k-skeleton. Then one can homotop f to have all point inverses "homotopy equivalent to" this fiber through dimension k, if $2k+2<m$.*

Homotopy equivalence is here meant in the sense of a limit of open sets containing the point inverses. The point inverses themselves can be quite pathological. The reader should examine what this says for P of

[37]Sadly, these lines were first written in July 1992.

larger dimension than M, and for the case where M and P are spheres of the same dimension (but the map is not degree one) – see also [Wal, Bes] for earlier work.

I would like to close with a number of conjectures that seem plausible to me.

MODIFIED RESOLUTION CONJECTURE. *Every ANR homology manifold is resolvable by a unique ANR homology manifold with the disjoint 2-disk property.*

HOMOGENEITY CONJECTURE. *Any ANR homology manifold with the disjoint 2-disk property is homogenous. (More speculative would be that they have locally contractible homeomorphism groups.)*

S-COBORDISM CONJECTURE. *S-cobordisms that satisfy the disjoint 2-disk condition are products.*

These conjectures lead one to believe that there is a whole new set of local types as beautiful as Euclidean spaces. Spaces modeled on these will be the best homology manifolds and should resolve arbitrary homology manifolds. Surgery theory for classification up to homeomorphism will then be as correct for these as for manifolds. However, for the most perfect view of surgery, it is best to consider as a whole the aggregate of spaces with all the local types together.

Finally, I mention one last conjecture:

DIMENSION CONJECTURE. *The theory described above is correct for $n \geq 4$. For $n = 3$ all homology manifolds are resolvable, at least modulo the Poincaré conjecture.*

REMARK. We don't yet have a good substitute for the disjoint disk condition for $n < 5$. Consequently, resolutions of those lower dimensional objects would only immediately imply higher dimensional results, e.g. that after $\times\mathbb{R}^2$ they were (or weren't) manifolds.

10 Proof of Main Theorems in *Top*

In this chapter I will complete my discussion of the proof of the main theorems (6.2) in the topological case. I will not give complete arguments ab initio but instead describe where they differ from the PT discussion given in chapter 8. We've already seen that controlled topological technique suffices to reduce to the PT case whenever holinks have vanishing algebraic K-theory, so it should be of no surprise that the same technology can be applied to deal with the general case.

10.1. The h-cobordism theorem

We recall the statement:

THEOREM ([Q2,St]). *Let* $\mathbf{Wh}^{Top} X$ *be defined by the following fibration:*

$$\mathbf{Wh}^{Top} X \to \mathbf{H}(X; \mathbf{K}_0^{BQ}(\textit{neighborhood})) \to \mathbf{K}_0^{BQ}(X).$$

$\pi_0\, \mathbf{Wh}^{Top} X$ *is in a* $1-1$ *correspondence with h-cobordisms on* X *provided that all strata of* X *are of dimension at least four.*

Actually, the statements in [Q2] and [St] are slightly different and make use of the general splitting that we've seen in chapter 7 of Browder-Quinn K-groups into the K-groups of the closed pure strata (or more precisely, the fundamental groups of the interiors of the pure strata). This means that $\mathbf{Wh}^{Top} X$ decomposes into a sum, one piece for each pure stratum. Let's write this decomposition out and see what it says about homotopy groups:

$$\begin{aligned} &\bigoplus H_0\big(X_i; Wh(\text{local fundamental group})\big) \to \bigoplus Wh\big(\mathbb{Z}\pi_1(X^i)\big) \\ &\to Wh^{Top} X \to \bigoplus H_0\big(X_i; K_0(\text{local fundamental group})\big) \\ &\to \bigoplus K_0\big(\mathbb{Z}\pi_1(X^i)\big). \end{aligned}$$

where sum is taken over all strata. Note also that on $X^i \subset X_i$, the local fundamental group is trivial, so these homology groups are supported on the union of the lower strata.

Let us rephrase it one more time:

$$Wh^{Top} X \cong \bigoplus Wh^{Top}(X_i \text{ rel } X_{i-1})$$

and

$$\ldots \to Wh\left(\mathbb{Z}\pi_1(X^i)\right) \to Wh^{Top}(X_i \text{ rel } X_{i-1})$$
$$\to H_0\left(X_{i-1}; \tilde{K}_0\left(\mathbb{Z}\pi(\text{holink})\right)\right) \to \tilde{K}_0\left(\mathbb{Z}\pi_1(X^i)\right)$$

where the holink is taken in X_i.

This is the form in which Quinn and Steinberger phrase their theorems. It is also a form that is more convenient for the proof.

Once we see the description of $Wh^{Top}(X_i \text{ rel } X_{i-1})$, the fact that $Wh^{Top}(X)$ has a map into the product will be quite clear; one defines directly an obstruction to h-cobordism that lies in a group isomorphic to this one (via the description of $Wh^{Top}(X_i \text{ rel } X_{i-1})$ as the controlled Whitehead group for a certain map). It is then a formality that the map $Wh^{Top}X \to \bigoplus Wh^{Top}(X_i \text{ rel } X_{i-1})$ is injective.

Surjectivity is also easy for the piece corresponding to the top stratum. Unfortunately, the literature does not explain how to extend h-cobordisms from a lower closed stratum. This is rather delicate and can be done with more advanced end machinery (a future paper promised by Quinn), the ϵ-surgery of Ferry and Pederson [FP1], or as a consequence of the teardrop neighborhood theorem (see 10.3.A). In any case, I will not discuss this issue in this section.

So let's move on to the calculation of $Wh^{Top}(X_i \text{ rel } X_{i-1})$. For convenience of exposition, let's suppose that X_i is a PT space, although this makes no serious difference. (We would otherwise have to deal with relative finiteness obstructions rather than absolute ones.)

The map $Wh^{Top}(X_i \text{ rel } X_{i-1}) \to H_0(X_{i-1}; \tilde{K}_0(\mathbb{Z}\pi(\text{holink}))$ is the obstruction to taking an h-cobordism on X_i rel X_{i-1} and extending the mapping cylinder structure around X_{i-1} to a neighborhood of $X_{i-1} \times I$. (The obstruction group should be $H_0(X_{i-1} \times I; \tilde{K}_0(\mathbb{Z}\pi(\text{holink})))$ but homotopy invariance reduces it to what is written.) In other words, it is an end obstruction.[1]

The end obstruction vanishes by the time we go to $\tilde{K}_0(\mathbb{Z}\pi X^i)$ since this relaxation is a Siebenmann end obstruction, which always vanishes in the K-group of the interior of the manifold (1.4). That is, if one has an open manifold W, then, by definition, the end obstruction in $\tilde{K}_0(\mathbb{Z}\pi')$ vanishes by the time we pass to $\tilde{K}_0(\mathbb{Z}\pi)$ (where as usual π and π' denote the fundamental groups of the interior and end respectively).

If the controlled end obstruction itself vanishes, then we can put a (controlled) boundary on this complementary part of the h-cobordism.

[1] In the previous chapter we've discussed the obstruction when the holink has constant fundamental group. This case requires the general stratified system of fibrations generalization from [Q2, pt. II]. It can be dealt with by the same formal steps as the constant case.

Then consider the torsion of the inclusion of the closed complement on the boundary into our newly constructed closed complement. This is the lift to $Wh(X^i)$.

This lift is not quite well defined. It depends on which mapping cylinder structure we choose. These structures are h-cobordisms controlled with respect to the retraction to the singular set X_{i-1}. These are classified by a controlled Whitehead group (see chapter 9), which, in turn, is a cosheaf homology group, as described. Thus, a change of the controlled boundary will kill this obstruction if (and only if) it lies in the image of the assembly map $H_0(X_{i-1}; Wh(\mathbb{Z}\pi(\text{holink}))) \to Wh(\mathbb{Z}\pi_1(X^i))$.

EXERCISE ([Q3]). The text only sketched the fact that an h-cobordism is a product if the topological torsion vanishes. Prove the realization of all elements of $Wh^{Top}(X_n \text{ rel } X_{n-1})$ as h-cobordisms, for X_n the top stratum.

NOTE. Similar arguments to those we've just given produce an end theorem for stratified spaces with obstructions in $\bigoplus \tilde{K}_0^{Top}(X_i \text{ rel } X_{i-1})$.

10.2. Stable surgery

Now I want to explain the proof of the stable classification theorem using controlled surgery. The reader can appreciate why we have to stabilize in the stratified situation but not in the classic one, because of the main theorem of controlled L-theory (9.3). Only stable surgery is a homology theory. For the manifold case, the normal invariants involve controlled L-theory with a local fundamental group with no K-theory, so stable = unstable, but, in general, we are forced to deal with the stabilized version.

There are several different approaches to the classification theorem possible, and these are somewhat interrelated.

One way is to redo the arguments from the PT category using controlled surgery to replace blocked surgery and making use of the result that stable controlled surgery is a homology theory [FP1].

Another way would be to map the structure space into the fiber of the surgery map and inductively use the picture given in the appendix to this chapter to verify stable isomorphism.

The way that I would like to describe is a little different. While it is not true that every manifold stratified space has a PT structure, even stably, one can produce an analogue of a PT structure stably.

In unpublished lectures and manuscripts I have called these structures STIBBs for stable transfer invariant block bundles, and their first mathematical application was my work with Rothenberg on the stable existence of equivariant Lipschitz structures for certain group actions [RtW]. I do not really want to maintain the name in perpetuity, since it is an awkward notion whose time probably has both come and gone, but it is quite intuitive.

In any case, the idea of the proof is this. To begin with, let's think about a space X with an isolated singularity. This corresponds to the one point compactification of a manifold with a tame end. The lack of a PT structure is determined by Siebenmann's end obstruction. However, after crossing with a circle, this completion exists.

This is not of much help in understanding our stratified space, however, because the one point compactification of this manifold is a singular space with a single point as singularity, not the circle that we need.

The way to achieve this is to improve the control on the boundary in the S^1 direction. It turns out that there is a canonical boundary on products with S^1. It is obtained as follows (compare 1.6): if V is any sufficiently small closed neighborhood of the singular point, and V' is any smaller neighborhood, there is an isotopy (constructed by engulfing using tameness) F_t of V within X that starts embedding V as V and ends up embedding it inside V'.

Now use this isotopy to embed the mapping torus of this self-embedding in $S^1 \times X$. This is a manifold with boundary, and the boundary is actually a completion of $S^1 \times X - S^1 \times \infty$. I leave the verification to the reader.

(By the way, I have found that this makes for some beautiful pictures that the reader should draw or model with clay. A nice case is to consider the neighborhood in $S^1 \times D^2$ of S^1 produced by starting with an annular neighborhood of a point of D^2. After unraveling it, you will of course see a torus.)

Figure 18 is a simple picture of the process.

The controlled boundary is independent of all choices as well (because there is a contractible choice of isotopies). Consequently, it is the same as its twofold cover (the TI part of STIBB).[2] Taking iterated covers one can improve the control in the circle direction as much as one wants (or improve it as much as one needs in order to squeeze). The result of the transfer invariant gluings gives a good neighborhood of the circle. (Compare 5.3.)

Now, if the singularity is not just a point, then one can do this to get a mapping cylinder structure after crossing with a circle. Recall from 9.4 that the difficulty in putting a block structure on a neighborhood is the result of solving a series of end problems. Solve these, after crossing with more and more circles, as necessary.

The end result is that for each simplex Δ of the original stratum one obtains a "block" that looks like a transfer invariant structure on $\Delta \times T \times$ holink and that is then canonically glued in over $\Delta \times T$. Such things can

[2]For our purposes, it suffices to argue using $\times \mathbb{R}^n$ and working boundedly over $\mathbb{R}^n$ as opposed to crossing with tori and working transfer invariantly.

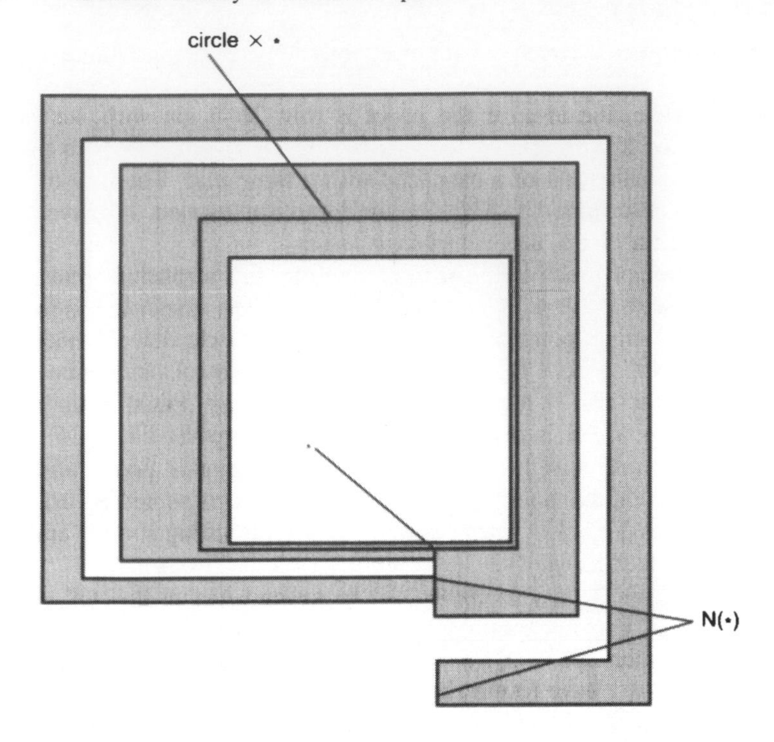

Figure 18. The neighborhood of a circle in annulus. Taking the mapping torus of a self-embedding of a (nonregular) neighborhood of a point produces a regular neighborhood of a circle (after crossing with a circle).

be classified by a (negatively decorated) version of blocked surgery, and one can now repeat verbatim the proof from chapter 8.

With which we close our sketch of the stable theorem.

REMARK. This repeated crossing with tori to obtain the analogue of block structures is quite similar to two other similar discussions in the literature that were discovered independently. (1) Ferry and Pederson [FP1] prove that the stable bounded surgery theory is a suitable homology theory by repeated splitting. However, this is obstructed by a bounded K-obstruction. By stabilizing, they kill this obstruction. (2) Weiss and Williams [WW] are interested in analyzing the topological space $BTop(M)$ which classifies fiber bundles with fiber M. They show that stably this space coincides with the classifying space of block bundles with fiber $M \times \mathbb{R}^i$ bounded in the Euclidean direction. Weiss and Williams were also interested in the destabilization problem, and they, too, relate it to Tate cohomology of an appropriate spectrum, but this is the topic of the following section.

10.3. Destabilization

I now would like to sketch a bit of what is involved in the destabilization theorem. Actually, we will prove a little less. The theorem asserts that there is a fibration

$$\mathbf{S}^s(X) \to \mathbf{S}^{-\infty}(X) \to \mathbf{H}^*\left(\mathbb{Z}_2; \mathbf{Wh}^{Top}(X)^{\leq \mathbf{0}}\right).$$

We will only sketch a looped version of this:

$$\mathbf{H}^{*+1}\left(\mathbb{Z}_2; \mathbf{Wh}^{Top}(X)^{\leq \mathbf{0}}\right) \to \mathbf{S}^s(X) \to \mathbf{S}^{-\infty}(X).$$

The delooping is a bit indirect. The current, weaker, destabilization suffices for the embedding theory developed in 11.3. There, following [CW3] for manifolds, we give a geometric form of Siebenmann periodicity, which puts a self four-fold loop space, and hence ∞ loop space, structure on the whole theory, which in particular deloops the fibration here.

Now, as for the fibration, one should consider the following diagram, where the horizontal rows are fibrations and the vertical arrows are homotopy equivalences:

$$\begin{array}{ccccc}
\mathbf{H}^{*+1}(\mathbb{Z}_2; Wh^{Top}(X)) & \longrightarrow & \mathbf{S}^s(X) & \longrightarrow & \mathbf{S}^h(X) \\
 & & & \swarrow \cong & \downarrow \\
\mathbf{H}^{*+1}(\mathbb{Z}_2; \tilde{K}_0^{Top}(X)) & \longrightarrow & \mathbf{S}^s(X \times \mathbb{R}, bdd) & \longrightarrow & \mathbf{S}^h(X \times \mathbb{R}, bdd) \\
 & & & \swarrow \cong & \downarrow \\
\mathbf{H}^{*+1}(\mathbb{Z}_2; K_{-1}^{Top}(X)) & \longrightarrow & \mathbf{S}^s(X \times \mathbb{R}^2, bdd) & \longrightarrow & \mathbf{S}^h(X \times \mathbb{R}^2, bdd) \\
 & & & & \downarrow
\end{array}$$

Of course, *bdd* denotes bounded. The isomorphisms of various h-structure sets with bounded s-structure sets of bounded objects are completely analogous to the theorem of Chapman referred to in 9.4. The horizontal fibrations are Rothenberg sequences and are a formal consequence of the (full) h-cobordism theorem (i.e. including realization of obstructions) as in the classic case (i.e. as in 2.4.A or [Sh1]). (We do not need Milnor duality here, because we define the duality by turning h-cobordisms upside down!)

One can see that the spaces $\mathbf{H}^{*+1}(\mathbb{Z}_2; K_{-i}^{Top}(X))$ are in fact Eilenberg-MacLane with the expected homotopy groups. This gives a spectral sequence for computing the homotopy group of the fiber of the stabilization map.

Tate cohomology of spectra (see [GrM, WW]) behaves well with respect to fibrations, so that at least one formally obtains the same spectral sequence[3] for computing the fiber and computing the Tate cohomology. (This spectral sequence is described for general Tate cohomology of spectra in [GrM]. This paper contains much information regarding its convergence, or failure of convergence, as well as a number of very interesting calculations.)

It is not impossible to make these approximating fibrations compatible with each other and deduce the theorem on the space level. In other words, one can compare the structures with differing amounts of stabilization inductively, and then take the direct limit, recognizing the fiber (of the limit = the limit of the fibers) as the Tate cohomology.

A nice apparatus for doing this type of argument is given in [WW], where they have to do considerably more work than is necessary here because their Tate cohomology involves the whole Whitehead spectrum, i.e. the higher as well as the lower algebraic K-groups. We will leave the details of the interactions between these Rothenberg sequences out, since they are rather complicated and combinatorial in nature.

REMARK. At the present there is no known group for which $K_i(\mathbb{Z}\pi) \neq 0$ for i below -2. Thus, the same lack of knowledge prevails for stratified spaces, so the spectra we are taking Tate cohomology of never have more than three nontrivial homotopy groups, and the homotopy groups of the Tate cohomology are thus of exponent 8.

10.3.A. The structure of neighborhoods

This book would not a description of be complete without a discussion of what neighborhoods really look like: not a description of what they look like stably and an obstruction theory for how to destabilize them. One knows, by the tubular neighborhood theorem, that (germs of) neighborhoods of manifolds in manifolds are the same thing as vector bundles, and in the PL case, they are the same thing as block bundles. What do the neighborhoods of strata in a manifold stratified space look like?

The answer, in [HTWW, Hu3], involves manifold stratified approximate fibrations (MSAFs). These are analogous to the one stratum case and they have a similar general theory (calculation as sections of a fibration). This is proven by induction on the number of strata simultaneously with the following (answer) theorem:

[3]This spectral sequence is described for general Tate cohomologies and analyzed in some cases in [GrM].

THEOREM. *There is a homotopy equivalence of appropriate spaces*

$$|\textit{Germ neighborhoods of } B \textit{ as a stratum of a manifold stratified space}| \leftrightarrow |\textit{MSAFs over } B \times \mathbb{R}|.$$

The map from the right-hand side to the left is a nice construction, called the "teardrop". Let $f\colon W \to B \times \mathbb{R}$ be an MSAF with k strata. $TD(f) = W \cup B$ topologized by a basis of open sets consisting of open sets of W, and the union of open sets O in B with the inverse image of sets of the form $O \times (i, \infty)$. In other words, for a sequence of points outside B to converge to a point in B they must lie in a certain teardrop-shaped region of this space. The MSAF condition on p guarantees that this is an MSAF with $k + 1$ strata.

The main geometry involved in showing that every neighborhood is a teardrop is a controlled stratified engulfing. The reader is invited to play with the construction in 10.2 and for MAF squeezing (9.4) to build the MAF associated to a stratified space with an isolated singularity.

This theorem allows an analysis of manifold stratified spaces that is inductive in the same way that the definition in the PL category can be formulated.

REMARKS. 1. Bruce Hughes has extended this theorem to an analysis of germs of embeddings of stratified spaces in one another.

2. That one looks at MSAFs over $B \times \mathbb{R}$ is useful for applications to h-cobordisms. After all, all h-cobordisms become products after crossing with a copy of $\mathbb{R}$. (This observation suffices for proving the realization of torsions in the s-cobordism theorem.)

3. The theorem can be used to provide an alternative approach to the proofs of the main theorems in the topological category. The idea for this is to induct on the number of strata, as in the PL case, and use the ideas of [Ch2, Hu1, HTW1,3] for the beginning of the induction, i.e. for providing a sufficient analysis of the space of MAFs. (See also 9.4.2)

PART III: APPLICATIONS

This part is devoted to applications of the theory developed in Part II to specific stratified spaces, and occasionally to obtain some deeper understanding of manifolds.

Chapter 11 reviews some aspects of manifold and embedding theories from the stratified point of view and extends the main embedding theorem to stratified spaces. In addition, we prove some interesting results on immersions of stratified homotopy equivalent spaces. The embedding theorems have a general theoretical importance because they are used for putting an infinite loop space structure on the microsurgery sequence.

Chapter 12 is devoted to a deeper understanding of singular spaces that have simply connected links and that somehow resemble algebraic varieties. Intersection homology is an important tool in this development, so this chapter reviews some of that theory as well. The theory here is very computable and closely resembles the theory for manifolds. There are a number of applications here to questions about eta invariants of manifolds and to singular algebraic varieties.

Chapter 13 contains the deepest applications to date of the theory and is devoted to many aspects of G-manifolds of group actions. We shall examine some of the foundational aspects of the theory, as well as compute in many cases an equivariant surgery exact sequence. The results here also show how to modify given actions to produce new ones with exotic fixed point sets purely from the point of view of the general theory. We will also give the first substantial results on topological actions of compact Lie groups. Here the theory does not seem to resemble anything else that is more or less standard.

Chapter 14, the final chapter, is mainly about conjectures. These are extensions of the standard Novikov and Borel conjectures (see 4.6 and the 9.4.A and B) to the singular case. I will try to make these extensions plausible and show what additional information they entail. They also lead to new information about the geometry and topology of manifolds, especially with fundamental group which is not torsion free. This final chapter will not go to any great lengths to bring the reader to the state of the art on proofs of special cases of these conjectures. The front is moving day by day, but the picture of the situation outlined in chapter 14 seems to be more or less stable.

11 Manifolds and Embedding Theory Revisited

This chapter contains the simplest illustrations of the general theory. In the first two sections nothing new will be proven; rather we will consider manifolds with boundary and with isolated singularities from the stratified point of view to compare with what should be familiar to the reader from Part I. In the third section I will turn to embedding theory and simultaneously re-prove and generalize the codimension three embedding theory for manifolds (see 4.4) to stratified spaces. In the topological case this, inter alia, re-proves the homotopical characterization of local flatness due to Ferry [Fe1] in codimension one, Bryant in high codimensions, and Quinn [Q2, pt. I] (see also [Ch2]). The final sections deal with immersions and with low codimensions.

The reader who has read chapters 5 and 6 should have no problem following the treatment here.

11.1. Manifolds with boundary

I will not bother writing down theorems. Presumably, the reader knows what I'm trying to do.

There are two questions to be addressed. Firstly, what does the classification theorem say, and secondly, what is the classification theorem classifying?

In the *PL* case, the theorem is at first glance not saying anything since we are violating the codimension five assumption about strata. As pointed out in 6.1 this condition is sometimes avoidable by special low dimensional considerations. The point for us is that if we decide that we are studying manifolds with boundary, then we already know what the link of each simplex on the boundary is. The holink is a point, which has the same geometry as predicted by surgery theory! In other words, surgery predicts that its structure set should be contractible and therefore so should spaces of point block bundles, but this is clearly the case by inspection.

A similar discussion is necessary for proving the results of low codimensional embedding theory (4.4).

Now, the Whitehead groups of the links are trivial, so simple homotopy transversality is just homotopy transversality; i.e. we are after stratified homotopy equivalence, which is equivalent to just a homotopy equivalence of pairs (see 5.2).

Wh^{PL} is just the sum of the Whitehead groups of the strata. Note however that the involution is unusual, i.e. does not preserve the sum decomposition but must be modified by the map on Whitehead groups induced by the inclusion (see 7.1).

The L-group is clearly the usual relative L-group.

The cosheaf of L-spectra is interesting here. There are two kinds of points, interior and boundary points. For an interior point, one has the usual $\mathbf{L}(\mathbb{R}^n)$, which is[1] a shifted $\mathbf{L}(e)$. At boundary points, one has the cosheaf $\mathbf{L}(\mathbb{R}^n_+, \mathbb{R}^{n-1})$ corresponding to the closed upper half plane. The $\pi - \pi$ theorem (2.4) asserts that this is contractible. Thus, the cosheaf homology is simply relative $\mathbf{L}(e)$ homology.

Note that this is Poincaré dual to absolute $\mathbf{L}(e)$ cohomology = maps into F/Top, the form we had in 2.4 for the surgery sequence (aside from Kirby-Siebenmann difficulties (2.5.B)).

Also note that had we done rel ∂ classification, we'd have obtained the $\mathbf{L}(\mathbb{R}^n_+, \mathbb{R}^{n-1} \operatorname{rel} \mathbb{R}^{n-1})$ cosheaf, which is isomorphic to $\mathbf{L}(\mathbb{R}^n)$, and the whole cosheaf homology would be absolute $\mathbf{L}(e)$ homology, which, again, Poincaré duality identifies with the sequence in 2.4.

The topological case is a little different. The codimensional issue does not arise. $Wh^{Top} = Wh^{PL}$ in this case because the links are simply connected. A little thought then identifies the topological and PL structure sets (modulo the Kirby-Siebenmann obstruction; actually, the PT structure set would be more appropriate; see chapter 8).

It is more interesting to think about what is being computed: Rather than manifolds with boundary what one gets a priori are manifold stratified spaces with two strata and holink homotopy equivalent to a point. Given that one has the same answer as for manifolds with boundary, one recovers the following result of Ferry:

THEOREM [Fe1]. *A space is a manifold with boundary if it is a manifold stratified space with two strata with contractible holink.*

This can be viewed as a local homotopical characterization of local flatness in codimension one.

11.2. Isolated singularities

Again, this section doesn't really contain anything new; it is here for illustration purposes. Suppose that X is a space with an isolated singularity. In the PL (or PT) case this means that X is given the structure of a manifold with boundary with the cone on the boundary glued on.

[1] The reader should think through this identification; it makes use of an orientation. For unorientable manifolds one gets twisted homology.

Then the simple homotopy transversality implies that any equivalences we consider will restrict to simple equivalences on the boundary components, and Wh^{PL} is simply the Whitehead group of the closed pure stratum.

For simplicity, let's only describe the rel singularity theory. The L-group is the ordinary absolute L-group of the complement. In the classical theory, however, the L-group would be the relative L-group, because the boundary is not fixed by a mere simple homotopy transversality. This difference is accounted for by the difference in the normal invariant terms.

Classically, the singular point doesn't contribute at all and we have the relative $\mathbf{L}(e)$ homology of the complement. From the stratified point of view the cosheaf homology has different stalks at the manifold points and at the singularity: at the manifold points we get the usual $\mathbf{L}(e)$, but at the singular point we get a $\mathbf{L}(\pi)$ where π is the groupoid of the boundary, which is precisely the difference in the surgery terms between the classic and our sequence. This is summarized in the following diagram:

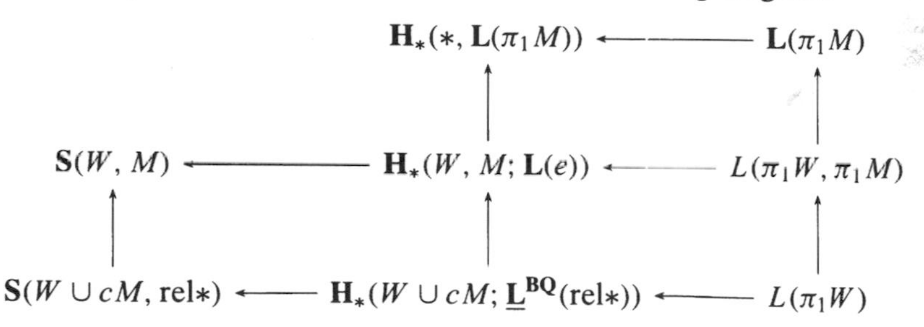

Note also that here we can see quite clearly the need for the codimensional assumption; without it, we'd be claiming surgical classification of low dimensional manifolds. (Note that in the PL case, modulo simplicity issues, a direct argument shows that $S(cM \times R^i) = S(M)$.)

The topological case corresponds to one point compactification of a noncompact manifold with tame ends. The Whitehead group is the proper Whitehead group of the top open stratum.

The surgery sequence that comes out of the stratified theory is somewhat more complicated than that described in 2.4.A because we have to stabilize and destabilize. In this example one can see the difficulty in just describing the structure set as the fiber of an assembly map: it is akin to trying to describe the L-spectrum decorated by a relative K-group as the fiber of some map between appropriately decorated absolute L-spectra. (It would involve h in the interior and p at the end, but this doesn't work because in the $\pi - \pi$ case one does get vanishing.) Thus, the fact that there is a nice conventional surgery theory in this case seems obscured by our trying to force it to be a global (i.e. no points excluded) fiber of a spectral cosheaf assembly map.

The sequence is simply the one obtained by stabilizing and destabilizing the sequence of 2.4.A. By the time we have removed all the K-theory, the relative L-group (spectrum) is the fiber of the map of corresponding absolute groups (spectra), and we measure the deviation in terms of a global object: Tate cohomology of a truncated global Wh^{Top}.

11.3. Embedding theory

Let us now return to the results of 4.4. We will ignore the low codimensional PL difficulties, since they are of the same nature as those of the manifold with boundary case.

DEFINITION. *An embedding $i : X \to Y$ is said to be* **pure** *if for each i there is a j such that $i(X^i) \subset Y^j$. In other words, pure strata are mapped into (not necessarily corresponding[2]) pure strata.*

DEFINITION. *A pure embedding is* **locally flat** *if restricted to each pure stratum it is a locally flat embedding into the corresponding pure stratum.*

DEFINITION. *A pure embedding has* **codimension at least c** *if the inclusion of each pure stratum in the stratum in which it lies has codimension at least c. Note that not every pure embedding has a codimension, although it is sometimes useful to view the codimension as a multi-index or to restrict attention to either the top stratum or the lowest codimensional strata or similar intrinsic combinations of strata.*

REMARK. Local flatness for codimension at least three embeddings of manifolds can be checked by verifying that the local fundamental group at each point (i.e. the fundamental group of the holink) vanishes. This follows from the material in 9.4, e.g. by putting a block bundle structure on the neighborhood or classifying an associated MAF (see [Ch2, Q2, pt. II]). In codimension two, one has to specify the local homotopy types; the cone of a nontrivial knot whose complement has fundamental group $\mathbb{Z}$ shows that local flatness cannot be characterized in terms of fundamental group alone.

DEFINITION. *A Poincaré embedding of X in Y consists of a Poincaré stratified object Z (see 5.1) where X is stratified homotopy equivalent to a closed union of strata, such that Z has a "coarsening" that is stratified homotopy equivalent to Y.*

Recall that a coarsening of a stratified space V is just another stratified space where strata are unions of old strata. It is not always possible to coarsen a stratification, but in the situation of an embedding it is possible to just ignore the subobject.

[2] Assuming the index sets are the same.

A Poincaré embedding is always pure in the above definition. Each pure stratum of X Poincaré embeds into a pure stratum of Y. If each of these has codimension at least c, then we say the Poincaré embedding has codimension at least c.

THEOREM. *Locally flat embeddings of $X \subset Y$ of codimension at least three are in 1-1 correspondence with Poincaré embeddings of $X \subset Y$.*

Note that when X and Y are manifolds this is the result of 4.4 (the Browder-Casson-Haefliger-Sullivan-Wall theorem).

COROLLARY. *If X and Y are respectively stratified homotopy equivalent to X' and Y', then there is a 1-1 correspondence between codimension ≥ 3 embeddings $X \subset Y$ and those of $X' \subset Y'$.*

This corollary can be used to give a pleasant description of an analogue of Siebenmann periodicity (3.4) for structure sets.

Here's the construction (see [CW3]) of a map $S(X) \to S(X \times D^4 \operatorname{rel} \partial)$ which is an isomorphism modulo the component $\mathbb{Z}$'s (see 6.2). Let X' represent an element of $S(X)$. Using the corollary embed it in $X \times D^3$; now take the pullback over the complement of X' of the Hopf bundle $(S^1 \to S^3 \to S^2)$ via the natural map to S^2. Make the radii of the circles shrink as you approach X' and glue X' in. This is a **branched $\mathbf{S^1}$-fibration.** This gives an element of $S(X \times D^4 \operatorname{rel} \partial)$.

It is not hard to parlay the argument given in [CW3] into a calculation of this map in our case; we will not do this here for two reasons. Firstly, the main application we had in mind was to produce circle actions on certain spaces; these results fall out much more simply from the results of the next chapter. Secondly, a more interesting view of the calculation will be described in 13.1 ("exotic products").

REMARK. This periodicity is a little surprising in that there are no K-theory difficulties. In terms of MAFs (see 9.4.B) and the geometry of germ neighborhoods (10.3.A) what is involved here is an isomorphism of homotopy groups:

$$\pi_{i+4}\left(\mathbf{MAF}(E \times \mathbb{R}^4 \to \mathbb{R}^{n+4})\right) \cong \pi_i\left(\mathbf{MAF}(E \to \mathbb{R}^n)\right)$$

at least in a stable range, where E is produced out of the local germ along a singular stratum. Thus the periodicity of structure sets is not quite the reflection of a periodicity of homotopy groups of a space of MAFs (which in fact only have the periodicity away from the prime 2, because of K-theoretic difficulties).

This geometric periodicity also geometrically gives the abelian group structure on structure sets and is therefore a useful thing to have around.

Another interesting example of the corollary is the case of group actions. The unequivariant case of the embedding theorem, proven in 4.4, depends on F_c/Cat_c stability. This is not true equivariantly (see 13.1), so it is pretty remarkable that the embedding problem remains homotopy theoretic.

Back to the proof of the theorem. Since one has a coarsening available, one can map $F(Y, X) \to F(Y) \times F(X)$ for any of the functors of stratified spaces we've considered, i.e. Wh in its BQ group and cosheaf versions and similarly for the various L's. In general one has a fibration $F(Y - X \text{ rel end}) \to F(Y, X) \to F(X)$ (see 7.1) so that the codimension assumption (together with the way these functors depend on fundamental group data) shows that the map $F(Y, X) \to F(Y) \times F(X)$ is an equivalence. Therefore the induced map on Wh^{Top}, its Tate cohomology, and the stable structure sets are all equivalences, and therefore the map

$$S(Y, X) \to S(Y) \times S(X)$$

is an equivalence as well. This is precisely the statement of the theorem except for the local flatness of the embedding.

The local flatness follows from the manifold case of the theorem. In that case, one argues that (see 8.1) $S^{PT}(Y, X) \to S^{Top}(Y, X)$ is an equivalence, and that the domain, built up out of block bundles, corresponds to topologically locally flat embeddings.

REMARK. There is an interesting point here. The forgetful map built by surgery $S(Y, X) \to S(Y)$ is geometrically interesting: it produces a manifold from a weakly stratified space with "homotopically mild" singularities (this means that all the holinks are homotopy spheres) along X. One can see that the manifold produced by the forgetful map "resolves" the original stratified space (this means that there is a map to the stratified space where the point inverses are contractible in any small neighborhood of themselves). In our codimension three setting, Edwards's theorem [Dvr] shows that any resolution can be approximated by homeomorphisms, so that, in particular, the original stratified space is a manifold. (We essentially used local simple connectivity to compare to the PT category and avoid Edwards's result.)

The reader might want to first think through the case where X is just a point. Then what has occurred is that the most natural forgetful map has led us to a "slightly singular manifold", i.e., a manifold with an isolated singularity whose holink is a sphere. Removing the point, one obtains a manifold with a tame end. It is simply connected (since the sphere is), so one can complete it by Siebenmann's thesis. The completion is a homotopy sphere, and therefore a sphere, so that the original singular space was a manifold.

We will use versions of this argument extensively in chapter 13.

See [Q5] (or our discussion in 9.4.D) for the general proof that an ANR homology manifold has a resolution if some open subset does, even if the singularities do not form part of a smaller stratified space.

11.4. Immersions

Classical immersion theory of Smale, Hirsch, Haefliger-Poenaru [HP], Lees, etc. reduces the problem of immersing one manifold in another to bundle theory. (An immersion is a map that is locally a locally flat embedding.) As a consequence of the stability theorems (4.4), one sees that homotopy equivalent manifolds simultaneously immerse in homotopy equivalent manifolds in the PL and topological categories (although not in the smooth category).

I do not know how to produce immersions via stratified space theory from bundle reductions, but below I sketch a more refined version of the homotopy invariance of immersion. As with the embedding theory, it also applies to general codimension three pure immersions, and hence applies to situations where bundle stability fails.

With a good solution to the following problem, one should be able to deduce the stratified homotopy invariance of immersion from the results of the previous section.

PROBLEM. Construct a stratified immersion theory for pure immersions.

THEOREM. *Consider an immersion $i : M \to W$ that is transverse to itself (i.e. in general position) and of codimension at least three. Then,*

$$S\big(W, i(M), \text{rel sing } i(M)\big) \cong S(W) \times S(M).$$

In particular, anything homotopy equivalent to M immerses in anything homotopy equivalent to W, and furthermore, these immersions have homeomorphic double, triple, and quadruple point sets, etc. in exactly the same configurations. (See also [Lvt2] for related results.)

That $S\big(W, i(M), \text{ rel sing } i(M)\big) \cong S(W) \times S\big(i(M), \text{ rel sing } S(M)\big)$ follows from the previous section. Now, we need to see that the "resolution of singularities map"[3] (blowing up!) $S\big(i(M), \text{ rel sing } i(M)\big) \to S(M)$ is an isomorphism. Using the codimension three assumption, Van Kampen implies isomorphism on Whitehead and global surgery theoretic terms.

To complete the argument one observes a quasi isomorphism of cosheaves $i_*\underline{\mathbf{L}}(M)$ with $\underline{\mathbf{L}}\big(i(M) \text{ rel sing}\big)$, which follows from examining how $i^{-1}(U)$ is a union of sheets, for any open set U.

[3]There is no general resolution of singularities in the theory of singular spaces. Here one is just associating the normalization which consists of pairs (x, g) where x lies in X and g is a "sheet" of X at x, suitably topologized. Each d-tuple point has d inverse images.

11.5. On codimensions one and two

The methods of this chapter do not directly apply to the situations of codimension one and codimension two submanifolds. However, here are some remarks on these submanifolds.

Codimension one.

I leave the following to the reader:

PROPOSITION. *The forgetful map* $\mathbf{L}^{BQ}(\mathbb{R}^n, \mathbb{R}^{n-1}) \to \mathbf{L}^{BQ}(\mathbb{R}^n)$ *is a homotopy equivalence.*

As a consequence, the normal invariant part in computing $S(W, M)$ for M of codimension one is, independently of M, the conventional normal invariant of W!

COROLLARY (See [Wa1]). *There is a fibration* $\mathbf{L}(\Phi) \to \mathbf{S}(W, M) \to \mathbf{S}(W)$*, where* Φ *is the diagram of fundamental groups of* M*,* $W - M$*, and* W.

One just has to identify this L-group of a diagram of groups with the fiber of the map $\mathbf{L}^{BQ}(W, M) \to \mathbf{L}(W)$.

The issue is then how to compute this L-group. The discussion varies according to whether or not the normal bundle of M in W is trivial. In the trivial case, Cappell's codimension one splitting theorem is equivalent to the following (see [Ca1-4], 4.6.A):

THEOREM. *If* $M \subset W$ *is a codimension one submanifold with trivial normal bundle, whose fundamental group injects, then the fiber* $\mathbf{S}(W, M) \to \mathbf{S}(W)$ *has 2-torsion homotopy groups. Furthermore, if the fundamental group of* M *includes that of* W *in a square root closed fashion (for* $g \in \pi_1 W$*,* $g^2 \in \pi_1 M$ *implies* $g \in \pi_1 M$*), and assuming some algebraic K-theory conditions, then this map is a homotopy equivalence.*

The algebraic K-theory conditions can be obviated through careful choices of decorations. Indeed Cappell introduced decorations to specifically handle this difficulty!

If the submanifold has nontrivial normal bundle, then the fiber can indeed have highly nontrivial homotopy groups. For the classical case of $\mathbb{R}P^n \subset \mathbb{R}P^{n+1}$, the analysis was given by Browder and Livesay [BLi]. The general case is discussed in [CS2].

Codimension two.

Again we have a local calculation:

PROPOSITION. *The forgetful map* $\mathbf{L}^{BQ}(\mathbb{R}^n, \mathbb{R}^{n-2}) \to \mathbf{L}^{BQ}(\mathbb{R}^n)$ *is a homotopy equivalence.*

This leads to a similar corollary regarding classification of embeddings in a given Poincaré embedding class. However, this is not such an interesting viewpoint for this problem, in general. It ignores the whole existence of knot theory, a large classical and still thriving subject!

A development of knot theory and more general codimension two embedding theory from the point of view of Poincaré embeddings appears in [CS1]. They also apply their methods beautifully to *PL* nonlocally flat embeddings in [CS4]. The basic idea is that codimension two embeddings naturally give rise to surgery problems in which one only tries to achieve homology equivalences. (After all, for knots in the sphere, all of the complements have the homology of a circle, but only the unknot's complement has the homotopy type of a circle.) Then, codimension two can be approached in a formally analogous way to the higher codimensions, with just other calculations entering. An important feature is a comparison theorem between homology surgery groups and conventional ones, which leads to qualitatively different results in odd and even dimensions. (See [CS8] for a survey of some of this work. It would be useful for someone to write a more modern survey of this approach to codimension two phenomena.)

Yet more recently, le Dimet [lD] has shown that it is convenient to weaken further the notion of Poincaré embedding for codimension two problems (or more precisely, lump many Poincaré embeddings together and classify them homotopy theoretically). An important application of this idea to link theory is given in [CO]. (See also [GiL].)

I believe that these ideas can be incorporated into the general framework of this book, at least in the *PL* case. (Cappell and I re-proved the nonlocally flat embedding theorem of [CS4] using ideas similar to those used for the replacement theorems in chapter 13.) The topological nonlocally flat theory seems harder in that controlled homology surgery must be developed, and there are certainly some new wrinkles. (Ferry and I have some tentative results for the odd dimensional case.) Le Dimet's idea can also probably be adapted to the more general homotopical framework of Poincaré stratified homotopy types. Whether or not this can be done efficiently and fruitfully is yet to be seen and seems to present a nice group of problems.

12 Supernormal Spaces and Varieties

In this chapter we will give a more substantial application of the theory of the previous chapters to some special classes of stratified spaces. We will assume that the spaces are **supernormal**, i.e. that the local holinks of all strata are simply connected and, for the more global results, in addition, that the space has only even codimensional strata. Algebraic varieties over $\mathbb{C}$ are important examples of spaces all of whose strata are even dimensional. It turns out that for these spaces one has a very close analogue of surgery theory for manifolds. Also, intersection homology fits into the picture very nicely, and we will be able to distinguish some spaces that very closely resemble varieties from varieties.

I would like to point out that most of the main results of this chapter are joint work with Sylvain Cappell and preceded the general theory. (It might therefore not be such a surprise that they fall out so easily from the general theory!) Many of the remaining results are due to the author, Cappell, and Julius Shaneson in various combinations.

12.1. Supernormal spaces

DEFINITION. *A manifold stratified space is* **supernormal** *if all the local holinks (see 5.1) of each stratum in all higher strata (i.e.* $X^i \subset X^i \cup X^j$*) are simply connected.*

This terminology is suggested by familiar definitions in algebraic geometry. One can always normalize a variety to obtain a normal one, which has the property that the link of each simplex in the singular set is connected. One can elaborate this to make higher connectivity assumptions. Sometimes one can deduce super(duper) normality from a codimensional assumption for a singular variety; see the "theorems of Barth type" [Ful1, GM2].

THEOREM ([CW2]). *If X is an n-dimensional supernormal space, $n \geq 5$, with singularities S, then* $\mathbf{S}(X, S) \cong$ *fiber* $\mathbf{H}_n(X; L(e)) \to \mathbf{L}_n(\pi_1 X)$. *In other words, the rel S structure set is given by the same homological description as for manifolds.*

This is because the rel S sheaves are just $\mathbf{L}(e)$. Note also that one only needs the simple connectivity of the holinks in the top stratum.

I would like to sketch an example from our original paper that means a lot to me.

EXAMPLE. If M and N are simply connected manifolds, the join $M * N$ is a supernormal space. There is a fibration

$$\mathbf{S}(M * N \operatorname{rel} M \cup N) \to \mathbf{S}(M * N) \to \mathbf{S}(M) \times \mathbf{S}(N).$$

We know how to calculate all the terms but the middle. However, there is a section to this fibration, given by joining! Thus one has

$$\mathbf{S}(M * N) \cong \mathbf{S}(M * N \operatorname{rel} M \cup N) \times \mathbf{S}(M) \times \mathbf{S}(N).$$

This is quite odd from the stratified point of view (6.1 and 6.2). The L-spectra around the singular components depend strongly on the homotopy types of M and N. This collapse of the fibration is an example of an important principle: **Trivial constructions in some special situations give rise to difficult calculations of the surgery sequence**. (If one is very lucky, these calculations can then be applied to other situations where the "trivial construction" is unavailable.) In our case, joining is a construction that the general stratified theory finds hard to assimilate. If the signature of M and/or N are/is 1, then this collapse is related to Siebenmann periodicity (3.4). If the signatures are otherwise, then this is a sort of periodicity with coefficients.

If M and N are complex projective spaces, then the structures on the quotient are the isovariant (see 13.1) structures on the linear action of S^1 on $\mathbb{C}P^n$. This calculation, by a much more awkward method, was one of the main results of the 1987 CBMS lectures on group actions on manifolds ([CW1]).

EXERCISE ([CW1]). What relationships are there between the splitting invariants of the fixed sets, the normal representations (i.e. the local holinks of the quotient) at the various components, and the invariants of the ambient homotopy $\mathbb{C}P^n$? (See 4.1 for splitting invariants of homotopy $\mathbb{C}P^n$'s.)

In particular, there are examples of locally linear S^1 actions on fake $\mathbb{C}P^n$'s with exotic Pontrjagin classes. A well-known, much studied conjecture of Petrie [Pe1] asserts that this cannot happen in the smooth case. These locally linear examples were first constructed in [CW3] as an application of the geometric construction of Siebenmann periodicity (see also 11.3). By Petrie's work, none of these are smoothable.

12.2. Intersection homology

Homology is, of course, the homology of the chain complex built up out of all chains on a pseudomanifold[1] X. If one only allows chains that are transverse to the singular set, i.e. that intersect any stratum of the singular set in a subset of the codimension of that stratum, then one obtains a complex that computes (according to an old theorem of McCrory) the cohomology (in the dual dimension) of X. In a manifold, there is no difficulty in moving all chains to be transverse, and the identification of these groups is Poincaré duality.

The failure, then, of Poincaré duality is related to the difficulty in making chains transverse to the stratification. Goresky and MacPherson [GM1, pt. I] had the bold idea of introducing many more homology groups, based on different perversities, i.e. failures of transversality. These groups interpolate between homology and cohomology. The groups in complementary dimensions with opposite perversities are dual as a type of Poincaré duality that is true for all pseudomanifolds.

More precisely, a **perversity** is a nondecreasing function $p : 2, 3, \ldots, n \to \mathbb{N}$ such that $p(2) = 0$ and $p(n+1) \leq p(n)+1$. There are two extreme perversities: the **zero perversity**, $0(c) \cong 0$; and the **total perversity**, $t(c) = c - 2$. Perversities m and n are dual if $m + n = t$.

A k-chain will be said to be p-transverse if its intersection with the codimension c stratum of X has dimension $\leq k - c + p(c)$ (and similarly for its *algebraic*[2] boundary). Closed k-chains modulo boundaries as usual form a homology group, denoted $IH^p(X)$.

THEOREM ([GM1, pt. I]). *$IH(X)$ is a topological invariant, independent of the choice of stratification. If we take field coefficients, $\mathbb{F}$, then the intersection homology groups in dual dimensions with dual perversities are paired perfectly by taking intersections of chains.*

More explicitly, chains of complementary dimensions with complementary perversities can be isotoped slightly to intersect[3] in a finite number of points, whose sum with sign $\in \mathbb{F}$ is well defined, yielding a pairing $IH_k^p \otimes IH_l^q \to \mathbb{F}$ when $k + l = n$, $p + q = t$. Furthermore, this gives an isomorphism $IH_k^p \to \mathrm{Hom}(IH_l^q, \mathbb{F})$.

For many spaces, it is unnecessary to give all values of the perversity. After all, if there is no codimension c stratum present, then $p(c)$ is irrel-

---

[1]A pseudomanifold is a polyhedron with codimension two singularities whose complement is a dense open set.

[2]The fact that we use algebraic boundary leads to a lack of identification of $IC(\ \ ; R)$ with $IC(\ \ ) \otimes R$, so that, in general, one does not have a universal coefficient sequence in IH. See [GS] for cases when the sequence is valid for more subtle reasons.

[3]More generally, one can intersect chains of different perversities to obtain a chain of the right codimension with coefficients in the sum of the perversities (assuming this is a perversity).

evant. For instance, if X has only even codimensional strata, then if we consider the two middle perversities m and n, $m = \{0, 0, 1, 1, 2, 2, \ldots\}$, $n = \{0, 1, 1, 2, 2, 3, \ldots\}$, then $IH^m(X) = IH^n(X)$.

COROLLARY. *If X has only even codimensional strata, then the middle intersection homology groups (with $\mathbb{F}$ coefficients) have a "Poincaré" self-duality.*

If we write an intersection homology group without specifying its perversity, then we will always mean with middle perversity.

REMARK. It is not true that any X with even codimensional singularities has a $\mathbb{Z}$ self-duality. The case of isolated singularities is instructive (for seeing this as well as many other things about IH). Below some dimension (determined by the perversity), chains and their boundaries are not allowed to touch the singular points, so that the homology is ordinary homology (with compact supports) of the complement. In high dimensions (again determined by the perversity), everything will be allowed through the singular points, so that one obtains the ordinary homology of the space = locally finite homology (= rel ∂ homology) of the (closed) complement of the singularities. In the critical dimension, one has chains in the complement with coboundaries allowed to extend through the singular set, which can be described as the image of an ordinary group into a relative one.

Also, it is very useful to extend, following Siegel [Sg] and Cheeger [Che1,2], to a larger class of spaces, the **Witt spaces**. One allows there to be some odd codimensional strata but demands that their middle dimensional middle perversity IH vanish. If you think about it, you'll see that the open cone on such a space (thought of as having an isolated singularity of a new type) satisfies Poincaré duality, so that these are the IH analogues of homology manifolds in the usual theory: they are the spaces which have self-duality for a local reason.

THEOREM ([Sg, Che2, Gm1, pt. II]). *Witt spaces have self-dual IH.*

Isomorphic (or, better, dual) cohomology groups were discovered entirely independently at around the same time by Cheeger [Che1,2] in the course of other, more analytic investigations. Sullivan, having spoken to both sets of workers and heard that they had discovered groups that satisfy self-duality for spaces with even codimensional strata, immediately conjectured that the groups are the same; a proof of this appears in [GM1, pt. II] as a consequence of verifying that both satisfy a certain characterizing set of axioms.

I will be brief with my description of Cheeger's work since I have nothing to add to his papers. He considers spaces which have locally conical metrics.[4] For an isolated point this means that the metric looks like the metric one would put on a Euclidean cone. Then for a product of such a cone with a manifold, one uses the product metric. All pseudomanifolds are built up out of these.

Then he considers the L^2 DeRham complex on the incomplete manifold $X - S$, where S is the singular set. Happily, there is a Hodge decomposition for Witt spaces given such a metric, and the $*$-operator induces the Poincaré duality ([Che1]). An immediate consequence of this is:

THEOREM ([Che2]; KUNNETH FORMULA). *If X and Y are Witt spaces, then $IH(X \times Y) \cong IH(X) \otimes IH(Y)$.*

It would be a real challenge to prove this from the combinatorial point of view. (An axiomatic treatment is given in [GM1, pt. II].)

Furthermore, there is a version of the Atiyah-Patodi-Singer [APS] index theorem; see [Che2]. The main application that Cheeger gives to his ideas is a local formula for L-classes.[5] In terms of putting piecewise flat metrics on the simplices of X one gets a formula

$$L_{n-4k} = \sum \eta(\text{link } (\Delta))\Delta$$

where the sum is taken over all Δ of dimension $n - 4k$. One way to prove this is to observe that this formula is correct for $n = 4k$. This is the content of an Atiyah-Patodi-Singer formula. Then one observes that this expression behaves correctly with respect to taking products with $\mathbb{R}^i$ and to taking open subsets. The method of Thom-Milnor (see [MS]) for defining L-classes then shows that the L-class agrees with any other definition.

Once we have gotten to this point, we can define the expression $\sum \eta(\text{link } (\Delta))\Delta$ for any Witt space and see that it's independent of choices of metrics and the like. Alternatively, and this is done in [GM1, pt. I], one can follow the Thom-Milnor approach directly using IH signature, its cobordism invariance, and a little transversality to define L-classes for Witt spaces. And, of course, the results agree.

Now, I shall turn to the "Deligne construction" [GM1, pt. II]. It is defined sheaf theoretically and therefore has the advantage of being definable, for instance, in characteristic p. One needs only a few formal operations on sheaves and a decent stratification theory to define it.

In addition to its theoretical advantages, it is also incredibly useful for formal manipulations and calculations. We shall be using this approach below.

[4]One of the remarkable new emphases suggested by his work was to focus attention on what happens for other classes of metrics. This issue is at the core of many important phenomena; see e.g. [CGM, SS, Lo].

[5]In certain cases, [CS5] extends the sheaf-theoretic approach to give local formulae for characteristic classes.

In [GM1, pt. II] sheaf-theoretic ideas are applied to proving the Kunneth formula and a Lefshetz hyperplane section theorem for singular varieties. In [GS] the Deligne construction is used for describing when integral Poincaré duality holds and for analyzing the universal coefficient theorem. Also, I cannot imagine how the proofs of the formulae in [CS5,6] would go without the formalism of the derived category.

Before explaining Deligne's construction and how this leads to a proof of duality, I must describe the derived category, its various derived functors and pushforwards, and Verdier duality. Useful references are [GM1, pt. II, Bo, Iv]. For us, **all sheaves are henceforth assumed to be constructible** (i.e. locally cohomologically constant with respect to some stratification).

Two *bounded* complexes of sheaves $\underline{C}$ and $\underline{D}$ are equivalent in the **derived category** if there is a third complex $\underline{E}$ with maps $\underline{C} \to \underline{E} \leftarrow \underline{D}$ which are **quasi isomorphisms**, i.e., which induce homotopy equivalences on every stalk. A morphism in the derived category is represented by a morphism of complexes of sheaves up to a quasi isomorphism. In the derived category, one replaces complexes of sheaves by injective resolutions, so that the homological algebra becomes nicer. The topologist might find it useful to think about the analogous homotopy category of spaces over X (without control, i.e. morphisms must strictly commute with the map to X).

Verdier introduced an interesting complex of sheaves D_X, called the **dualizing complex**, with the property that its stalk cohomology at x is the local cohomology of X at x, $H^i(X, X - x)$. (In fact, D_X is equivalent to the local singular chain complex; for an open U we get $C_p(X, X - U)$.) Global cohomology of X with coefficients in the dualizing complex is ordinary homology. Geometrically, this is dual to the fact we started this section with: homology defined using transverse chains is cohomology. Dualizing is closely related to local Spanier-Whitehead duality and defining homology as the cohomology of the Spanier-Whitehead dual. If $\underline{E}$ is in the derived category, we define $D\underline{E} = R\mathrm{Hom}\ (\underline{E} : D_X)$ and it is called the **dual of $\underline{\mathbf{E}}$**.

We can define some interesting functors between different derived categories on different spaces associated to a map. If $f : X \to Y$ is a continuous function, then there are at least the following induced maps on derived categories. We can take the derived functors of pushforward and pullback Rf_* and Rf^*, and we can take the analogous proper analogues $Rf_!$ and $Rf^!$. These latter have nice descriptions in terms of dualizing: $Rf_! = DRf_*D$. For proper maps, $Rf^* = Rf^!$ and for closed inclusions $Rf_* = Rf_!$, but in general these notions disagree. **Verdier duality** asserts that

$$Rf_* R\mathrm{Hom}(\underline{A}, Rf^!\underline{B}) \cong R\mathrm{Hom}(Rf_!\underline{A}, \underline{B}).$$

Verdier duality implies that the induced pairing of hypercohomology groups $H^i(\underline{A}) \otimes H^{-i|}(D\underline{A}) \to H^0(D_X) \to \mathbb{F}$ is perfect.

It is quite feasible to do homological algebra in the derived category; indeed it is built for that. The reader should realize that it is not an abelian category, so that one cannot take kernels and cokernels. What replaces this is the **distinguished triangle**. It is the obvious extension of the situation one gets by considering a mapping cone, where one gets in addition to the usual degree zero maps of complexes a degree one map from the cone to the domain of the map.

The final ingredient in the construction is the collection of **truncations** of complexes.

$$(\tau^{\leq p}\underline{A})^n \begin{cases} = \underline{A}^n & \text{if } n \leq p-1 \\ = \underline{\ker} \, d^n & \text{if } n = p \\ = 0 & \text{if } n \geq p+1. \end{cases}$$

There is a similar definition for **cotruncation**. Furthermore, one can truncate over a closed subset by sheafification of the result of truncating only on the open sets that touch the closed set. All of these constructions pass to the derived category.

THEOREM ([GM1, pt. II]). *Let X be a stratified space and let $U_k = X - X_k$ be a filtration by open sets, and i_k the inclusion $U_k \to U_{k+1}$. Let*

$$\underline{P} = \tau^{\leq p(n)-n} Ri_{n*} \ldots \tau^{\leq p(2)-n} Ri_{2*} \mathbb{F}_{X-S}[n].$$

The final $[n]$ is a shift by n. Then $\underline{P}$ is quasi-isomorphic to $IC^p(X)$.

The proof, in the spirit of Eilenberg and Steenrod, is axiomatic. There are various axiomatizations of intersection homology with the following ingredients: constructibility, normalization ($\mathbb{F}$ on the nonsingular part, i.e. the analogue of the dimension axiom), a lower bound axiom for the vanishing of homology, and an upper bound axiom. One also has some choice. One can assume an attaching axiom that tells how the pieces are connected. Alternatively, one can use support and cosupport conditions restricting the dimensions of pieces with high dimensional local homology and cohomology. For the purposes of these notes we do not need any specific characterization, so I will not quote any.

From this point of view Kunneth becomes quite easy; one simply verifies the axioms for the $IC(X) \otimes IC(Y)$. This is typical of the way in which the abstract uniqueness theorem for IH becomes a powerful computational tool.

I will not explain how one proves these uniqueness theorems in detail. Roughly speaking, one does obstruction theory in the derived category. The basic lemma for producing maps is the following:

LEMMA ([GM1, pt. II]). *Suppose that $\alpha : \underline{A} \to \underline{C}$ and $\beta : \underline{B} \to \underline{C}$ are morphisms in the derived category, that $H^i(\underline{A}) = 0$ for $i \geq k+1$, and that β is a cohomology isomorphism for $i \leq k$; then α has a unique lift $\underline{A} \to \underline{B}$.*

The proof is not that hard and resembles in an algebraic language the proof of the analogous results for extending maps between spaces with low dimensional cofibers to spaces with highly connected fibers.

The duality of IH^p and IH^q is now deduced by verifying the q axioms for the dual of IC^p.

12.3. Characteristic classes of self-dual sheaves

The sheaf-theoretic construction of the Poincaré duality in IH emphasizes the significance of self-duality of IH on the sheaf level. I would like to continue the development by explaining how to associate a characteristic class to any self-dual sheaf, so that the L-classes obtained via the Cheeger-Goresky-MacPherson procedures are equal to the Pontrjagin character of a K-homology class associated to the IC sheaf. In an appendix we will find some further applications of this additional generality.

The idea of this section is joint work with Cappell and Shaneson [CSW] (restricted to the case of trivial group). Although I will be working with fields throughout, if one assumes a torsion-freeness condition as in [GS] one can work over more general rings. These results apply equally well to equivariant situations and so will be of some use in the following chapter.

DEFINITION. *A sheaf $\underline{E}$ on X is self-dual if the following data exist: A quasi isomorphism $\varphi : \underline{E} \to D\underline{E}$, a homotopy from φ to $D\varphi$ (identifying $DD\underline{E}$ with $\underline{E}$), a homotopy of this homotopy to its dual, etc.*

REMARK. If the field $\mathbb{F}$ has characteristic $\neq 2$, then one just needs a homotopy from φ to $D\varphi$.

PROPOSITION. *If X is a Witt space, then $\underline{IC}(X)$ is self-dual.*

The higher coherencies are produced using the obstruction theory lemma from the previous section.

PROPOSITION. *There is a functor from self-dual $\mathbb{F}$ sheaves on $X \to$ controlled visible algebraic Poincaré $\mathbb{F}$ complexes on X.*

We will not go through the necessary visible theory here; see [Ws]. If we invert 2, we can use the pairing $L^*(\mathbb{F}) \otimes L_*(\mathbb{Z}) \to L_*(\mathbb{F})$ and the quadratic theory we discussed in chapter 9. In the PL case one can use transversality and the methods of 2.5 and its appendices.

COROLLARY. *One can associate a symmetric signature of a Witt space in $L^*(\mathbb{F}\pi)$. It has all of the usual cobordism invariance properties.*

COROLLARY. *We can associate to any self-dual sheaf on X a characteristic class in $H_*(X; \mathbf{VL}^{*-\infty}(\mathbb{F}))$. For $IC(X)$ this hits the class of the previous corollary under assembly.*

This enables one to redo many of the arguments of manifold theory for Witt spaces (away from 2 and for Goresky-Siegel spaces even at 2). For some first steps in this direction, see [Cu]. As another corollary one sees that these characteristic classes are topological invariants.

12.3.A. Applications of Witt spaces

The Witt spaces that have been introduced as the IH homology manifolds also have some remarkable applications to the parts of topology that involve the signature operator (which is a large chunk).[6]

One should start with the first theorem of this sort:

THEOREM (SIEGEL [SG]). *The map sign : $\Omega_*^{Witt} \to L^*(\mathbb{Q})$ is an isomorphism (above dimension zero).*

COROLLARY.

$$\Omega^{Witt}(Z) \otimes \mathbb{Z}[1/2] \cong KO(Z) \otimes \mathbb{Z}[1/2].$$

$$\Omega_n^{Witt}(Z) \otimes \mathbb{Z}_{(2)} \cong H_n(Z; \mathbb{Z})_{(2)} \oplus H_i(Z; L^{n-i}(\mathbb{Q}) \otimes \mathbb{Z}_{(2)})$$

where the sum of the right is taken over $i < n$.

The corollary follows from the methods of 2.5 and 2.5.A. Siegel's proof of the theorem is beautifully geometric.

Recently Goresky and Pardon [GP] have computed additional bordism of interesting singular spaces. Pardon [Pa] has shown that the bordism of the spaces introduced in [GS] defines $\mathbf{L}^*(\mathbb{Z})$-homology. These results have a useful consequence:

COROLLARY ([RSW4]). *There is a natural transformation*

$$\bigoplus H_{n-4i}(X; \mathbb{Z}_{(2)}) \to K_n(X) \otimes \mathbb{Z}_{(2)}$$

such that if X is a manifold there is a natural refinement of the class of the signature operator to homology at 2.

---

[6] I hate to suggest this, but it might require another book to do this topic justice.

In that paper, Rosenberg and I also identify the preimage in terms of the characteristic class of topological manifolds introduced in [MoS]. This class enters in proper integral forms of the Novikov conjecture (see 4.6.A). This also requires the identification of the class of the signature operator with what is produced by PL techniques.

Cappell and I noticed that one can apply this cobordism result to re-prove, without serious analysis, the disproof of the integral Hodge conjecture given in [AtH1]. From the fact that algebraic varieties are Witt, one sees:[7]

COROLLARY. *Any algebraic cycle in* $H_*(Z; \mathbb{Z}[1/2])$ *comes from* $KO(Z) \otimes \mathbb{Z}[1/2]$.

You should look at the original paper to see how this restricts cohomology operations on algebraic cycles and how to construct the counterexample varieties and classes from this criterion. The [AtH1] result at 2 is an immediate consequence of resolution of singularities and the results in [Th]. In fact, one sees that for any algebraic cycle, any Steenrod operation Sq^I with any odd index vanishes (e.g. anything like Sq^9 and $Sq^3 Sq^2$ must vanish). See [Ful2] for more information.

There have been some additional applications of this cobordism calculation to do interesting calculations of signatures and peripheral invariants. One of the nicest is in the paper [Sg]. Using the "pinch cobordism" he proves Novikov additivity for the signature of the union of two manifolds along a boundary component. (The pinch cobordism is obtained by taking the mapping cylinder of the collapse map collapsing the boundary component to a point, and adding a collar to the target; see fig. 19.) A more complicated version gives the formula for the deviation from additivity when one glues along a piece of boundary.

THEOREM ([WEI6]). *The invariant* $\eta_\rho(M)$ *associated to an odd dimensional manifold and a unitary representation* $\rho : \pi \to U(n)$ *by [APS] (see also 4.7.A) is a homotopy invariant for any group* π *satisfying the Borel conjecture (rationally) or its* C^**-algebra K-theory analogue.*

See 4.6.A for the Borel conjecture. The free abelian case[8] and, more generally, the case of the Cappell-Waldhausen class are verified in that appendix. Farrell and Jones have recently announced that it holds for at least all lattices in connected real Lie groups. For more information and applications of this theorem see [Wei6]. In the exercises in 4.7.A we have already seen that for familiar groups with torsion this invariant is not a homotopy invariant. However, I recently combined the following

[7]For this argument one does not need an isomorphism of Witt bordism to K-theory, just the existence of a map related to signature.

[8]In this case Neumann [Neu] has given not only a proof of this theorem but also a homotopy invariant formula for the invariant.

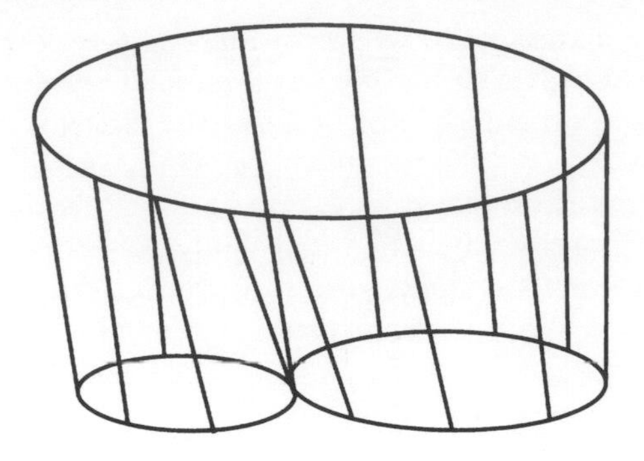

Figure 19.

argument with work of Farber and Levine to show that for homotopy equivalent manifolds M, M', $\eta_\rho(M) - \eta_\rho(M')$ is rational.

The argument goes like this. The Novikov conjecture asserts that for homotopy equivalent manifolds the higher signatures agree, and therefore homotopy equivalent manifolds are Witt cobordant (rationally). By [APS], extended by Cheeger [Che2] to Witt spaces, the difference $\eta_\rho(M') - \eta_\rho(M) = \text{sign}_\rho(X)$ where X is this cobordism. By Novikov additivity, this signature does not change if we glue the ends together by the homotopy equivalence to obtain a closed object. This is a bad object since it now has codimension one singularities. Nonetheless, if we use the ordinary constant sheaf away from the old singularities and the Deligne sheaf at those, we obtain a closed $\mathbb{C}\pi$ Poincaré complex. According to (C^*-algebra K-theory) Borel, the (signature of the) underlying algebraic Poincaré complex is cobordant (rationally) to one coming from a smooth closed manifold. For these, the Atiyah-Singer theorem gives the vanishing of sign_ρ.

REMARK. The method of this example applies as well to other secondary (peripheral) invariants.

EXERCISE. Show that the same argument can be made for rational equivalences. Deduce a vanishing theorem (for certain fundamental groups) for $\eta_\rho(M)$ if M supports a free action of a finite group such that $M \to M/G$ splits on fundamental groups, and the action of G on twisted homology is trivial.

EXERCISE. Show that all of the above results hold true for Witt spaces and homotopy transverse maps.

EXERCISE. Using the Goresky-Siegel structure on $c\mathbb{C}P^{odd}$ show that if M and M' are homotopy equivalent and $E \to M$ is a block fibration with $\mathbb{C}P^{odd}$ as fiber, then the map $E \to M'$ is also homotopic to a block fibration. (See also 3.3, 4.4, and 13.4.)

EXERCISE (1) ([GM1, pt. II]). Show that if $f : X \to Y$ is a small resolution of singularities, which means, by definition, that for each r, codim $\{y | \dim f^{-1}(y) \geq r\} \geq 2r+1$, then $IH(X) \cong IH(Y)$. (Hint: Prove that $Rf_* \underline{IC}(X) \cong \underline{IC}(Y)$.)

(2) (Cappell, Shaneson, and Weinberger) Show that $f_* L(X) = L(Y)$ (and similarly for Sullivan classes). In particular, a variety with a small resolution has L-class in the image of intersection homology.

REMARK. In 12.4.A we will extend this final exercise to restrict L-classes of arbitrary singular varieties.

12.4. Spaces with only even codimensional strata

Now let's apply the general theory to supernormal spaces with only even codimensional strata.[9] We have the following result:

PROPOSITION (SEE [CW2]). *If X is a supernormal space with only even codimensional strata, then $L^{BQ}(X) \otimes \mathbb{Z}[1/2] \cong \bigoplus L(\mathbb{Z}\pi_1(X_i)) \otimes \mathbb{Z}[1/2]$. Furthermore, the same is true on the sheaf level.*

Note that supernormality implies that $\pi_1(X_i) \cong \pi_1(X^i)$. The previous section produces a map $L^{BQ}(X) \to L(\mathbb{Q}\pi_1(X))$. Remember that $L(\mathbb{Q}\pi) \otimes \mathbb{Z}[1/2] \cong L(\mathbb{Z}\pi) \otimes \mathbb{Z}[1/2]$. Thus we get a map $L^{BQ}(X) \otimes \mathbb{Z}[1/2] \to L(\mathbb{Z}\pi_1(X)) \otimes \mathbb{Z}[1/2]$. By restriction we get a map into the right-hand sum. Now, thinking about how $L^{BQ}(X)$ is built up out of the $L(\mathbb{Z}\pi_1(X^i))$ makes the result clear.

EXERCISE. Verify that $Wh^{top} \cong \bigoplus Wh(X_i)$.

COROLLARY ([CW2]). *Away from the prime 2, if X is a supernormal space with only even codimensional strata, then*

$$S(X) \otimes \mathbb{Z}[1/2] \cong \bigoplus[\textit{fiber } H(X_i; \mathbf{L}(e)) \to \mathbf{L}(\mathbb{Z}\pi_1(X_i))] \otimes \mathbb{Z}[1/2].$$

This means that one can vary the L-classes (or the Δ-class in $KO[1/2]$) of such spaces at will only subject to the condition that these give rise to the symmetric signature after assembly, just as for manifolds.

One also can then phrase analogues of the Novikov conjecture for stratified homotopy equivalences, and the like.

REMARK. The proposition in this section shows that the universal characteristic class explained in chapter 6 (both sections) just reduces to the direct sum of standard Δ-classes for the various closed strata.

---

[9] Actually, we could work with Witt spaces all of whose strata are Witt, or work relative to strata that are not.

We also see that for some of the joins of the projective spaces considered in 12.1 (when both projective spaces are odd complex dimensional) the classification is in terms of a priori invariants, since these are supernormal Goresky-Siegel spaces. It still is a little mysterious to interpret the surgery theory of that example in the more general case, since the links have nonvanishing signatures, so that it seems to take a stratified homotopy equivalence to *define* the characteristic classes involved.

12.4.A. The BBDG decomposition theorem and its application to characteristic classes of singular varieties

In this appendix, I will try to describe some implications of algebraic structure for the characteristic classes considered in the previous sections. I believe that we are just seeing the tip of the iceberg, so I will be content with just indicating the direction.

We have seen that intersection homology has for certain nonmanifolds such as varieties many of the same properties that ordinary homology has for manifolds. In addition to Poincaré duality, we have added the tools of surgery theory and indices of the signature operator.

It is not all that (conceptually) difficult to add to the list the Lefshetz hyperplane section theorem. In [GM1, pt. II] a proof is given by the sheaf-theoretic techniques, and in [GM2] a proof is given by a stratified extension of Morse theory and the argument given in [Mi8].

The consequences of Hodge theory have proven more resistant. Hard Lefshetz is due to A. Beilinson, J. Bernstein, P. Deligne, and O. Gabber and appears in [BBD]; Hodge structure, first conjectured in [CGM] was proven by M. Saito [Sai] by D-module techniques.

The method in [BBD] is via an analysis of the category of algebraically constructible **perverse sheaves** (these are constructible sheaves that have slightly worse connectivity and coconnectivity properties than the intersection chain sheaves) and a comparison with characteristic p. Their theory has many more remarkable consequences for the topology of algebraic maps. (See [GM3] for some of these: collapse of spectral sequences, generalized invariant cycle theorems...) Their main result is a decomposition theorem for perverse sheaves:

THEOREM ([BBD]).

a) *The category of perverse sheaves on X is an artinian abelian category whose simple objects are the complexes $IC(V; L)$ where V is an irreducible subvariety of X and L is an irreducible local system on V.*
b) *One has all of the usual pushforward and ! maps defined for algebraic maps.*
c) *$Rf_*IC(X)$ is actually a sum of simple objects.*

Furthermore, they show that in part c the sum looks like a Hodge decomposition to some extent. This "refinement" includes the hard Lefshetz, so there is certainly much important information that I am leaving out.

Cappell and Shaneson [CS6] have proven a very general decomposition theorem for self-dual complexes of sheaves (under an appropriate cobordism relation) under arbitrary stratified maps. (To contrast, [BBD] only deals with algebraic maps.) Their theorem could be used just as effectively for most of the following arguement. In fact the restrictions given below on characteristic classes for algebraic varieties were first obtained in that paper. The only novelty in this appendix lies in showing how much of this follows from the formal structure of the theory together with [BBD].

To reiterate the theorem, if $W \to X$ is a resolution of singularities, then one can write

$$H_*(W) \cong IH_*(X) \oplus \Sigma\, IH(V; L)[?]\otimes ?$$

Here we are denoting all the singular strata of X by V_p and L is some local system on V. Now combining the decomposition theorem with resolution of singularities, one inductively sees that $L(X) = L(W) + \Sigma$ correction terms associated to the V's. (This is proven in [CS6] with an explicit formula[10] for the corrections.) The $L(W)$ piece, according to [BBD], comes from the $IH_*(X)$ piece. Therefore, we see that the L-classes of an arbitrary variety are a sum of pieces each of which lifts to an intersection homology group. In particular:

COROLLARY. *If X is a variety, and ∂ is the boundary of the regular neighborhood of the singular set, then its L-classes pushed into $H(X)$ lift to $IH(X)$.*

This can be applied to all of the strata. It is asserting a kind of vanishing theorem for L-classes as one approaches singularities. (Just consider what is asserted about ∂ when the singularity set consists of a point.) Using the material of the previous section one can now produce polyhedra with almost complex structures on their pure strata that are stratified homotopy equivalent to varieties but are not themselves varieties.

REMARK. In [CS9] Cappell and Shaneson have described a more algebraic-geometric version of their decomposition theorem (although the relation between the two versions and their relation to [BBD] are not entirely clear) and applications of this to the theory of characteristic classes of singular varieties. This leads to a formula for Todd classes of

[10] The relation between their explicit formula and [BBD] seems somewhat unclear, at least to me. Maybe the later work of Cappell and Shaneson mentioned at the end of the section will shed some light on this connection.

toric varieties. (Note that we have by now left the topologically invariant setting of L-classes.) This, in turn, by work of Danilov, yields a formula for the number of lattice points inside a convex polytope whose vertices lie in the lattice. All in all, this work shows, in an extraordinary way, the calculational feasability of computation by the "decomposition method".

13 Group Actions

Problems concerning group actions on manifolds are among the most interesting and important in topology and are studied by a wide variety of different techniques. In this chapter I will explain some of the applications of the ideas of this book to various sorts of topological and PL actions. The reader will need to have read chapter 6 to understand this chapter (although occasionally some tricks are used that have been discussed in chapters 11 and 12).

The first section deals with basic properties of topological G-manifolds like handlebody structures, transversality, and the like. It turns out that these do not hold equivariantly the way they do for manifolds. As a consequence, it is not such a good idea to try to develop the theory by equivariant analogy with usual topology. This was, in fact, a source of much consternation until recently. However, as we will see further in the chapter, it is possible to get around these difficulties and still obtain profound geometric results.

Section 2 develops an equivariant characteristic class for finite group actions with manifold fixed sets, extending the (Sullivan orientation) class Δ that has played such a fundamental role in, say, 2.5.A and in the previous chapter. It calculates the relevant cosheaf homology for us, away from the prime 2.

This class was first constructed for odd order group actions by Madsen and Rothenberg [MR] as a consequence of equivariant transversality, more or less along the lines of Sullivan's original PL construction. Because of the failure of transversality for even order groups, it was necessary to develop different techniques. Rothenberg and I [RtW] extended the class to arbitrary compact Lie groups acting locally linearly by producing stable Lipshitz structures and appealing to the work of Teleman [Te]. (See also [Ros1, RsW3] for the necessary analysis.) Subsequently, Cappell, Shaneson, and I have given a purely topological approach [CSW] that works for all finite groups and applies as well to group actions on Witt spaces. The approach I will take here mediates between these latter two approaches.

I will also describe in section 2 what is known about equivariant Siebenmann periodicity. This topic falls between equivariant and stratified views of the subject of group actions. For this reason, it is also a useful technical device for solving problems of one sort by the methods of the other. (13.6 gives an example of this.).

The next section explains what we can say about compact Lie groups. It turns out that for actions that are locally free, i.e. that have finite isotropy, the analogue of Δ plays exactly the same role in classification, and we have a nice theory. If some points have positive dimensional isotropy, it is clear that different phenomena occur and examples of this will be studied in both this section and section 6. Hopefully, a new paradigm that shifts away from equivariant K-theory will develop by thinking through the theory in this case.

Section 4 specializes the previous material to the "nonlinear similarity problem". This is the question of when two linear representations are topologically conjugate. Much is known about this question, but it has not been completely resolved. I will not prove the state of the art theorems but will just show how some of the main results of the subject arise from our point of view. (An interesting exercise or problem would be to prove the results of [CSSW, CSSWW1,2] from the point of view of this book.)

Sections 5 and 6 study in a little more detail questions about fixed sets of group actions that resemble some standard given ones. We prove theorems that enable one to modify the fixed sets to be related manifolds, or to obtain nonmanifolds as fixed sets. These are subjects with long involved histories, but I should mention the early work of L. Jones here in this introduction; he studied these problems in the case of actions on the disk in an acutely insightful way (at least implicitly using a number of techniques that were completely elucidated only much later).

I also should say that many of the results presented in this chapter are not the last word on the problems they confront; there are often extra hypotheses that need removing. These often concern fundamental group restriction, parity of group order, finiteness of groups, etc.

We delay discussion of the equivariant Novikov conjecture and related rigidity phenomena to the next chapter. However, a "higher G-signature theorem" is presented in the exercises in 13.2. More detailed histories will appear in the sections that discuss the individual problems.

13.1. Remarks on the foundational theorems of manifolds

As I stated before, the early theory of G-manifolds was developed from the point of view that all manifold notions should have G-analogues. The two most serious problems stemmed from the failure of there to be G-handlebody structures (which interfered with the analysis of h-cobordisms) and the failure of G-transversality (which made cobordism calculations of all sorts impossible). Although interesting, the results of this section will not be used in any of the later sections.

These problems were, for the most part, systematically analyzed by Steinberger and West and Quinn and by Madsen and Rothenberg respectively. I should mention that these authors work in the locally linear topological category. This means that each orbit has a smooth equivariant neighborhood. (Thus, for instance, arbitrary free actions on topological manifolds are automatically locally linear: the manifold structure gives appropriate coordinate charts.)

EXERCISE. Prove that a manifold stratified action (i.e. an action where all fixed sets are locally flat submanifolds), abbreviated MS, is locally linear iff it is locally linear at some point in each component of each pure stratum.

Therefore, we view the issue of local linearity as something artificial to demand in a construction. If one is interested in it, one checks it at the very end.[1]

The issues about handlebodies are most easily understood. The difficulty stems from the fact that we are in a manifold stratified category, not a PT category. Nonuniqueness of handlebody structures arises from PL nontrivial h-cobordisms that are topologically products (e.g. 5.1). Very natural examples of the failure of uniqueness occur in the nonlinear similarity problem. DeRham had shown that PL conjugate representation spheres are linearly conjugate (see [Rot, Luck1] for modern versions). However, Cappell and Shaneson gave examples that show that different representation spheres can be homeomorphic (see [CS3,7] and 13.4 below).

Here is a very simple example of a locally linear G-manifold with no equivariant handle decomposition:

EXAMPLE. Suppose that G acts on $\mathbb{R}^n$ semifreely[2] and smoothly so that its fixed set is a punctured manifold, whose complement has a finitely dominated quotient with nontrivial Wall finiteness obstruction (1.1). We will produce such shortly using the material from 4.5. Then one point compactify to produce an MS action.

If this did have an equivariant handle decomposition, one could fit the pieces containing parts of the fixed set together and thus produce an equivariant closed complement. The quotient would then have the homotopy type of a finite complex since it would be homotopy equivalent

[1] This is parallel to our point of view on using homology manifolds with nontrivial resolution obstruction for doing manifold classification. The "right" thing to do in the equivariant setting is to deal with homology-manifold stratified actions and check (manifoldness and) local linearity at the very end, if one is interested in these features.

[2] Recall that a group action is semifree if each point is fixed either by the whole group or just by the identity element.

to a compact manifold (the quotient of the closed complement), which violates our assumption on the Wall finiteness obstruction.

Now to produce the action. Start with your favorite submanifold of S^n which is a mod 2 homology sphere for which the Swan obstruction (4.5) is nonzero for, say, Q_8 (the quaternion group of order eight). It is more elementary to perform this construction with a submanifold with trivial normal bundle; using a lens space we can arrange for the normal bundle to be trivial but for the Swan obstruction to not be so. The trivial normal bundle allows one to build a semifree action on a neighborhood of the submanifold. Let's puncture the submanifold and try to extend to the complement in $\mathbb{R}^n$. If we remove an open disk and try to do this PL, we run into the Wall obstruction, as in 4.5. By puncturing, we don't, because we can do a proper surgery version of the argument in 4.5 to produce the extension.

REMARK. In 13.6 we will remove the condition of triviality of the normal bundle.

The question of whether or not a locally linear G-manifold has an equivariant handle structure has a very nice answer:

THEOREM (STEINBERGER-WEST [STW]). *If M is a locally linear G-manifold, there is an obstruction in $H_0(M/G; \tilde{\mathbf{K}}_0^{BQ})$ which vanishes iff M has an equivariant handle structure.*

Steinberger and West write their answer in a form that makes the equivariant nature of the obstruction clearer, but I will leave it as I've stated it. This obstruction is just the obstruction to controlled finiteness. The proof of this uses the same ingredients as the proof of equivariant h-cobordism [St, Q3].

The issue of transversality is more complicated. One must be careful about what one means. (See [Pe2, CoW].) We cannot make a map from $M \to \mathbb{R}$ transverse to the origin if M has a trivial action and $\mathbb{R}$ has the flip $\mathbb{Z}_2$ action. However, in the smooth category, if the domain has a very large equivariant tangent bundle in comparison to the range (in terms of the dimensions of the subbundles associated to the various irreducible representations), one can do the transversality.

THEOREM (MADSEN AND ROTHENBERG [MR]). *For odd order G-manifolds stable G-transversality holds, but for $G = \mathbb{Z}_2$ it does not.*

The proof for odd order groups is very long and complex. It is not so hard to reduce the problem to analyzing the stability properties of spaces of equivariant homeomorphisms for representations $Top(V) \to Top(V \times W)$. (We suggest the reader think through how smooth transversality to a submanifold with nontrivial normal bundle is proven.) These results of [MR] can be obtained from the surgery results of the following section.[3]

I should remark that stable transversality is enough for certain applications, like reducing the calculation of normal invariants to an equivariant cohomology theory. This is part and parcel of the [MR] approach.

The failure for $G = \mathbb{Z}_2$ comes quite easily out of the work of Browder and Livesay [BLi] (see the exercise in 4.7). Cappell and I noticed this failure independently in [CW5]. Madsen and Rothenberg were motivated by the observation that their perspective on nonlinear similarity related nontrivial topological similarities between representations to failures of stable transversality, and for cyclic groups of larger order, nonlinear similarities had already been produced in [CS3].

EXERCISE. Fill in the details of the following argument. Use Wall realization of an element of $L_{n+2k}(\mathbb{Z}_2)$ that defies desuspension (4.7) on the boundary of the regular neighborhood of the interior of a top simplex of the fixed set of the involution on $S^n \times D^{2k}$ which is trivial on the first factor. This will still be locally linear after coning, by an Eilenberg swindle (see 5.3). This yields a new action on $S^n \times D^{2k}$. However, show that this cannot be made transverse to the trivial flip action on $S^n \times \mathbb{R}$.

ALTERNATIVE EXERCISE. Show that the forgetful map from equivariant to nonequivariant block bundle theory $B\widetilde{PL}_k^{\mathbb{Z}_2} \to B\widetilde{PL}_k$ is a $\mathbb{Z}[1/2]$ equivalence for k even but not for k odd. Deduce the failure of equivariant stability from the validity of unequivariant stability!

13.2. Equivariant surgery for finite group actions

Before discussing the main substantive theorems, I would like to get one thorny problem out of the way.[4] A map is equivariant if it commutes with the group action. Such maps give filtered maps of the quotient spaces. To obtain stratified maps of the quotients, one needs isovariant maps, i.e. maps with $G_{f(x)} = G_x$ for isotropy groups. The following result relates these notions in an important special case:

PROPOSITION ([Br3]; SEE ALSO [Do]). *If M and N are G-manifolds satisfying the (large) gap hypothesis that* $\dim M^H \geq 2 \dim M^{H'} + 1$ *whenever*

[3] Again, I think it is an interesting exercise to re-prove all of the results of [MR] from the point of view of this book; one discovers that a large chunk of their theory follows formally from the machinery and ideas here, but some of their more precise results still require the same sorts of difficult calculations that they do. The results on higher homotopy groups of homeomorphism spaces still require the kind of geometrical analyses that they do or, alternatively, something like the machinery of 10.3.A.

[4] The entire discussion in this section applies, with only the slightest changes, to proper actions of discrete groups. One must work with "proper equivariant K-homology". See [BC] and [Phi].

$M^{H'} \subset M^H$, $M^{H'} \neq M^H$, then any equivariant homotopy equivalence between M and N can be equivariantly homotoped to an isovariant homotopy equivalence; i.e. the quotient spaces are stratified homotopy equivalent.

This is rather more challenging to show than the parallel result in embedding theory, i.e. that homotopic inclusions of low dimensional manifolds in a much higher dimensional one (large gap condition) are stratified homotopy equivalent; one knows by a simple general position argument that they are even isotopic! As an exercise, you should try to prove these results by hand. If you despair, see Dovermann's paper [Do] for a surgical proof of a closely related result, and then try again!

This result is ideologically a cornerstone to our conception of the theory. We will not make any essential use of the equivariant category anywhere. All equivariant classifications that I know of can be obtained from isovariant ones, either through this proposition or through arguments special to that situation (as in the results of 13.6).

It is a fundamental problem to make our theory more equivariantly natural. For instance, are equivariant (or isovariant) structure spaces equivariantly natural? Some pieces of the theory are natural, but I do not know if the whole theory is. (See [CWY] for our current partial result.)

This would even have interesting *computational* significance.

In any case, if we, say, invert 2, the key issues surround the calculation of the L-cosheaves (because we will see that the surgery obstruction group always decomposes into a sum of ordinary L-groups), which we do below in three steps. We decompose the cosheaves into simpler ones, geometrically interpret these simpler cosheaves, and, finally, use the signature operator to build a morphism from these geometric versions of the homology to K-theory. This morphism is an equivalence.

PROPOSITION. *Away from 2, for G finite, there is a decomposition of the L-cosheaf $\underline{\mathbf{L}}^{\mathbf{BQ}}(M/G)$ as a sum $\bigoplus \underline{\mathbf{L}}(M^H/NH$, rel sing) where H ranges through the subgroups and NH is the normalizer of H in G, assuming that G acts with small gaps, i.e.,* $\dim M^H \geq \dim M^{H'} + 3$ *whenever $M^{H'} \subset M^H$, $M^{H'} \neq M^H$.*

In other words, the cosheaf (and the L-spectra) breaks up away from 2 into a sum of pieces, each of which is an L-cosheaf supported on a closed stratum.

The proof, as in the Witt case of the previous chapter, relies on Ranicki's localization result that working rationally in the coefficient ring does not change an L-group (or spectrum or cosheaf) *away from the prime 2*. We thus have to produce a map $L^{BQ}(M/G) \to L(\mathbb{Q}G)$ naturally, and the result will follow, on localization. Think of $L^{BQ}(M/G)$ as being built out of (certain) equivariant surgery problems. Unfortunately there are fixed points for various subgroups so that the chain complexes are only $\mathbb{Z}$ free, not $\mathbb{Z}G$ free, but by the time we pass to the rationals all finitely generated modules become projective and we get a well-defined $\mathbb{Q}$-surgery obstruction. It is now not hard to see that the sum of the maps associated by restriction to each stratum with these rationalizations gives an isomorphism $\otimes\mathbb{Z}[1/2]$ (using Ranicki and the fibrations in 6.1, 6.2).

These "simple" cosheaves were actually used in [CSW] prior to the invention of the surgery theory to study certain analytically defined topologically invariant characteristic classes[5] purely topologically. Here I will reverse the process and use surgery to interpret these cosheaf homologies geometrically and then analytically define classes which calculate the homology.

REMARK. Recently Cappell, Yan, and I extended the calculation of equivariant L-groups in [LM] to the cosheaf level, showing the existence of a similar splitting of structure sets for odd order locally linear group actions into pieces corresponding to individual strata [CWY]. These integral splittings are not compatible with the $\mathbb{Z}[1/2]$ splitting just discussed!

In any case here is the result regarding these simple pieces:

THEOREM ([CSW]). *There is an isomorphism* $H_0(M/G; \underline{\mathbf{L}}(M/G, \text{rel sing})) \otimes \mathbb{Z}[1/2] \to KO_n^G(M) \otimes \mathbb{Z}[1/2]$.

COROLLARY. *One can topologically intrinsically define a class* $\Delta(M) \in KO_n^G(M) \otimes \mathbb{Z}[1/2]$ *associated to any ANR G-manifold.*[6]

This class, when M is a smooth G-manifold, is the class of the equivariant signature operator and can be viewed as more or less computable by classic index theory [AS].

The corollary follows from the theorem via an identification of the homology as controlled (as in chapter 9), visible (see [Ws]), equivariant algebraic Poincaré complexes (as in chapter 3) extending the nonequivariant recognition theorem discussed in chapter 9. The class in the corollary is associated to the equivariant self-dual sheaf on M given by the singular chain complex of M. This corollary was first proven for odd order G in [MR] and even order G in [RtW].[7] The map that forgets control

$$H_0(M/G; \underline{\mathbf{L}}(M/G, \text{ rel sing})) \otimes \mathbb{Z}[1/2] \\ \to \mathbf{L}(\mathbb{Z}\pi_1(\mathbf{M} - \text{singularities})/\mathbf{G}) \otimes \mathbb{Z}[\mathbf{1/2}]$$

[5]The paper [CSW] also develops the characteristic classes more generally. The additional generality stems from the remarks on Witt spaces in 12.3 with calculation that we redo here of the cosheaf homology.

[6]Or, following [CSW], any G-ANR Witt space.

[7][MR] show that in the locally linear setting this class is actually an orientation for $KO^G \otimes \mathbb{Z}[1/2]$; this is not in general true for even order groups or for odd order groups if the action is not locally linear.

takes the characteristic class to an equivariant symmetric signature of M; this statement can be viewed as a (higher) G-signature theorem (see [RsW5, CSW, RsW2,1].

EXERCISE. Verify that this is a G-signature formula when M is simply connected by using the multisignature (see 4.7) to map $L(\mathbb{Q}G) \to RO(G) \otimes \mathbb{Q}$. In the nonsimply connected case, show that if the action on fundamental groups is unextended as in the exercises in 12.3.A, then one obtains the higher index theorem of [RsW5].

EXERCISE ([RsW2]). The target of the forgetful map contains an invariant that is invariant under equivariant maps that are homotopy equivalences (called pseudoequivalences in [Pe2]). Use this to reduce an equivariant Novikov conjecture to an injectivity statement. We will discuss this in more detail in the next chapter.

To prove the theorem, first one reduces to the construction of the natural transformation from homology to K-theory in the case where M is a PL manifold and G acts on it by PL transformations. By crossing with D^3 and applying the $\pi - \pi$ theorem, this homology now corresponds to equivariant PL structures $\otimes\mathbb{Z}[1/2]$. Now we can use [Che2] to produce an equivariant signature operator on domain and range. To our original homology class we assign the difference in $KO^G \otimes \mathbb{Z}[1/2]$ of these signature operators. Then one checks that this element is independent of choices.

How does one see that this map is an isomorphism?[8] One has to check that the induced map on homotopy groups is an isomorphism at the stalks around orbits. This is basically a consequence of the G-signature formula, the stratified structure sequence, and the calculations of KO_n^H(point) and $L(\mathbb{Q}H)$, both away from 2, for various subquotients of G.

REMARK. This construction is related to the original construction Rothenberg and I used. We produced stably (after crossing with tori) equivariant Lipschitz structures on $M \times T$ and then used [Te] to produce a signature operator. The stable construction makes use of the geometric picture of stable structures described in the proof of stable surgery in 10.2. Here we've assembled the ingredients differently. We've used the structure of stable structures to show that there's little difference from the PL case and then used Cheeger's rather simpler signature operator. Along the way, we've picked up the benefit of the calculation of an interesting controlled L-group ($\bigotimes \mathbb{Z}[1/2]$).

REMARK. We also have computed, inter alia, the "universal signature class", also denoted Δ, away from 2 mentioned at the end of 6.1 and 6.2.

[8]One does not need this for many of the applications, e.g. to nonlinear similarity. However, it is critical for classification results.

It is just the direct sum of these (NH/H)-equivariant classes associated to all of the fixed sets of all of the subgroups $H \subset G$.

Putting everything together we obtain:

EQUIVARIANT SURGERY EXACT SEQUENCE. Suppose that G is a finite group acting orientation preservingly on a manifold M with small gaps and with all fixed point sets locally flat submanifolds. Suppose also that all fixed sets have dimension at least five.[9] Then we have an exact sequence for isovariant structure sets $\otimes\mathbb{Z}[1/2]$:

$$\begin{aligned} \dots \to \oplus \mathbf{L}_{\dim M^H+1}\big(\mathbb{Z}\pi_1(M^H - \text{singularities})/(NH/H)\big) \otimes \mathbb{Z}[1/2] \\ \to S^{G-\text{iso}}(M) \to \oplus KO^{NH/H}(MH) \otimes \mathbb{Z}[1/2] \\ \to \oplus \mathbf{L}_{\dim M^H}\big(\mathbb{Z}\pi_1(M^H - \text{singularities})/(NH/H)\big) \otimes \mathbb{Z}[1/2]. \end{aligned}$$

REMARK. For odd order groups acting locally linearly this was obtained in an equivalent form by Madsen and Rothenberg [MR]. The homology term was replaced by an equivariant normal invariant which was viewed, via transversality, as maps into an equivariant version of F/Top which was computed to be a sum of equivariant BO's. The Poincaré duality needed to identify their cohomology with the homology here is given by the fundamental class Δ. If we are not locally linear or G is even order, then there are two failures: firstly, Δ is harder to define and is in any case not an orientation, and secondly, there is no transversality so we cannot get a cohomological description of normal invariants. Happily, our technique solves these two problems at one shot.

REMARK. In [CWY] the normal invariants are computed at 2 for odd order locally linear group actions. See also [Na] for the odd order abelian case, proven using the Madsen-Rothenberg technique. These results provide generalizations of Sullivan's calculations of unequivariant normal invariants (2.5). For even order groups, the calculation, at the prime 2, doesn't parallel Sullivan's.

EXERCISE. Prove the Wall desuspension theorem (4.7) by relating structures on a lens space to those on a representation space and by applying Atiyah's proof of equivariant Bott periodicity [A1]. (Note that we are using the signature operator, not the Dirac operator. This is responsible for the failure of desuspension for even order groups.)

REMARK. One can extend this to arbitrary linear representations of an odd order group and nonsemifree "suspension". In a stable range, this is due to Madsen and Rothenberg [MR].

[9] As usual, one can work relative to low dimensional strata and also can handle some four dimensional difficulties using [FrQ].

EXERCISE ([CSW]). If G acts on an odd dimensional simply connected manifold with simply connected fixed point sets, then analyze the surgery exact sequence associated to the cone of M viewed as a Witt space.

PROBLEM. Compute the normal invariants for actions that do not preserve orientations.

Note that away from the prime 2, both normal invariants via their identification with equivariant K-theory and surgery obstruction groups via their decomposition into L-groups of pure strata suggest that there is a form of Siebenmann periodicity true for some nontrivial representations. (For $4 \times$ trivial representation, this follows from the stratified Siebenmann periodicity proven in 11.3.) The following was proven by Min Yan:

THEOREM ([Yn]). *Let G be an odd order group. If $V = 4 \times$ permutation representation, and if $M \times V$ and M have the same isotropy structure,*

$$S^{G\text{-}iso,-\infty}(M) \cong S^{G\text{-}iso,-\infty}(M \times V),$$

except for at worst one $\mathbb{Z}$ per component of pure stratum, as usual.

In particular, one can stabilize by $4 \times$ regular representation. Ultimately Yan deduces his theorem from the fact that for arbitrary G-sets, S, crossing an equivariant surgery problem with $\times \mathbb{C}P^2$ where one takes the product of S copies and the action is given by permutation of coordinates induces a periodicity of L-groups. (See also [DoS].) In [WY] there is the following improvement in a special case:

THEOREM ([WY]). *For G an abelian group one has a(n unstable) periodicity for any V which is twice a complex representation:*

$$S^{G\text{-}iso}(M) \cong S^{G\text{-}iso}(M \times V).$$

I do not want to go into the detailed proofs here. For the second theorem [WY] it was necessary, for reasons related to equivariant bordism theory, to define some new products in stratified surgery. These seem worth exposing here:

CONSTRUCTION (EXOTIC PRODUCTS). *Suppose that X is a stratified space and S is a closed union of strata such that $S = \partial T$ is a stratified $\pi - \pi$ situation. (That is, each pure stratum of S is in the closure of exactly one pure stratum of $T - S$, with the corresponding union of pure strata actually* *a manifold with boundary of π – π type.) Then given any Y, these data define a product map*

$$L(Y) \to L\big(Y \times (X - S)\,\mathrm{rel}\,\infty\big).$$

To see what's involved in this, realize that to most straightforwardly define a product map one would need a coboundary (over its own fundamental group) of the holink of S. *If the holink fibration extended over T*, then the above data would easily produce this, but we have not assumed this.

An important example is $X = \mathbb{C}P^2$, $S = \mathbb{C}P^1 = S^2$, and $T = D^3$. Then the associated exotic product induces $L(Y) \to L(Y \times D^4\,\mathrm{rel}\,\partial)$ and is responsible for the usual Siebenmann periodicity.

It is in fact critical that S bound over its own fundamental group and not over anything smaller, because of phenomena like nonmultiplicativity of signature in bundles [A2]. One cannot see that a signature of a bundle vanishes just from knowing that the signature of the base is trivial. Atiyah has shown that the deviation of such a vanishing statement is a characteristic number involving the monodromy map of $\mathbb{Z}\pi$, so that if one knew that the base bounded over its own fundamental group, even if the bundle didn't extend, one would get vanishing (or, in general, multiplicativity).

What is implicitly involved in these products is an extension to all sorts of signatures, and not just for manifolds but also for stratified spaces. Nonetheless, the proof that exotic products exist is quite direct.

EXERCISE. Using the definitions of BQ-groups in chapters 6 and 7, give a three line construction of the exotic products.

EXERCISE (REGARDING CHARACTERISTIC NUMBERS OF BUNDLES). Show that no characteristic numbers besides the higher signatures can be multiplicative in arbitrary (block or fiber) bundles with connected structural group. Show that these in fact are multiplicative, assuming the Novikov conjecture.

(Hint: This is not so easy. The first part for block bundles uses blocked surgery. Then one must produce fiber bundles: [WW] is convenient for this. The converse is based on facts about transfer; if you get stuck, the material in 13.5 should help. This converse fact, for higher signatures of manifolds with free abelian fundamental group that fiber over a 2-connected manifold, was first proven in [Lus].)

REMARK. The starting point for all of the recent work in elliptic cohomology is that there are additional genera that are multiplicative for finite dimensional connected structural groups. The $\hat{A}$-genus is one such. These genera have analogues of many of the familiar formulae for signature. Thus, for instance, the formulae for fixed point sets in the next section should be true for smooth actions; [CS9] have formulae for some such genera in algebraic-geometric situations. The above exercise shows that for PL actions and stratified maps, there can be no such similar theory.

EXERCISE ([WY]). Exotic products can be used to kill products as well as define them. To make $\times A$ trivial, one need not find a stratified object with boundary A which is stratifed $\pi - \pi$. All one needs is a stratified coboundary that is "exotically" $\pi - \pi$ (i.e. there can be extra strata in the coboundary, but they must be glued together to form a singular substratified space with contractible $\mathbf{L}^{\mathbf{BQ}}$). Use this to show that $\times \mathbb{R}P^{4k+3}$ always induces 0 on surgery groups. (Its symmetric signature does not always vanish!)

EXERCISE. Deduce from Yan's theorem and from the stable transversality of Madsen and Rothenberg that $\Omega^V F/Top \cong F/Top$ for $V \cong 4\times$ permutation representation of an odd order group. (Here F/Top denotes the equivariant version of this space.)

13.3. Locally free compact actions

Everything done in the previous section applies equally well to actions of compact groups where the isotropy groups are finite. Such actions are called **locally free**. The key is simply to use the equivariant signature operator.

THEOREM. *If G acts locally freely on a manifold, then one has the universal class* $\Delta(M) \in \oplus KO^{NH/H}(M^H) \otimes \mathbb{Z}[1/2]$. *The difference of these classes detects the stratified normal invariant away from 2.*

This is proven by the same technique as the result of the previous section. This means that we have completely analyzed (away from 2) the surgery sequence for locally free actions.

EXERCISE. Let S^1 act on S^{2r-1} (viewed as the unit sphere in $\mathbb{C}^r$) by multiplication by $\big(\exp(2\pi i a_1 t), \exp(2\pi i a_2 t), \ldots, \exp(2\pi i a_r t)\big), 0 \leq t \leq 1$. What is the rank of the equivariant structures of this space?

However, for general actions of compact Lie groups the signature operator does not seem to play the same role: equivariant K-homology does not seem to be the correct measure of tangential data. What replaces this I cannot yet guess. I have done some calculations in special cases but have not yet discerned a pattern.

The following theorem is the simplest result showing why things are different in general:

THEOREM.[10]

a) Suppose $G = S^1$ *or* $SU(2)$ *acts semifreely*[11] *on a manifold* M*, and* F *is a component of fixed set of codimension an even multiple of* $\dim G + 1$*, then stably*

$$S^{G\text{-}iso} \cong S^{G\text{-}iso}(M \operatorname{rel} F) \times S(F).$$

If F *is of another codimension, then*

$$S^{G\text{-}iso}(M) \cong S^{G\text{-}iso}(M - F).$$

b) For such an action one has an equality of higher signatures:

$$f_*(L(M) \cap [M]) = \Sigma(fi)_*\big(L(F) \cap [F]\big) \in H_*(B\pi; \mathbb{Q}),$$

where $f : M \to B\pi$ *classifies the fundamental group,*[12] *and the sum is taken over the components of codimension* $\equiv 0 \bmod 2(\dim G)$.

The two parts of the theorem are obviously related. The first part says that in appropriate codimensions, the complement of the action determines what the fixed set is and how it is glued in. For the other codimensions, the fixed set and the total stratified space are as independent as possible. The second part describes the relationships of characteristic classes: the higher signature of the manifold acted upon is precisely that of the "flexible" components of fixed set. Replacing a fixed point set will change the ambient manifold. Alternatively, the manifold acted upon forces some conditions on fixed points.[13]

The universal characteristic class encodes all relations among the L-classes of the various closed strata, and then encodes whatever secondary classes are implicit in the universality of the relations among these L-classes.

EXERCISE. Prove the theorem. (Hint: Observe that all quotients are supernormal. In the rigid codimensions, verify that the local stalks of the L-cosheaf are contractible, and compute L^{BQ} to deduce the result. In the other dimensions observe that one has a Goresky-Siegel space, so that

---

[10]Strictly speaking this theorem is only correct with Siebenmann-periodic or homology-manifold structure theory. Otherwise the right-hand side might be one $\mathbb{Z}$ larger for each component of fixed set–the missing fixed set can be a nonresolvable homology manifold.

[11]Recall that a group action is semifree if each point is fixed either by the whole group or just by the identity element.

[12]This formula presupposes that $F \neq \emptyset$. Otherwise, one has to take instead $pi/\langle orbit \rangle$. Remember that the orbit of any point gives a central loop in the fundamental group which is independent of the point chosen. This formula is also valid, at least in the circle case, for actions that are not semifree; see [Weil1, pt. II].

[13]These conditions on higher signature can be refined to an integral statement even in cases where the higher signatures are <u>not</u> homotopy invariant.

the machinery of the previous chapter gives the independence. For part *b*, try to use the action to build an appropriate cobordism between M and a projective bundle around the fixed set, and use the exercise from the previous section on multiplicativity of higher signatures in bundles.)

EXERCISE. Compute the L-cosheaf homology $\otimes\mathbb{Z}[1/2]$ for an arbitrary S^1 action on a manifold using the material of this section. Interpret the universal class. (Hint: It is important to separate the elements of finite order in whose fixed sets the fixed set of the group has codimension $2 \bmod 4$ from those in which the codimension is $0 \bmod 4$.)

13.4. Nonlinear similarity

The nonlinear similarity problem is to determine when two orthogonal representations V and W of a finite group are topologically conjugate.[14] This is a problem which is still the object of considerable research. Much of the material in this book was developed for proving and understanding the theorems I will sketch in this section. Not surprisingly, it turns out that we can now describe the essential features of the current state of the art more efficiently.

After original pioneering work of DeRham (the PL case), the first progress was due to Schultz [Scz1] and Sullivan independently, who proved that topological and linear conjugacy are identical for odd p-groups.

EXERCISE (DERHAM; SEE ALSO [Rot, Luck1]). Prove DeRham's result using Reidemeister torsion and the method of calculation for lens spaces (see 1.2.A).

EXERCISE. Prove the theorem of Schultz and Sullivan in the following steps. First reduce to the case of a cyclic group. Then show that one can assume that the difference between the two representations is in primitive eigenvalues. Show that the corresponding lens spaces are normally cobordant. By taking products with semifree representations show that the lens spaces remain normally cobordant after any number of suspensions. (Compare 4.7.) Now show that homotopy lens spaces with odd p-group fundamental group are homeomorphic if all suspensions are normally cobordant. (Hint: Use the connection between normal invariant and ρ-invariant; Atiyah's theorem on $K(B\pi)$ is useful but not essential.)

The subject can really be said to have become nontrivial with the breakthrough by Cappell and Shaneson [CS3], who showed that nonlinear similarities exist for $\mathbb{Z}_{4k}$ for $k \geq 2$; that paper also gave a complete

[14] Actually, it is a very interesting question to study this for noncompact groups of linear transformations as well. The history in the case of individual matrices as well as the situation for compact groups is surveyed in [Sh2], which, although somewhat outdated, remains useful, especially for its overview of related problems.

classification of certain very special representations. Over the following few years, the main results were:

(ODD ORDER GROUP) THEOREM ([HP] AND [MR]). *For odd order groups, topologically conjugate linear representations are linearly conjugate.*

(TOPOLOGICAL RATIONALITY) THEOREM ([CS7]). *Let K be the real subfield obtained as the extension of $\mathbb{Q}$ by all odd roots of unity. Then letting $R^{Top}(G)$ denote the Grothendieck group of linear representations modulo topological equivalence, one has*

$$R^{Top}(G) \otimes \mathbb{Z}[1/2] \cong RK(G) \otimes \mathbb{Z}[1/2].$$

(ALGEBRAIC) EXERCISE ([CS7]). Deduce the odd order group theorem from the topological rationality theorem. Furthermore, deduce which groups have nontrivial topological similarities. Use a density argument and continuity of characters to work out the compact Lie case.

These theorems are not that hard to understand from the point of view of the G-surgery results of 13.2. A concrete description of what the Δ-class means for linear representations was discovered in [CSSWW1,2] (see also [RtW] for the essential point):

PROPOSITION. *The renormalized Atiyah-Bott numbers (or, better, characters) are topological invariants of linear representations.*

The **renormalized Atiyah-Bott numbers** are defined as follows for a group element g. If $V^g = 0$, define $AB(g, V)$ by the formula in [AB]. If $V^g \neq 0$, let $V \cong V^g + W$, and define $AB(g, V) = AB(g, W)$.

The way one proves the result on Atiyah-Bott numbers is to first do the discrete case. The identity follows from the topological G-signature formula (of 13.2), and the localization formula in equivariant K-homology. (The Atiyah-Bott numbers from the smooth theory are given a purely topological interpretation by this formula: they compute the contribution of fixed points to g-signature, just as in the smooth case [AS, pt. III].) If the fixed set is not discrete, then one knows that there will be a homeomorphism $\mathbb{R}^n \times W \cong \mathbb{R}^n \times W'$ which is the identity on the $\mathbb{R}^n$ factor (why?). This implies that W and W' are homeomorphic after crossing with a torus. (This follows from the argument in 10.2, for instance.) Now the topological "higher G-signature formula" gives the result.

Now, we return to the odd order group theorem. The Franz independence lemma (used in the Reidemeister torsion argument for classic lens spaces) is used to show that inequivalent representations have different renormalized Atiyah-Bott numbers. It takes a little more care to show that $RK(G) \otimes \mathbb{Z}[1/2]$ is a lower bound for the size of $R^{Top}(G) \otimes \mathbb{Z}[1/2]$. (What happens is that when one has a group element with -1 as an eigenvalue, the Atiyah-Bott number vanishes and we lose precious information about the other eigenvalues of the element.)

To show that $R^{Top}(G)$ is not larger than this, one has to produce some explicit nonlinear similarities. The simplest examples I know are those in [CSSW]; they deal with a case where one can see a priori that there will be no 2-torsion in the calculation, so the Δ-class actually detects. (Of course, they necessarily have to argue more ad hoc.) The onus of the paper [CW7] is now to show that what is obtained by induction and restriction from these examples actually cuts $RO(G)$ down to the stated size, and this is done purely algebraically.

EXERCISE (FOR THOSE WITH A LITTLE HANDS-ON COMPUTATIONAL KNOWLEDGE). Verify the example in [CSSW] by filling in the details in the above sketch. Show that if we take two representations of $G = \mathbb{Z}_{4k}$ for $k > 2$ as the sum of 2 copies of rotation by $2\pi a/4k$ (or $2\pi(a+2k)/4k$), the one dimensional nontrivial representation of G and the one dimensional trivial representation, they are topologically conjugate. (Hint: First show that any two homotopy equivalent 3-manifolds are normally cobordant.[15] To compute an inevitable surgery obstruction, make use of the fact that for cyclic groups L^p is torsion free except for Kervaire invariant and codimension one Kervaire invariants. These finite order obstructions can be eliminated because they are in the image of the assembly map, or by comparing $\mathbb{Z}_{4k}$ to $\mathbb{Z}_{2k}$. The nontorsion part of the obstruction can be computed from the ρ-invariants, which are given by Atiyah-Bott numbers.)

The most precise information that we now have on this problem is a complete calculation for cyclic 2-groups [CSSWW1]. The reader can find there some information on the normal invariants at 2 and how to calculate these. Using this, one can find elements of R^{Top} with arbitrarily high exponent (at 2, of course).

In [CSSWW2] the complete low dimensional situation is worked out. This is done using the renormalized Atiyah-Bott numbers and a number/algebraic K-theoretic argument based on examination of the proper Whitehead group of the top pure stratum (which turns out to be $\cong Wh^{Top}$ in all the relevant cases). The result is that the only six dimensional examples are the ones of the previous exercise (and that there are no examples in any lower dimensions).

It seems that we can reasonably expect to someday have, at least stably (with respect to trivial representations), a complete topological classification of linear representations of cyclic groups. Unstable and other calculations are apt to involve number theoretic and algebraic K-theoretic

[15] Hint for this step: The normal invariants of a 3-manifold are determined by its fundamental group, and $L_3(\mathbb{Z}_2) \cong \mathbb{Z}_2$ is determined by a codimension one Kervaire element.

difficulties. (I am not sure whether the successful resolution of these issues announced in [CSSWW1] can be extended beyond the situation of 2-groups.)

13.5. Replacement theorems

Replacement theorems describe situations where one manifold can be replaced by any other in a certain class as the fixed set of a group action.

EXERCISE. Suppose that G acts smoothly, PLly, or MSly on a simply connected manifold M with fixed set F, and suppose that F' is h-cobordant to F. Then there is a concordant action of G on M with F' as fixed set. In fact, one only needs a semi-h-cobordism from F' to F. (A semi-h-cobordism is a manifold with boundary where the inclusion of one of the boundary components is a homotopy equivalence.)[16] What if M is not simply connected?

Recall that actions are concordant if there is an action on $M \times I$ that restricts on the boundary components to the given actions. This exercise introduces our theme: under certain circumstances one can take a given action with a given fixed point set and automatically find another action with a related manifold as the fixed set. Now, I'd like to study the replacement problem within a given equivariant simple homotopy type.

THEOREM ([CW8]). *Suppose that G is an odd order group acting PL locally linearly and smoothly on a neighborhood of an equivariant 1-skeleton, satisfying the small gap hypothesis on a manifold M; then any PL manifold simple homotopy equivalent to the fixed set is in fact the fixed set of another PL locally linear G-action on M that is equivariant simple homotopy equivalent to the initial action.*

ADDENDUM. The theorem remains valid for G abelian if the smoothness assumption is dropped. This assumption can also be dropped if the sphere bundle of the normal representation satisfies the "large gap hypothesis".

In this theorem, the new actions are strong replacements of F' for F. **Strong replacement** means that there is an isovariant homotopy equivalence of the group actions that is a homeomorphism in the complement of a small neighborhood of the fixed set and is nonequivariantly homotopic to a homeomorphism relative to this complement, as an ordinary unequivariant map between manifolds.

To describe some obstructions that arise for even order groups, we need some characteristic classes. The Kervaire classes of a homotopy equivalence can be viewed as the composite of the associated normal

[16] This is a useful notion. For instance, every PL $\mathbb{Z}$-homology sphere is semi-h-cobordant to the sphere.

invariant $F \to F/Top \cong \mathbf{L(e)}$ with the natural map $\mathbf{L(e)} \to \mathbf{L}(\mathbb{Z}_2, -) \cong \prod \mathbf{K}(\mathbb{Z}_2, \mathbf{2i})$. (These can be normalized to be trivial for i even.)

THEOREM ([CW8]). *Let G be an even order group that acts semifreely and PL locally linearly on a PL manifold with the small gap hypothesis and with simply connected fixed set. It is possible to strongly replace the fixed set iff the Kervaire classes of the homotopy equivalence vanish.*

We conjecture that this criterion applies to all even order group actions. There is some evidence for this in the work of [HMTW] in surgery theory.

As the name "strong replacement" suggests, there are weaker types of replacement possible, and they can be approached via the classification theory as well. For instance, decomposition theorems for structure sets imply that one can often replace a fixed set, but at the cost of changing the manifold acted upon! [CW8, CWY] give some examples where this weaker type of replacement is possible, but strong replacement is not. (An example of this is implicit in the results of 13.3 for semifree S^1 actions.) One might also want to analyze what happens when the map isn't an equivariant homotopy equivalence, only a pseudoequivalence: this can be handled (for certain actions of p-groups) through a merger of the techniques of this and the following sections.

PROBLEM. Work out the replacement theory for circle actions using the results of the previous section. There should be no difference between the pseudoequivalence version and the equivariant homotopy equivalence version.

To get a quick idea of how obstructions to replacement arise, let's suppose we could replace. In that case, since the complements are the same, the quotient of the boundary of the regular neighborhood must block fiber over the new fixed set. If we were dealing with circle actions, the putative fiber would be a complex projective space. To fiber, the transverse inverse image of a submanifold must have signature = signature(submanifold) × signature(fiber) (ignoring monodromy issues). Since it already fibers over the original fixed set, we obtain a condition–if the signature of the fiber is nontrivial.

Even assuming we can change the base of the fibration, we then have to deal with the difficulty of identifying the manifold obtained by pasting in the new fixed set, which can lead to more obstructions.

I will now indicate how the proof goes in a little more detail. This is mainly a reorganization of part of what's in [CW1,8] in light of better technology; the main advantage is that some conditions which seemed useful as sufficient conditions for performing the constructions in those papers can be shown to be necessary using the present treatment.

For simplicity, I would like to start off by describing the semifree PL locally linear case and then describe the modifications necessary for the general case.

To deal with strong replacement it helps to have an auxiliary construction, called the **bubble quotient**. This consists of the identification space made by identifying to points all orbits that do not get too close to the fixed set. Figure 20 illustrates a bubble quotient for the flip involution.

Now we are interested in the issue of **base change** for the bubble fibration over F. In other words, the bubbled neighborhood of F is a stratified block bundle over F, and we would like to know whether the map to F' can be homotoped to be a stratified block fibration as well.

In the simply connected case, the point is that the inclusion of the "open local holink" into the whole holink is a stratified $\pi - \pi$ open inclusion, which induces an isomorphism of L-spectra, which trivializes the L-cosheaf. In general, we use a trick to trivialize the cosheaf of spectra that relies on the reduction of transfers to analysis of monodromy. The following is the relevant assertion at the level of surgery obstructions (see [LuR, CW8]):

LEMMA. *If a fibration is trivialized over the 1-skeleton of a manifold, and the fundamental group of each pure stratum is isomorphic to the product of the fundamental groups of the base and the corresponding pure stratum of the fiber, and these isomorphisms are compatible with the trivialization over the 1-skeleton, then the transfer equals the product.*

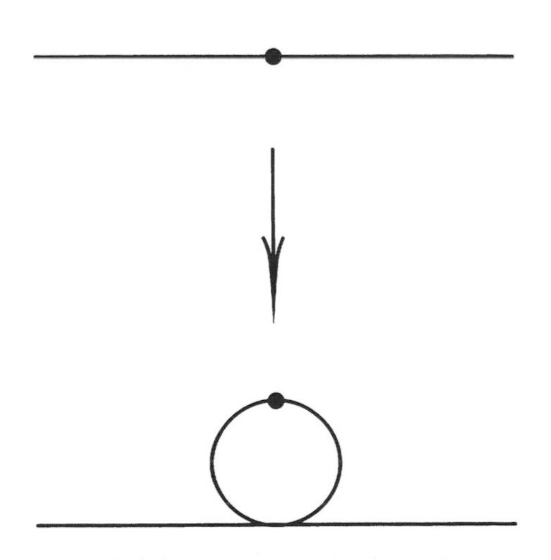

Figure 20. A bubble quotient

Now, the base change obstruction is the transfer of the obstruction to fibering $F' \to F$, i.e. the structure this simple homotopy equivalence is. This transfer is trivialized (on the cosheaf level) assuming trivial monodromy and vanishing "stratified symmetric signature".[17] This is sufficient to prove the first theorem and the addendum in the stable case. The fact that the fibers bound in a relevant sense is proven by a bordism argument (see [LM] for a slightly less refined version).

The difficulty in general is that the monodromy is an isovariant self-homotopy equivalence, and these are hard to analyze without a large gap hypothesis. What one does in the general case of odd order abelian groups is apply the periodicity theorem of [WY], described in 13.2, to reduce to the stable case. There Browder's theorem (see 13.2) is available and reduces isovariant monodromy considerations to equivariant ones.

The result for even order groups is done by analyzing characteristic classes (systematizing the heuristic given above) and following them around as one transfers from F/PL to other L-spectra. According to [Wa1], the product of the Kervaire problem with an even projective space has nontrivial obstruction, so we see that we cannot get rid of the primary obstruction to blocked fibering (3.2) unless the Kervaire classes vanish. Accordingly, strong replacement is then impossible (since either the boundary of the regular neighborhood or the total space of the bubble has a cover which is an even projective space). Conversely, if these vanish, then the product obstruction vanishes as well, by direct calculation.

In the simply connected case we can avoid the secondary obstruction to block fibering by puncturing the manifold, producing a $\pi - \pi$ situation, and then verifying that the action is locally linear on the boundary, using the techniques of, say, [Wei3].

REMARK. For many problems in surgery theory, and especially replacement problems, the following principle is very useful and is implicit in the lemma above:

FLATTENING OF L-COSHEAVES. Any L-cosheaf associated to a stratum of a stratified space is trivial over a two-connected base. If the fundamental group of the local holink maps injectively into that of the holink, then it is trivial assuming that a monodromy vanishes. More generally, assuming a π_2 condition, there is always a fibration of L-spectra over $B\pi$ from which the given bundle can be pulled back.

PROBLEM. Work out a good replacement theory for even order groups. This is especially interesting for nonsimply connected manifolds.

---

[17]This notion does not quite exist as a useful computational tool, but we'll use it for heuristic purposes.

EXERCISE. As a direct consequence of the results of 13.2, prove a general replacement theorem $\otimes \mathbb{Z}[1/2]$ for all orientation-preserving actions of finite groups.

PROBLEM. When can one replace strata above the fixed set? This includes bottom strata, i.e. fixed sets of subgroups that do not contain fixed sets of any larger group, and also involves understanding what happens above the bottom. Some results on weak replacement in these situations follow from the decomposition theorem [CWY].

REMARK/PROBLEM. I have verified (in old unpublished joint work with Bruce Williams) that the odd order group replacement theorem remains for topological semifree actions. I believe that all of the theorems in this section actually hold in the category of MS actions. However, I have not verified this (although I have convinced myself of the locally linear case.)

13.6. Semifree actions

By considering the quotient space, semifree group actions give a much studied class of stratified spaces which, with typical assumptions, have two strata. From the point of view of group action theory these form an interesting special case and are often the first stage in an induction. Here I will consider just some classic results on the disk and sphere. For some history, the reader might see [Wei2,3]. Cappell and I have combined the methods of the previous section with those that follow to obtain a homological replacement theory for semifree actions on more general manifolds.

Before discussing the delicate geometry of locally linear actions, I would like to sketch a result first announced (in a special case) long ago in [J2]. Indeed, the proof in broad outline involves many of the same steps as Jones's.

THEOREM. *Suppose that G is an even order group that can act freely on S^{2d-1}, and K is a subcomplex of the disk of codimension $2d$, $d \geq 2$; then there is a PL semifree G-action with fixed set K iff*

1) $H_*(K; \mathbb{Z}G) = 0$,
2) *K is a $\mathbb{Z}/G$ homology manifold properly embedded, and*
3) *the Swan element of every local link and for K vanish (see 4.5).*

The necessity of the first two conditions is due to P.A. Smith and is the first result of what is nowadays called Smith theory. See [Bre] for a textbook reference. The third condition is essentially due to Assadi; it arises for the same reason as the condition arises in 4.5.

The sufficiency goes like this. Using the homotopy theory of 4.5 locally (on holinks) and globally (on the complement of the putative fixed point set), one can construct a Poincaré object in the stratified category of the correct type. (Note that if the polyhedron is not a manifold, this object can have quite a large number of strata.) We'll puncture first and work on the disk. The total surgery obstruction (rel singularities) then lives in $KO^G_{n-1}[1/2]$ since we are mod 2 acyclic by condition 1 and the calculation in 13.2 of the cosheaf homology. We'll first concentrate on the neighborhood of the bottom stratum. (One can just as well make use of 3.3.A.)

Now the local surgery obstruction (i.e. in $L(\mathbb{Z}[G])[1/2]$) can be computed by working rationally (i.e. with the group ring $\otimes G$) and viewing this as a calculation for the original embedding problem $\times BG$ (which is a $\mathbb{Q}G$ Poincaré complex, and $S \times BG$ is rationally equivalent to the local stratified Poincaré space). This suffices because of Ranicki's localization theorem [Ra3]. This local obstruction vanishes (in L^p, which is good enough for us, since 2 is inverted),[18] since the obstruction to embedding vanishes, as we are specifically dealing with embedded polyhedra! One can now use the extension theorem from 4.5 to extend to an action on the disk, which can then be coned to produce an action on the sphere.

EXERCISE. Prove the parallel theorem in the topological case for MS semifree actions without condition 3.

REMARK. I expect that the work of [FP1] yields a proof of the topological analogue without assuming MS; i.e., arbitrary ANR $\mathbb{Z}/G$ homology manifolds will be made into fixed point sets.

REMARK. For odd order groups the above proof naturally leads to obstructions that are mod 2 homology classes. (In fact one can explicitly identify these mod 2 obstructions in terms of the exact sequence of a localization in L-theory (spacified, of course). This leads to the remaining results in [J2].)

EXERCISE. Give another argument using the fact that one can view the local problem as the obstruction to controlled surgery on a certain Poincaré complex, and that it is a controlled *closed manifold* problem, so that away from the order of G, the obstruction can be lifted to the simply connected case. (See e.g. 3.4; this is because obstructions for closed manifolds are in the image of the assembly map, whose domain is a homology theory.) The simply connected obstruction vanishes by the $\pi - \pi$ theorem or, better, because one started off with K embedded.

REMARK ([CW1]). This second proof works just as well for PL locally linear actions on manifolds and combines to show the vanishing of the primary obstruction away from the order of G. At the order of G, Smith

---

[18] We are led into projective L-theory because BG is a finitely dominated $\mathbb{Q}G$ Poincaré complex but has nontrivial finiteness obstruction.

theory shows that for pseudoequivalences one gets a homology equivalence of fixed sets, so one can use the base change ideas from the previous section to get pseudoequivalence replacement theorems for locally linear actions for semifree G-manifolds.

For PL or topological locally linear actions, the previous exercises give an argument that shows that any manifold satisfying condition 1 (and 3 for the PL case) can be made into a fixed set. These were first proven in [CW5] and [Wei2]. The same classification also applies to the topological locally linear case on the sphere [Wei2] by one point compactifying and using the homogeneity of manifold stratified spaces [Q3]. The proofs in [CW5, Wei2] make use of Atiyah-Singer classes which for the more subtle odd order group case are refinements of the class Δ from 13.2.[19] This is related to subtle divisibility questions regarding the G-signature of PL locally linear manifolds. For instance, there is a nontrivial restriction on the G-signature of a PL locally linear manifold for $G = \mathbb{Z}_p$, p prime, iff the class number of p is even,[20] but there is never any restriction on the G-signature in the topological case.

Semifree PL locally linear fixed sets on the sphere and their classification were achieved by quite ad hoc arguments (equivariant rather than stratified in some places) in [Wei3] (after the case of p-groups was settled in [CW5]). I believe that it is important to understand these results more intrinsically before progress can be made on the issues regarding the secondary obstructions in these types of replacement problems. The results are:

THEOREMS.

a) *A PL locally flat submanifold Σ^n of S^{n+k} for $k > 2$ is the fixed set of an orientation preserving semifree PL locally linear G-action on S^{n+k} iff Σ is a $\mathbb{Z}/G$ homology sphere, $\mathbb{R}k$ has a free linear representation of G, and certain purely algebraically describable conditions hold for the torsion in the homology of Σ.*

b) *Two orientation-preserving semifree PL locally linear G-actions on S^{n+k} with Σ as fixed set differ by equivariant connected sum with a semilinear sphere[21] iff the equivariant Atiyah-Singer classes for the two actions coincide.*

To make these theorems more explicit one needs some definitions. We denote by $\tau(\Sigma)$ the product $\prod |H_{2i}(\Sigma\text{-point})| / \prod |H_{2i+1}(\Sigma\text{-point})|$. This

[19] These Atiyah-Singer classes enter also in relative versions of the pseudoequivalence replacement problem. For instance, in extending the extension across homology collars result of 4.5 to the semifree case, these classes enter [CW5]. In fact it was from this "hard extension" theorem that the PL locally linear classification on disks first arose.

[20] See [Wei7] for this and deeper results regarding the composite case.

[21] A semilinear sphere is a sphere with group action with the property that the fixed set of each subgroup is a sphere.

is a multiplicative analogue of the usual Euler characteristic. Poincaré duality implies that it vanishes (i.e. equals 1) if n is even. In this case, one defines $\tau_{1/2}(\Sigma)$ to be the same product but with indices only running through integers at most half the dimension of the manifold. Part a asserts that the question of whether or not Σ is a fixed set can be determined purely algebraically from τ and $\tau_{1/2}$. For odd order groups the condition is that the Swan homomorphism applied to τ is trivial.

For even order groups, in addition to the Swan condition, we must make use of some constructions of Jim Davis [Da2]. The numerical surgery element is defined depending on the dimension as follows. If Σ is even dimensional, say dimension $2k$, one takes $\tau_{1/2}(\Sigma)$ and applies the Swan homomorphism to this element to get an element of K_0 which defines an element in Tate cohomology and thus in L-theory via the Rothenberg sequence. In dimension $4k+1$ one observes that $H_{2k}(\Sigma)$, in virtue of its possession of a skew-symmetric linking form, has square order (since it's odd order). Consequently $\tau(\Sigma)$ is a square, and one applies the same process to its square root to get an element of L-theory. Finally, in dimension $4k+3$, following Davis [Da2], one must be more devious. Davis first shows that in dimension three, $L^h(\mathbb{Z}G) \to L^A(\mathbb{Z}_\Gamma G)$ injects, where L^A denotes an intermediate Wall group (see 2.4.A) and A is the image of $K_1(\mathbb{Z}G)$. Thus to determine an element of $L^h(\mathbb{Z}G)$ I must specify an element of $L^A(\mathbb{Z}_\Gamma G)$, and this is determined by viewing $\tau(\Sigma)$ as an element of $K_1(\mathbb{Z}G, S)$ (the relative term in Bass's localization sequence [Ba]) and, as usual by now, pushing forward in an appropriate Rothenberg sequence. (This element only depends on $\tau \bmod 4\Gamma$.) These elements are defined so that if one has certain sorts of surgery problems with finite surgery kernels of order prime to Γ, then the surgery obstruction is the given element. Finally the condition on even order groups is that the image of the numerical surgery element of $L^h(\mathbb{Z}G)$ associated to $H_*(f)$ is zero in $H^*(\mathbb{Z}_2; Wh(\mathbb{Z}G))$.

The proof of this has a number of steps and is quite complicated and, unlike the earlier results, is quite closely geared to the specific question it addresses (i.e. uses properties of homology spheres). (See [Wei3].) For further progress on pseudoequivalence replacement questions beyond cyclic groups in the PL locally linear category, it will be necessary to get a more conceptual hold on this characterization theorem.

14 Rigidity Conjectures

This chapter discusses the question, which stratified spaces are rigid? In other words, for which stratified spaces does stratified homotopy equivalence imply homeomorphism? We shall see that this is quite an important problem with many implications. However, the chapter is mainly about conjectures and the area is a fast-moving one.

Indeed, we shall see that much information about K- and L-theories for group rings is encoded in rigidity conjectures, and such information has great impact on our picture of arbitrary (nonrigid) manifolds. *Philosophically*, every manifold seems to best approximate some rigid space, and then the deviation of homotopy to homeomorphism is identical to the difference between the manifold and the rigid model.

On the other hand, it does not seem to me that the rigidity conjectures have the same direct global implications for the study of stratified spaces (in general)—nonetheless, one can often use rigidity to indirectly get information even about stratified spaces which have no rigid approximation.

This chapter is divided into four sections. The first gives some differential geometric motivation for rigidity. The second discusses topological forms of rigidity conjectures motivated by section 1, and gives algebraic reformulations of these. We also give counterexamples to "wrong" topological versions of the geometric rigidity conjectures. Section 3, perhaps the most tentative in this book, is devoted to presenting some evidence for (perhaps slightly modified versions of) these conjectures. Undoubtedly, by the time you read it, it will be obsolete. The final section is devoted to pointing out some of the implications of the rigidity conjectures, when applied to special cases.

14.1. Motivation

The goal of this section is to motivate the topological rigidity conjectures. The most direct source of motivation is found in the beautiful rigidity theorems of Bieberbach, Mostow, Margulis, and others.

Tori are given by lattices in Euclidean space. Let T be the standard rectilinear torus. If we form the space

$$\mathcal{G} = \{Z, \varphi : Z \to T \mid Z \text{ is a torus of volume one, and } \varphi \text{ is a homotopy class of homotopy equivalences}\}/\text{isometry},$$

then $\mathcal{G} \cong SL(n, \mathbb{R})/O(n)$, which is abstractly a Euclidean space. According to Bieberbach, every flat manifold has a canonical cover which is in fact a flat torus. Thus, for instance, one can deduce that the space of flat manifolds with a homotopy equivalence to a given one is always diffeomorphic to Euclidean space. Since these are connected spaces, any two homotopy equivalent flat manifolds are diffeomorphic.

This analysis goes much further. For instance, the famous rigidity theorem of Mostow asserts that if we look at symmetric spaces of rank at least two or hyperbolic manifolds of dimension at least three, every homotopy equivalence is homotopic to a unique isometry.

In dimension two this fails, but the analogue of $\mathcal{G}$ still is a topologically Euclidean space, the so-called Teichmuller space.

As manifolds, the key feature that all of these manifolds share is that they are aspherical. Since aspherical manifolds with the same fundamental group are homotopy equivalent by obstruction theory, and since diffeomorphism is too strong a conclusion to ever expect from a homotopy assumption (since the sphere has different differential structures, and one can always connect sum with a sphere) Borel reportedly suggested the following question:

BOREL CONJECTURE. *Any two closed aspherical manifolds with the same fundamental group are homeomorphic. Indeed, the space* $\mathbf{S}(M)$ *for an aspherical manifold is contractible.*[1]

In 4.6.A we discussed this problem and showed that it is equivalent to the vanishing of the Whitehead group of any such manifold and to the statement that the assembly map in L-theory for the fundamental group of any such manifold is an isomorphism. These statements suffice to extend the Borel phenomenon to manifolds with boundary or noncompact manifolds relative to a homeomorphism at ∞.

In turn, this would suggest that there are (differential geometric) rigidity theorems for symmetric spaces with convex boundaries or rel ∞. I do not know much about this.

Now, let's turn to other rigidity theorems in differential geometry and see what kind of topological conjectures one can make parallel to the

[1] This form of rigidity excludes the sphere despite the generalized Poincaré conjecture's validity. A sphere $\mathbf{S}(S^c) \cong F/Top$ for $c \geq 2$. The nonrigidity is reflected in the existence of nonresolvable homology manifolds which are homotopy spheres. The Borel conjecture plausibly embraces homology manifolds with equanimity.

Borel conjecture. In the remaining sections of the chapter we will explore these conjectures and their consequences. A natural program that we do not touch on is to find geometric concomitants of the topological conjectures that seem to fit the context (like the above suggested generalizations to noncompact manifolds of the usual rigidity—which cause no pain to the topologist but are somewhat serious for the geometer).

The first generalization that one can study, which is no additional trouble at all, is equivariant rigidity.

PROPOSITION. *If G is a finite group of isometries of a compact hyperbolic manifold M of* dim ≥ 3, *or an irreducible symmetric space of rank* ≥ 2, *then the action of G on the fundamental group of M, i.e. the map $G \to Out(\pi)$, determines the action up to conjugation by an isometry.*

Actually, $\mathrm{Out}(\pi)$ is isomorphic to the isometry group of M. This is because $\mathrm{Out}(\pi)$ is the collection of autohomotopy equivalences of M up to (free) homotopy, and each such class contains a unique isometry.

As a consequence there is rigidity for actions on a hyperbolic manifold. The same rigidity holds for actions on symmetric spaces of higher rank. We leave the verification of differentiable rigidity for flat manifolds to the reader. (This is an interesting exercise, because the homomorphism to $\mathrm{Out}(\pi)$ does not determine the action.) This suggests:

EQUIVARIANT TOPOLOGICAL RIGIDITY CONJECTURE. *If (M, G) is an equivariantly aspherical manifold (i.e. arbitrary subgroups of the fixed set are unions of aspherical manifolds), then any (N, G) equivariantly homotopy equivalent to M is equivariantly homeomorphic to it.*

We will see in the next section that this conjecture is quite false. However, assuming the gap hypothesis (see 13.2) and avoiding certain torsion phenomena, it does seem to stand a chance. And, as in the classic manifold case, the rigidity conjecture suggests a "Novikov type conjecture," which has a great deal more validity.

Mostow's rigidity theorem can be rephrased as the assertion that isomorphisms between lattices in appropriate Lie groups extend to isomorphisms of the Lie groups. Margulis's theorem is an extraordinary generalization of this: it asserts that any homomorphism with sufficiently large image of a lattice to another Lie group extends to a homomorphism of the ambient Lie group. (See [Mar, Zi] for precise hypotheses, proofs, applications, and much fascinating discussion.)

This has a number of simple geometric implications:

COROLLARY. *The embeddings of a compact symmetric space in another are rigid.*

COROLLARY. *An exact sequence $1 \to \Gamma' \to \Gamma \to \Gamma'' \to 1$ with Γ', Γ'' lattices in higher rank symmetric spaces implies that Γ is itself a lattice in a Lie group.*

This second corollary means that one can inductively realize homotopy fibrations by manifolds; it is similarly related to rigidity of certain fibrations over aspherical manifolds.

This then suggests the notion of **cylindrical rigidity**. We say that M is cylindrically rigid if for any stratified space X, $\mathbf{S}(M \times cX \operatorname{rel} M \times X)$ is contractible; or better yet, if for any $p : E \to M$ a MSAF (e.g. block bundle; see 9.4), $\mathbf{S}(Cyl(p) \operatorname{rel} E)$ is contractible.

CYLINDRICAL BOREL CONJECTURE. *If M is an aspherical manifold, then it is cylindrically rigid.*

We warn the reader that cylindrical rigidity can fail because of Nil phenomena, so that it is perhaps best to consider the above statement as being asserted for $\mathbf{S}^{-\infty}$.

We will see in the next section how a simple variant of the rigidity phenomenon that follows quickly from surgery, modulo rigidity itself, when combined with cylindrical rigidity leads to an important class of *rigid stratified spaces*.

For now, we will content ourselves with some exercises and remarks. In what follows, we will always be dealing with stable structure sets (see 6.2 for the definition).

EXERCISE. Show that if M is cylindrically rigid, and the fiber of p is (cylindrically) rigid, then E is (cylindrically) rigid.

EXERCISE. Using Farrell's fibering theorem 4.6 and the classification theorem (6.2), show that a circle (and hence a torus, in light of the previous exercise) is cylindrically rigid.

Alternatively, re-prove the Farrell fibering theorem directly for stratified spaces, and then interpret this result as cylindrical rigidity.

Also, show that nilmanifolds are cylindrically rigid.

EXERCISE. Modify the arguments in 9.4.B to show that Bieberbach manifolds and infranilmanifolds are cylindrically rigid.

EXERCISE. Using Cappell's splitting theorem (or alternatively Mayer-Vietoris sequences) as in 4.6.A, show that surfaces are cylindrically rigid.

The following two exercises display somewhat different types of rigidity.

EXERCISE. If $p : E \to B$ has rigid fibers, then $\mathbf{S}(Cyl(p) \operatorname{rel} B)$ is contractible. Furthermore, if B has a stratification for which p is a stratified system of fibrations with rigid fibers, then the same holds.

REMARK. According to [CFG] this is the situation when Riemannian manifolds collapse. This suggests that there might be a theorem of [GPW] type (see 9.4.C) for other collapsing situations ([GPW] relying on the rigidity of a point!), with more sophisticated invariants entering to understand the topology of the collapse.

EXERCISE. The following total space of an MAF is always rigid, despite neither base nor fiber is aspherical: Let X be arbitrary and $p : E \to X$ be an MAF with fiber a simply connected manifold of signature 1. Then $(Cyl(p) \text{ rel } E)$ is rigid.

14.2. Variant forms of the conjecture

Last section we saw how the rigidity theorems of differential geometry lead one to more topological conjectures than merely the Borel conjecture. In particular, we saw that embeddings are rigid and that there are equivariant and cylindrical forms of rigidity.

In this section I would like to (1) make a stratified rigidity conjecture that includes all of those cases, (2) point out some of its drawbacks, yet (3) make explicit its implications for characteristic classes in certain cases, and (4) point out its analogues in algebraic and analytic K-theories. In the next section, I will discuss the evidence for it a little bit more and a way around some of the difficulties.

Before dealing with our conjectures, I would like to point out that equivariant rigidity is false, quite generally, because it is not properly stratified! We will follow [Wei12] and give a certain number of details, since the construction is slightly computational.

THEOREM (COUNTEREXAMPLES TO EQUIVARIANT BOREL CONJECTURE). *For every finite G there is an affine G-torus with an equivariantly homotopy equivalent manifold not G-homeomorphic to it or even isovariantly homotopy equivalent to it. These examples are stable in that crossing with Euclidean spaces does not change the situation.*

We consider a torus T with a G-action with isolated fixed points. We assume that T satisfies a very large gap hypothesis ($=$ large$+4$) and that the Borel conjecture is true for T.

LEMMA 0. *The space of equivariant homotopy equivalences* $\mathrm{Equi}(T)$ *has vanishing homotopy groups above dimension one.*

PROOF. This is immediate from equivariant obstruction theory and the fact that T is equivariantly Eilenberg-MacLane.

Let $C_k(T)$ denote the configuration space of k ordered points in the torus T.

LEMMA 1. *There is a natural map $\Omega C_k(T) \to$* $\mathrm{Aut}(T, k)$.

PROOF. A map $X \to \Omega C_k(T)$ gives k-sections of $T \times I \times X$ over $X \times I$. Remove these sections and view this complement as a fibration.

Note that if k is at most the number of fixed points, we can build a group action on $\Omega C_k(T)$, where the loops are based at a subset of the fixed set.

LEMMA 2. *The action of G on $\Omega C_k(T)$ leaves fixed nontrivial elements of π_i.*

We first prove this for $k = 2$. We compute $\pi_{n-1}(C_2(T)) = \mathbb{Z}[\mathbb{Z}^n]$, since (as usual) there is a fibration

$$(T - p) \to C_2(T) \to T.$$

The G-action corresponds to the action on homology (since we are dealing with a split crystallographic manifold). Since the G-invariants of $\mathbb{Z}[\mathbb{Z}^n]$ are nontrivial (e.g. $\Sigma g(e) \neq 0$ is invariant) we have verified the lemma.

Now, the same fibration for larger k shows that $\pi_{n-1}(C_k(T)) \to \pi_{n-1}(C_{k-1}(T))$ surjects, so Schur's lemma gives the result on fixed sets.

LEMMA 3.[2] *The image of such an element gives $\otimes\mathbb{Z}[1/G]$ a nontrivial element of $\pi_{n-1}\,\mathrm{Iso}^G(T)$.*

PROOF. First of all, away from G the homotopy groups of $\mathrm{Iso}^G(T)$ are the invariant elements of the homotopy of $\mathrm{Aut}(T, k)$ where k is the number of fixed points. This is immediate from the Federer spectral sequence [Fe] (which gives a method of computing homotopy of function spaces), the decomposition of the E^2 term of this sequence according to representations, and the existence of transfer yielding an isomorphism on the E^2 level of the invariant part of the spectral sequence for stratified self-maps of (T, k) and the whole E^2 term of $\mathrm{Iso}^G(T)$.

Next consider the fibration over the sphere associated to an element of homotopy. The fixed sets link each other. That is, if one considers the map of one sphere into the complement of the first, one sees that this is a nontrivial element.

LEMMA 4. *The map $T' \to S^{i+1} \to B\,\mathrm{Iso}^G(T)$ gives rise to an isovariant Poincaré G-space, equivariantly an affine torus, but not isovariantly an affine action.*

PROOF. A fibration over a Poincaré space with Poincaré fibers is Poincaré. Lemma 0 gives the equivariant result, and the proof of lemma 3 gives the isovariant one.

LEMMA 5. *This total space is isovariantly a G-manifold. Equivariantly, this manifold is an affine torus.*

[2]Since writing the first draft of this section, I received [DS], which begins the development of a systematic theory for analyzing the difference between isovariant and equivariant maps. Presumably their work will lead to a less ad hoc and more systematic approach to the issues considered here.

PROOF. By rigidity and gap hypotheses we know rigidity for T and $T \times$ low dimensional tori. This then implies that the inclusion $\widetilde{\text{Homeo}}(T) \to \text{Iso}(T)$ is a homotopy equivalence, since it is for the first four homotopy groups. (See the periodicity result of 11.3.) Consequently, the isovariant bundle can be realized by a geometric block bundle, whose total space is the desired G-torus. Since it comes from the fiber of $B\text{Iso} \to B\,Equi$, it is equivariantly linear.

These lemmas complete the proof of the theorem, since we have an equivariantly affine torus that is not isovariantly affine. However, the disproof shows that the whole problem stems from the fact that the equivariant conjecture is not appropriately stratified, which, consequently, suggests that the isovariant conjecture might be correct.

Even this is not correct, because of difficulties involving Nil and UNil. The Nil difficulty can be removed if (following a suggestion of Frank Connolly) we restrict attention to isovariant homotopy equivalences that are topologically simple (see 10.1). However, here is an example that shows that UNil difficulties actually arise:

COUNTEREXAMPLE (TO ISOVARIANT BOREL CONJECTURE), *If we take a torus T and let $\mathbb{Z}_2$ act on it by flipping an odd number of coordinates and being trivial on a few additional coordinates (three or more will certainly suffice), then the quotient is predicted to be rigid but is neither simply nor stably rigid. In other words, this involution on the torus is not equivariantly rigid.*

The construction makes use of a remarkable example of Cappell [Ca2] of a manifold W homotopy equivalent to $\mathbb{R}P^{4k+1}\#\mathbb{R}P^{4k+1}$ but not itself a connected sum. He detects this via mapping to a finite dihedral group and comparing various Kervaire-Arf invariants, i.e. by a stable technique.

What is remarkable about Cappell's example is that it is based on the failure of codimension one splitting in the non-square root-closed case (see 4.4 and 4.6.A). One would try to homotop the map so that the inverse image of the sphere in the #-decomposition of $\mathbb{R}P^{4k+1}\#\mathbb{R}P^{4k+1}$ is a homotopy sphere, and hence a sphere, decomposing W. The failure of this means that a Mayer-Vietoris sequence is not valid in L-theory (again, see 4.6.A). The fundamental groups of the top strata in these examples have a split surjection to the infinite dihedral group, so we can make use of this element.[3] This, in turn, is interpretable in terms of the surgery sequence, which, after all, describes structures as being the difference between genuine L-theory and the homology version of it (i.e.

[3]A more intelligent thing to do would be to consider the element of $S(\mathbb{R}P^{4k+1}\#\mathbb{R}P^{4k+1})$ and use an embedding into the quotient of the top stratum and functoriality (perhaps after crossing with some circles to get the dimmod4 correct) to get a structure of the top pure stratum rel its end, and therefore a structure of the top pure stratum.

the version satisfying some sort of Mayer-Vietoris and giving a nontrivial element in the structure set).

It is not that hard to fill in the details of this, especially given the discussion below of the algebraic reinterpretation of the conjecture, but even easier is to make use of the fact that Cappell actually produced an infinitely generated group of counterexamples. Consequently, inspection shows that the homology term in the surgery sequence is finitely generated, so there is certainly enough room to construct counterexamples.

PROBLEM (SUGGESTED BY CAPPELL). Simplicity is a property that just involves chain complexes and is therefore defined for homotopy equivalences between polyhedra. Is there some notion that involves duality that is only defined for simple Poincaré complexes?[4]

When this is done, is it the case that the obstruction to topological self-dual simplicity is the UNil contribution to the structure set?

This is probably a good place to mention a positive bit of evidence for our conjecturing (but which does not simplify by use of the framework described in the next section).

THEOREM [CK1]. *If G is an odd order group acting affinely on a torus T with small gaps,*[5] *then T is simply isovariantly rigid.*

REMARK. A similar result on the calculation of L-groups of crystallographic groups is due to Yamasaki [Ya]. The theory here deduces the L-result from the geometric rigidity. Quinn in [Q6] proves something quite like the algebraic K-theory component of this result. (He does not actually compute the Tate cohomology.) All of these results are proven by extensions of the techniques of [FH1]. The proof given in [CK1, pt. II] is especially elegant.

In 14.3, we will mention a suggestion of Farrell and Jones for dealing with Nil and UNil. For now, we will formulate conjectures modulo these contributions. (This will make some sense when we discuss algebraic reformulations of the conjectures.)

To continue, let us first generalize the Borel conjecture to a larger class of manifolds.

DEFINITION. *A compact manifold M is* **haspherical** *(= homology aspherical) if the map $M \to B\pi_1(M)$ is a $\mathbb{Z}$-homology isomorphism.*

---

[4]An answer to this is given by the theory of supersimple homotopy equivalences; the first one involving $H(\mathbb{Z}_2; Wh_2(\pi))$ was defined by Cappell (unpublished) and developed in general in [WW, III].

[5]Recall that this means that all fixed point sets that are included in one another differ in dimension by at least three.

We note that if A and B are manifolds and $\varphi : A \to B$ is an isomorphism on fundamental group and integral homology, then $\varphi_* : S(A) \to S(B)$ is an isomorphism. Thus, if $B\pi$ is a closed manifold for which the Borel conjecture is true, and if M is haspherical with fundamental group π, then M is rigid.

Haspherical rigidity is also equivalent to the isomorphism of the assembly map (as in 4.6.A and below). Thus, in some sense, this is as well founded as the usual Borel conjecture.[6]

DEFINITION, CONTINUED. *If M has boundary or is noncompact, then we say that M is* **haspherical** *if the proper map from M to the triad $(B\pi, B\pi(\partial), B\pi(\infty) \times [0, \infty))$ is a (locally finite) $\mathbb{Z}$-homology isomorphism in relative homology. There is a similar notion if we choose to work relative some boundary component or end as well; then we work with absolute homology in that direction.*

There are many more haspherical manifolds than aspherical ones, as we will see shortly.

PROBLEM. When is there a finite dimensional haspherical (Poincaré) complex with fundamental group π?

REMARK. We can make haspfericality yet more general by picking a coefficient system other than $\mathbb{Z}$. Other convenient choices are $\mathbb{Z}/2\mathbb{Z}$ and 0 (i.e. no condition). In the conjecture below, one should then only assume haspfericality with respect to coefficient systems arising from the local holinks. That is, some holinks have small L-theory, and one shouldn't then require $\mathbb{Z}$-haspfericality of the corresponding stratum.

STRATIFIED RIGIDITY CONJECTURE. *(Modulo Nils)*[7] *If X is a stratified space such that (1) all holinks map injectively on fundamental groups to the holinks in which they are included, and (2) all pure strata are haspherical, then X is rigid. We will call a space satisfying (1) and (2)* **crigid**.[8]

The problem that condition 1 avoids is the following:

EXAMPLE/EXERCISE. Suppose that $X = (B^3, I \subset \text{Int}\, B^3, \{0\} \cup [1\})$ and we consider X rel S^2. Verify that all pure strata are relatively haspherical, yet show that X is **not** relatively rigid. Compute the structure set.

---

[6]Note also that the Gromov-Lawson-Rosenberg conjecture would preclude haspherical manifolds from having positive scalar curvature.

[7]There are various ways to make precise "true conjectures" (i.e. conjectures that might turn out to assert true statements, as opposed to, say, theorems or statements known to be false but still worth trying to prove since although any such attempt must fail, interesting theorems will be proven along the way) out of this false one, especially on examining its algebraic counterparts below. One geometric way, which does not always make sense, is to assert that the rigidity is virtual, i.e. only starts on passing to a subgroup of finite index. Alternatively, one would expect that this is correct for structure sets $\otimes\mathbb{Z}[1/2]$.

[8]For conjecturally rigid.

The difficulty here is that the local holink at an interior point of I is a circle with fundamental group $\mathbb{Z}$, but it is included in the holink of the whole I stratum which is an S^2, a simply connected space.

REMARK. As in 9.4.A it is possible to combine this conjecture with quasi-isometric phenomena. I will not be explicit about this; the connection between some of these ideas will be discussed in the following section.

REMARK. There are some people who talk about the "existence part of the Borel conjecture." This would assert that if $B\pi$ is a Poincaré space, then there is a topological manifold realizing it. This is not as well founded as the Borel conjecture; but if one used homology manifolds (9.4.D), it would be. Then the haspherical version would also be as well founded, and one would assert the stratified version as well. To understand this, do the following:

EXERCISE (COUNTEREXAMPLE TO HASPHERICAL MANIFOLD EXISTENCE CONJECTURE). Start with M^{2k} the boundary of a regular neighborhood of the torus T^{2k+1}; it has fundamental group $\mathbb{Z}^{2k+1}$. Do a Wall realization on the element $T^{2k+1} \times (1 \in L_0(e))$. According to [CS1], we can surger this normal cobordism to be a $\mathbb{Z}$-homology h-cobordism. Attach this to the complement of the 2-skeleton and glue the other end to the neighborhood of the 2-skeleton to obtain a haspherical Poincaré complex. Show that it is not homotopy equivalent to a manifold but that, using 9.4.D [BFMW], it is homotopy equivalent to a homology manifold.

One big advantage of the stratified rigidity conjecture is that it enables one to move the rigidity framework to situations that have torsion in the fundamental group, unlike the unstratified version of 4.6.A.

To appreciate the conjecture one needs some examples of crigid spaces.

EXAMPLE 1. Mapping cylinders of MSAFs to crigid spaces rel the domain.

EXAMPLE 2. The inclusion of a pure embedding of one crigid space in another of codimension at least three.

EXAMPLE 3. If X is crigid, then so is X/Γ for any group Γ acting freely and properly discontinuously on X. This includes $K\backslash G/\Gamma$ for Γ a discrete subgroup of a Lie group G and for K the maximal compact of G. The second exercise below includes a more general statement.

EXAMPLE 4. Mapping cylinders of maps with crigid local fibers rel the base are crigid.

EXERCISE. Show that in the case of embeddings of symmetric spaces in one another, one has crigidity using example 3. (Hint: Excise.)

EXERCISE. Suppose that G acts by isometries on a manifold of nonpositive curvature; show that the quotient is crigid. (Hint: First observe that the stratified space obtained by labeling components of fixed set strata is crigid, and then use example 3.) More generally, if G acts properly discontinuously on a contractible manifold with all fixed sets contractible, then the quotient is crigid. Given an action on a locally finite simplicial complex with this property, one can equivariantly thicken the complex to obtain a manifold example.

Surgery theory, as in 4.6.A, "reduces" the stratified conjecture to the following two:

K-THEORY CONJECTURE. *Let X be crigid relS, then for any ring R,*

$$\mathbf{H}_0(X; \underline{\mathbf{K}}^{BQ}(rel\ S; R)) \to \mathbf{K}^{BQ}(X \,\mathrm{rel}\, S; R)$$

is an isomorphism.

L-THEORY CONJECTURE. *Let X be crigid relS, then for any ring R,*

$$\mathbf{H}_0\big(X; \underline{\mathbf{L}}^{BQ}(\mathrm{rel}\ S; R)\big) \to \mathbf{L}^{BQ}(X \,\mathrm{rel}\, S; R)$$

is an isomorphism.

Recall from 6.1 that one introduces R coefficients in Browder-Quinn spectra by formally using the spectra for $R\pi$ instead of $\mathbb{Z}\pi$ and then inductively building up, as before.

It is of course just the π_0 statements that carry directly geometric information.[9]

EXERCISE (USING 6.2). The π_0 statements of the algebraic conjectures are jointly equivalent to the stratified rigidity conjecture when we restrict to $R = \mathbb{Z}$.

If R is a $\mathbb{Q}$-module, then the K-theory version might be true as far as we can tell, and similarly, if $1/2 \in R$, the L-theory version also has a shot. In those cases, Nil and UNil vanish. There is also a similar C^*-algebra version, which I only see how to phrase rel the whole singular set; then the sheaves decay into being K-functors of the C^*-algebras of the fundamental groups of the local holinks. (This is closely related to the Baum-Connes conjecture [BC] when we restrict attention to the case of Γ acting properly discontinuously on a contractible space with all fixed sets contractible.)

[9]The higher homotopy groups of these conjectures are related to parametrized rigidity theorems, see [G1].

REMARK. We will see that these conjectures include the Mayer-Vietoris sequences of Waldhausen, Cappell, and Pimsner [Wald3, Ca4, Pi] in these various settings. They also contain formulae for K- and L-theory of twisted group rings (cross product algebras) by reflecting on the case of cylinders of bundles. Furthermore, in 14.4, we will see that for all groups of finite virtual cohomological dimension (or, stretching the imagination, all discrete groups) the conjectures give predictions of the algebraic and analytic K-theories as well as the L-theory.

STRATIFIED NOVIKOV CONJECTURE(S). *All of the maps described above are split injective over $\mathbb{Z}$.*

I'm emphasizing the over $\mathbb{Z}$ part because our earlier versions of Novikov were essentially rational; here the $\mathbb{Z}$ version does stand a chance of being true. All of the Nils and UNils that are known to occur naturally split off (and although it is hard to imagine why they must continue to do so in the course of an inductive argument, they nonetheless seem to!).

I want to close this section by returning to the equivariant situation. Despite our difficulties with equivariant rigidity conjectures, the equivariant Novikov conjecture is quite reasonable when thought about carefully and has some interesting additional aspects to it. Basically these come about because the problem natural to group actions is an equivariant one, and we have dealt only with functoriality associated to homotopy transverse maps.

CRUDE EQUIVARIANT NOVIKOV CONJECTURE. *If X is an equivariant $B\pi$, then for $f : M \to B\pi$ an equivariant map, the class $f_*(\Delta(M)) \in KO^G(M) \otimes \mathbb{Q}$ is an equivariant homotopy invariant.*[10]

For the finite case, an equivariant $B\pi$ means that all components of all fixed sets are aspherical. For the compact case, I recommend the appendix by Peter May to [RsW2].

This conjecture is crude for several reasons. Firstly, there is the shameful $\otimes\mathbb{Q}$. Also, the usual G-signature is invariant under a larger class of maps: those that are equivariant and a homotopy equivalence. Following Petrie [Pe2], we shall call such maps **pseudoequivalences**. Finally, for compact groups, we might want to use the really rather different theories discussed in 13.3. Even with all of this, the conjecture is incorrect: for $G = S^1$, the space ES^1 is an equivariantly aspherical space, and this conjecture contradicts the classification of homotopy CP^n's (4.1).

[10]We are tacitly assuming in all of this that G is compact. It only takes a small amount of reformulation to allow locally compact groups, but then one wants cocompact quotients (or more subtlety at infinity of the sort we've already seen e.g. in 9.9.A) and more importantly locally uniformly compact isotropy.

The solution I will adopt provisionally is to assume that $B\pi$ is a finite dimensional space. (However, this is not necessary; probably one wants the quotient to be a "filtered aspherical space", which is just equivariant asphericality, when the group acting is finite.) In that case, we can better use the universal classes of 6.1, 6.2; however, note that for finite groups the argument in 13.2 splits off the top piece of the homology theory away from 2; by using $L^{-\infty}$ and taking coefficients in $\mathbb{Q}$[11] (or even just $\mathbb{Z}[1/G]$) one can break this piece off, and ask that this top homology piece be a pseudoequivalence invariant.

In the compact Lie case, one still can ask that the equivariant signature operator give a pseudoequivalence invariant. This was studied in [FRW, RsW2]. I do not believe that this can play the same role in understanding the geometry of G-actions as it does in the finite case (because of the discussion in 13.3). In fact, in those papers, we essentially reduce to the finite case.

EQUIVARIANT NOVIKOV CONJECTURE. *If $B\pi$ is a finite dimensional equivariant aspherical complex, and $f : M \to B\pi$ is equivariant, then $f_*(\Delta(M)) \in H_*(B\pi/G; \mathbf{L}(\mathbb{Q}G_x))$ is a pseudoequivalence invariant (as is the class of the equivariant signature operator in $KO^G(B\pi)$).*

I should point out that there might be more than one $B\pi$ that one can map to, both with the same (unequivariant) fundamental group as M. With infinite dimensions, this is quite simple (restricting to finite groups, I see no reason not to make a $\mathbb{Q}$ conjecture allowing infinite dimensions). One can use a point or EG or even contractible spaces whose fixed sets are various unions of other aspherical spaces (like 2 points)!

These various conjectures can be related to the stratified conjecture by making use of the interpretation of the domain of the surgery map as being ($1/2|G| \in$ coefficients) the controlled equivariant visible algebraic Poincaré complexes over $B\pi$ (as in 6.2), a description which is explicitly functorial in the equivariant category, and observing that the splitting off of the top L piece (in the groups, spaces, and cosheaves) is dependent on only the pseudoequivalence type.

REMARK. It is often possible to use the techniques from harmonic maps to produce maps to specific equivariant $B\pi$'s (see [RsW2, ScY2] for a discussion). For instance, given an action on a manifold there is an affine action on a torus and an equivariant map $M \to T$ inducing an isomorphism on $H^1(\ ;\mathbb{Z})$. The proof is a pleasant exercise in the Hodge theorem on harmonic representatives of DeRham cohomology. Consequently, we see that the equivariant Novikov conjecture for affine tori implies restrictions on arbitrary G-manifolds.

[11] This still gives interesting information at the prime 2; I think that this theory still has a map to equivariant KO-homology integrally.

14.3. Evidence

Why should we believe anything like the previous conjectures? We've seen some evidence already in 4.6, 4.6.A, 9.4.A, 9.4.B, and at the end of the last section. In this section I would like to make the following argument. In light of Cappell's codimension one splitting theorem and our remarks on haspherical generalization of the Borel conjecture, stratified rigidity follows inductively from cylindrical rigidity of strata. As for cylindrical rigidity, this is as well founded a conjecture as rigidity in that the main methods for proving rigidity apply just as well to cylindrical rigidity. That this is correct for codimension one splitting techniques is essentially the exercise about surfaces in 14.2; we will here discuss the Novikov type methods of [FeW1,2] and the remarkable dynamical-foliation idea of [FJ1-3].

Note that even in this sketch-proof I show my cavalier approach to Nil and UNil. Cappell's splitting theorem is obstructed by a UNil obstruction. In addition, cylindrical rigidity is related to replacing $\mathbb{Z}$'s in the usual conjectures about group rings by $\mathbb{Z}\pi$'s, perhaps twisted by an action of the group Γ. Since $\mathbb{Z}$ is a regular ring, Nil vanishes for it, but this will not be the case for cylindrical rigidity when the fiber has a nonregular group ring. However, some of these objections can be handled by an idea developed in [FJ3, ConS, G1].

The first step in the alleged justification of the conjecture is reducing (modulo UNils) the conjecture inductively to cylindrical rigidity via the splitting theorem. The following example is instructive.

EXAMPLE (RELATION BETWEEN RIGIDITY AND MAYER-VIETORIS = SPLITTING).[12] Consider a manifold W with a two-sided codimension one submanifold V. We can form a stratified space by taking the mapping cylinder of a map $W \to [0,1]$ (or the circle if V is only locally two-sided) which projects a tubular neighborhood of V to [0,1] and collapses the complementary regions to the endpoints. If the fundamental group of V injects into that of W, we can conjecture that this space is rigid relative to $W \cup [0,1]$ (which is the singular set) or even just rel W.

If one thinks geometrically, one sees this statement as asserting that any other splitting of W along a manifold h.e. to V is the original splitting. Applying this principle to $W \times$ tori would also lead, via a periodicity argument, to the existence of splittings! (See 3.3 for the total surgery obstruction, which explains why existence and uniqueness questions are of the same sort in topological surgery theory.)

If we work algebraically using the surgery exact sequence of 6.2, the rigidity statement becomes equivalent to the statement that the global

[12] This is a somewhat sophisticated solution to a problem in 4.6.A.

L-term of the top stratum (which is L of an amalgamated free product) is identical to the homology term, which can be computed by Mayer-Vietoris sequence. The sequence looks like this:

$$\to L_n(W - V) \to H \to L_{n-1}(V) \to \dots$$

(This is a Mayer-Vietoris sequence because the $W - V$ term breaks up into two pieces, one for each component of this complement.)

It is interesting to note that this case shows that in general rigidity is false because of UNil. However, here the Novikov version is correct and corresponds to a theorem of Cappell [Ca3] that the UNil term splits off the L-theory of the amalgamated free product!

The analysis of this example shows how to inductively reduce the rigidity conjecture to other cylindrical rigidity conjectures. We try to use cylindrical rigidity to remove the bottom stratum, together with a teardrop neighborhood (10.3.A).

EXERCISE. Using the codimension one splitting theorem, 6.2, and the discussion above, prove a codimension one splitting theorem for stratified spaces.

When we remove the bottom stratum we now have a stratified space with fewer strata; the issue is to see its rigidity. Once it has been verified rigid, there will be a unique way to glue in the missing stratum because of its cylindrical rigidity.

EXERCISE. Show that conjecturally $S(X) \cong S(X - S)$ if S is cylindrically rigid.

However, by the conjecture, to see that it is rigid is a homological calculation comparing manifold homology to group homology. The desired result holds because group homology has excision for injective homomorphisms (see [Brnk]).

REMARK. L-theory also has the same excision, so one can reduce to the rel ∞ version for the space with fewer strata if desired.

In any case, let us return to methods for verifying Novikov and Borel conjectures, and discuss how well they behave under the extension of the conjecture to the cylindrical case.

14.3.1. The method of descent

This method is developed explicitly in [FeW2] and is implicit in one form or another in [GL3, Kas2, CoGM, Car, CP]. The idea is to deduce Novikov conjectures for a group Γ from bounded versions for the metric space Γ (see 9.4.A) via a *families bounded transfer*.

We will first sketch the proof of Novikov for manifolds with nonpositive curvature, where the method looks more geometric. Then we will rephrase it to apply to many assembly maps besides the $L(\mathbb{Z})$ one, and the consequence will be cylindrical Novikov theorems as well as an extension to many other groups (compactifiable groups).

REMARK. In [FRW] we explained that the geometric version of the method actually leads to cylindrical Novikov as well as equivariant Novikov theorems (by replacing the tangent bundle by the equivariant tangent bundle). The same method yields (integral) results for arbitrary groups with torsion, not only the virtually torsion free ones which [FRW] and [RsW2] were restricted to for notational simplicity. However, these integral results do not include integral injectivity of the conventional assembly map when the groups have torsion. Indeed, that is false![13] (See [BDO, Gn, Og1] for related results in the C^*-algebra context.)

The geometric version goes like this. We assume that $f : E'/\Gamma \to E/\Gamma$ is a homotopy equivalence of aspherical manifolds. Consider the map

$$E' \times_\Gamma E' \to E \times_\Gamma E'.$$

Actually, $E' \times_\Gamma E' \to E'/\Gamma$ is equivalent to the tangent bundle. (After all, what is the tangent bundle if not the best Euclidean approximation to the manifold at each point, but here we have a bundle whose fibers are rigged to be locally diffeomorphic to the manifold!) To show that f is tangential, we must move, in a parametrized way, the universal cover of f (based at various points) to be a homeomorphism $E' \to E$. Note that the map is bounded in the universal cover. If we knew the bounded structure set vanished, then we'd be done. (Note this is a question about the nature of the universal cover.)

If E' has nonpositive curvature, then the α-approximation theorem (9.4) gives the vanishing of the structure set, at least if we use the logarithm map (the inverse of the exponential map) to map the bounded structures of E over E to the bounded structures of E over Euclidean space, because Euclidean space has no intrinsic scale, so bounded = controlled. (Actually the original bounded structure set vanishes; see [HTW2, FeW2].) A little more care enables one to handle noncompact complete manifolds of nonpositive curvature.

In any case, let us reformulate the argument now. The α-approximation theorem for $\mathbb{R}^n$ is essentially equivalent to the statement that

[13] Recently, Farrell and Jones have given a calculation of the K- and L-groups of discrete subgroups of real connected Lie groups in terms of their finite subgroups and ideas similar to the ones that follow here.

$H(\mathbb{R}^n; \mathbf{L}(e)) \to \mathbf{L}^{bdd}(\mathbb{R}^n)$ is a homotopy equivalence. We argued in terms of the following diagram:

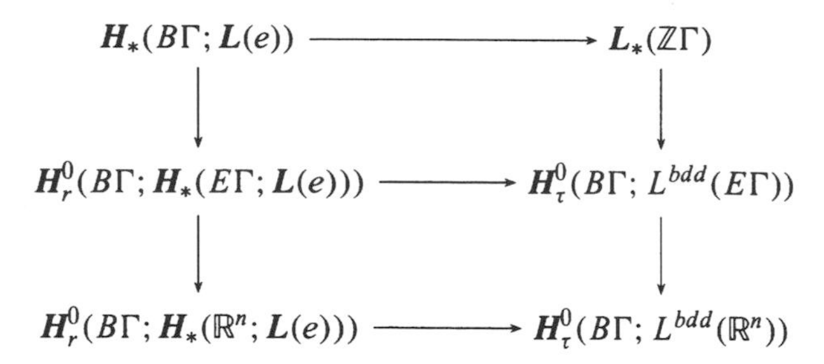

The first line is the usual assembly map. The second is the map on the level of space of sections of the family of assembly maps associated to the fibration $E \times_\Gamma E \to B\Gamma$.[14] The bottom line is the same sequence after identifying via the logarithm map $E\Gamma$ with $\mathbb{R}^n$. The first set of vertical arrows indicates family bounded transfers. Geometrically, a chain or surgery problem with target $B\Gamma$ gives rise, for each point of $B\Gamma$, to the transfer of that chain or problem to the copy of $E\Gamma$ based at a lift of that point.

The fact that the left-hand side vertical arrows are homotopy equivalences is due to a form of Spanier-Whitehead duality and the fact that $E\Gamma$ to $\mathbb{R}^n$ is a proper homotopy equivalence.

Putting together the maps $\mathbf{L}_*(\mathbb{Z}\Gamma) \to \mathbf{H}^0_\tau\ (B\Gamma; L^{bdd}_\tau(E\Gamma)) \to \mathbf{H}^0_\tau(B\Gamma; \mathbf{L}^{bdd}(\mathbb{R}^n)) \to \mathbf{H}^0_\tau(B\Gamma; \mathbf{H}_*(\mathbb{R}^n; \mathbf{L}(e))) \to \mathbf{H}_*(B\Gamma; \mathbf{L}(e))$, we have split the assembly map.

There are now several extensions possible of the argument. Firstly, we would not need to compare to Euclidean space if we knew that $\mathbf{H}_*(E\Gamma; \mathbf{L}(e)) \to \mathbf{L}^{bdd}(E\Gamma)$ was a homotopy equivalence. Another alternative (to the use of the logarithm map) would be if we knew that the map was split injective in a Γ-equivariant fashion by some other method.

Another point is that one can use other functors and assembly maps, like L of an arbitrary (twisted) ring or K-theory or the coefficient systems arising from an MSAF (which is why cylindrical Novikov follows with no further ado) or A-theory or pseudoisotopy theory or... This will be discussed further in [FeW2], but enough has been said for our purposes.

As for the verification of the bounded Novikov conjecture equivariantly or the bounded rigidity statement, for all functors this is possible with many different hypotheses. One alternative is an equivariant contractible compactification of $E\Gamma$ or an unequivariant Z-set compactification of $E\Gamma$ for Novikov and Borel respectively. Another possibility is an appropriate type of combing of the group; see [CEHPT] and see [FeW2] for more details.

---

[14]The notation H^0_τ is intended to capture the idea of "twisted cohomology."

But, in any case, this method implies all integral versions with arbitrary coefficients for all of the Novikov type conjectures for hyperbolic groups, discrete subgroups of real and p-adic Lie groups (or products of such), groups that act properly discontinuously on $Cat(0)$ spaces, and many others. However, I have had no success at all in promoting these (or the following) methods to "groups of infinite rank" and other seemingly ungeometric groups. I also should point out that bounded rigidity does sometimes fail for uniformly contractible manifolds [DFW].

14.3.2. The dynamic/foliation method

This method is quite simple in conception, difficult in detail, and has led to a number of really terrific results. I can't do it justice in a couple of paragraphs; the reader must consult the original papers and the further installments of the series as they appear.

For simplicity, let us consider the problem of showing that for an odd dimensional negatively curved manifold $Wh(\pi)$ has exponent 2. Consider an uncontrolled geometric module on M. Consider its transfer to the unit sphere bundle. Since the dimension is odd, this sphere has Euler characteristic two, so this step loses some information.

The sphere bundle has a geodesic flow on it, which then foliates it into one dimensional leaves. Applying the flow yields modules that are better and better controlled with respect to this foliation. (This is analogous to the gaining of control using self-maps of almost flat manifolds in 9.4.A.) Thus Farrell and Jones are led to a foliated control situation. It turns out that the lines contribute nothing unusual, but everything concentrates around the closed leaves, the circles which correspond to closed prime geodesics in the manifold. These behave like $R[t, t^{-1}]$ where t generates the geodesic.

This then gives a formula $K(R\Gamma) \cong H(B\Gamma; K(R)) \oplus \Sigma\ \mathrm{Nil}(R)$ where the sum is taken over the closed geodesics of Γ.

There are several difficulties to be dealt with in general; one must find good bundles to transfer to that have nice flows with dynamical properties that enable one to gain control without losing information. One must also prove foliated control theorems with respect to appropriate types of leaves to study the effect.

Farrell and Jones have been quite successful at this project and have used this technique to prove the Borel conjecture for all discrete subgroups of connected real Lie groups (the K-theory Borel conjecture follows from their work on pseudoisotopy according to Goodwillie [G1]).

Note that the method is built to give information about Nils; they come up here parametrized by conjugacy classes of elements of the group. This is quite similar to the description of cyclic homology of group rings (see [Bu]); this cannot be accidental, and the connection between cyclic techniques, K-theory, and these dynamical ideas is a fascinating subject. It is remarkable that cyclic homology, introduced for expressing index theorems as generalized traces, is entering as such a fundamental tool in the most geometric investigations.

Farrell and Jones have suggested ([FJ3] for pseudoisotopy, but it doesn't take much imagination to extend this to other functors) that there is a "circular assembly map" that gives the best approximation to $\mathcal{F}(B\Gamma)$ for a "continuous spectrum valued functor of spaces" from the knowledge of $\mathcal{F}(S^1)$. It is based on looking at the space of maps from S^1 into $B\Gamma$ and modding out by the action of $\mathrm{Aut}(S^1) =$ automorphisms of S^1. (This is slightly different from what is conventional in cyclic homology, where one considers the quotient under the rotational action on the circles, in that it involves power maps as well. This is somewhat similar to [BHM], which also seems to me quite thought provoking. . . .)

If one computes what this says about $\mathbb{Z}^k$, it yields for an arbitrary ring R a formula for all of the Nils of R in terms of the first one. This was discovered for $R = \mathbb{Z}\pi$, π finite, by Connolly and da Silva [ConS] by algebraic methods.

The fact that one works geometrically using flows (or if one were to algebraically rewrite their work using geometric modules as suggested above) allows one to see that the cylindrical setup is just a special case of the general expected perspective. In [FJ2] they prove the cylindrical version of rigidity for hyperbolic manifolds simultaneously with the ordinary rigidity.

REMARKS. 1) Oddly enough, despite the power of the Farrell-Jones methods there seems to be an advantage (besides generality) to descent: nowhere does one stabilize. Since unstable diffeomorphism and homeomorphism spaces are nowhere near algebraicization, we have no methods, as far as I know, for destabilizing results. In [FeW1] the A-theory assembly map was split by stabilizing an unstable result proven by the descent method. One shows that at least on the identity component, $\mathrm{Diff}(M) \to \mathrm{Homeo}(M)$ has a section (arbitrarily close to the identity) for nonpositively curved manifolds. Such theorems do not yet seem to fit into any general framework.

2) The other use of "spaces of circles" or, relatedly, cyclic homology in Novikov type problems is via the construction of trace maps; see [CM, BHM] for related implementations of this idea. This method essentially gives characteristic classes that can detect elements of K-theory (think of L-theory as a Hermitian analogue of K-theory, if you are optimistic). If one is successful, one can detect by such methods big parts of $K(R\Gamma)$; however, it is dependent on being able to find a "trace way" of detecting $K(R)$ to begin with. Roughly speaking, whatever one can detect of a

$K(R)$ via traces, one seems to be able to continue detecting (in "higher versions") after assembly in $K(R\Gamma)$.

Inversely, the fact that some assembly maps are not injective for finite groups somehow forces certain traces not to detect too much for the coefficient ring!

Finally, needless to say, this method does not adapt gracefully to cylindrical generalization.

14.4. Applications

Here I'd like to discuss some implications or applications of the conjectures discussed above. In the cases where these conjectures have been verified, the applications are quite genuine. I cannot claim, however, that the applications surpass the conjectures themselves in beauty.

14.4.1. Spaces with crigid holinks

The following is a calculation of the cosheaf homology:

THEOREM. *Let X be a stratified space with all holinks crigid. The rigidity conjecture for the holinks implies that*

$$H_*(X; \underline{\mathbf{L}}^{BQ}) \cong [X : \mathbb{Z} \times F/Top],$$

and of course the genuine manifold theoretic (i.e. no nonresolvable homology manifolds) normal invariants can be computed simply by removing the $\mathbb{Z}$.

As an example, if we study codimension two locally flat embeddings from the point of view of chapter 10, then one is led to a stratified space with S^1 holinks. The theorem then shows that the submanifold is irrelevant to the normal invariant theory, which goes a long way towards explaining the paucity of locally flat codimension two embeddings.

In addition, it asserts that the difference of whatever homology characteristic classes we have for homotopy equivalent objects has a natural lift to cohomology. It is indeed rare, however, for such characteristic classes themselves to be definable in cohomology. (For certain spaces with crigid holinks occurring in algebraic geometry, Goresky and Pardon have found such liftings.)

The theorem is quite believable. Intuitively, the rigidity of the links forces transversality in the traditional sense (7.1), which we've seen has conventional normal invariants. The actual proof is roughly speaking "Verdier duality for spectra over $L^*(\mathbb{Z})$" and then plugging in the algebraic form of the rigidity conjecture.

REMARK. As in manifold theory, this cohomological form of the normal invariant does not have good functoriality properties. However, it is useful for calculations, and the surgery sequence is then a sequence of group homomorphisms.

EXERCISE. What does this show about crigid spaces all of whose holinks are crigid? In particular, what are the L-groups of the top stratum?

14.4.2. Proper actions and groups with torsion

Let us consider a discrete group Γ acting properly and discontinuously on a manifold (although one could go further to Witt spaces for much of the discussion). All of the discussion regarding Whitehead groups and L-spectra in chapter 13 applies to this case because the quotient is an *orbifold*. The analogue of the statement that the homology is equivariant KO-homology uses the K-theory of proper actions [Phi] but the differences are insignificant.

In this setting Baum and Connes have written down a pretty Chern character [BC], at least in the complex case, but it is not too difficult to modify it for the real case.

If the action of Γ is on a finite dimensional contractible simplicial complex K with the fixed sets of all subgroups empty or contractible, then we can thicken the complex equivariantly to produce a manifold whose quotient is crigid rel ∞. If we work rel singularities, we obtain:

CONJECTURES.

$$H_*(K/\Gamma; \underline{\mathbf{K}}(\mathbb{Z}\Gamma_k)) \to K(\mathbb{Z}\Gamma),$$
$$H_*(K/\Gamma; \underline{\mathbf{L}}(\mathbb{Z}\Gamma_k)) \to L(\mathbb{Z}\Gamma), \text{ and}$$
$$K_*^\Gamma(K) \to K_*(C^*\Gamma)$$

are all isomorphisms.[15]

As usual, these conjectures must be modified because of the Nil type phenomenon; so one should, say, use $\mathbb{Q}$ coefficients in the rings, or demand only split injectivity, or one could utilize the suggestion of Farrell and Jones discussed in 14.3, and implicate the nearly cyclic subgroups of Γ.

Once one has gotten to here, it's possible to go further and ask the same thing when K is not finite dimensional but the action resembles one that is: namely that the fixed sets of the finite subgroups are nonempty (and contractible). This permits the general construction of a universal complex which conjecturally computes the various theories.

---

[15]Because of examples involving Kazhdan's property T one must insist here on using the maximal C^*-algebra [BC].

Baum and Connes were led to make the final conjecture in the C^*-algebra context just by thinking about proper actions, and for them K was $E'\Gamma$, the universal space for proper Γ-actions (which is unique up to Γ-homotopy equivalence). They also have an extension to foliations, which should be studied in the more topological settings as well. See [BC, Bau] for more discussion.

These maps can be proven to be injections quite often and can presumably be seen to be isomorphisms under more restrictive conditions (as discussed in the previous section). For instance, the ideas in [FJ1-3] suffice to prove the result for discrete subgroups of $O(n, 1)$. They have recently extended this to discrete subgroups of arbitrary connected real Lie groups.

COROLLARY. *If (the above conjecture is true for injectivity and) Γ acts properly discontinuously with compact quotient on M and N, and $f : M \to N$ is a Γ-map which is a homotopy equivalence, then the pushforward of the "signature classes" of M and N into K are the same.*

Note there are really two different corollaries depending on whether one is working in the analytic or L-theoretic context; both apply to the higher signature (although L is a bit more refined at 2), but the analytic approach has applications to other operators, as we have seen in 4.6.A.

14.4.3. $H\rho$

The fact that we conjecturally have a calculation of L-groups certainly has implications for ordinary (even smooth) manifolds. In [Wei4] I defined an invariant for homotopy equivalences of manifolds with fundamental group a discrete subgroup of a real semisimple group, but the construction applies whenever one knows the rational split injectivity of the maps of the previous section. I called it $H\rho$, for higher ρ, because it generalizes the ρ invariant we considered in thinking about lens spaces in 4.7. It should also be possible to use operator techniques to relate it to eta type invariants of a more subtle type than usual, but so far this has only been done for a few groups; see [Lo].

Unlike ρ this invariant needs some sort of acyclicity hypothesis for its definition. Again, [Wei4] discusses how to do this in some cases by making use of some of the theory of analysis on Witt spaces [Che2]. Here is the problem that requires the acyclicity:

EXAMPLE. If L and L' are homotopy equivalent three dimensional lens spaces, then $(S^1 \times L)\#k(S^2 \times S^2) \cong (S^1 \times \acute{L})\#k(S^2 \times S^2)$ for k sufficiently large. Therefore we cannot pick up the codimension one ρ invariant from the lens space for general $\mathbb{Z} \times \mathbb{Z}_p$ situations.

In any case, here is the idea. Suppose one knows the conjectures of the previous section $\otimes\mathbb{Q}$ for Γ. Then:

EXERCISE. One also has the split injectivity of the usual assembly map for Γ.

In fact let's choose a splitting. This then gives a map

$$S(B\Gamma) \to H_{*+1}\big(K/\Gamma; \underline{L}(\mathbb{Z}\Gamma_k, \mathbb{Z})\big) \otimes \mathbb{Q},$$

where $S(B\Gamma)$ is just the fiber of the assembly map (we aren't taking a geometric model of $B\Gamma$). Now given any element of $S(M)$ for a closed manifold with fundamental group Γ one can take the composition with this map to define $H\rho$. To be a little more explicit, one can combine this with the Chern character of [BC] to get an element of $H_{*+1}(\Gamma; F'\Gamma)$ where $F'\Gamma$ is the free $\mathbb{Q}$ vector space on nontrivial elements of finite order acted on by Γ by conjugation.

PROBLEM. Show how to associate an element of $H_{*+1}(\Gamma; F\Gamma)$ to a pair of metrics of positive scalar curvature on a manifold which gives an obstruction to them lying in the same component of the space of such metrics. Is there any realization of this invariant?

PROPOSITION. *If the $\mathbb{Q}$-isomorphism conjectures are true, then for any manifold, there are maps*

$$S(M) \to KO_{*+1}(B\Gamma, M) \otimes \mathbb{Q} \times H_{*+1}\big(K/\Gamma; \underline{\mathbf{L}}(\mathbb{Z}\Gamma_k, \mathbb{Z})\big) \otimes \mathbb{Q},$$

the first of which is a "solution to the Novikov conjecture" and the second is given by $H\rho$, which together yield an isomorphism.

I leave this as an exercise to the reader.

It is beautiful how the rigidity theorems for stratified spaces yield (and seem necessary for) conjectures on exactly how flexible arbitrary manifolds are. Unfortunately, there does not seem to be enough functoriality in our present theory to enable one to do the same for stratified spaces. (For orbifolds, there are reasonable conjectures one can make, but they are also tied to functoriality issues in equivariant topology.)

14.4.4. Approximate fibering

If we combine the rigidity conjecture with the picture of stratified spaces described in 10.3.A (and believe everything is true for families of spaces as well as individual spaces), we get:

CONJECTURE. *Suppose that M is a manifold and $B\Gamma = N$ is a closed aspherical manifold which has fundamental group a quotient of the fundamental group of M. If the induced cover has finite type, then modulo a Nil obstruction, the natural map $M \to N$ is homotopic to an MAF.*

Indeed, one would conjecture the same assuming that $B\Gamma$ was finitely dominated; then $B\Gamma$ would, conjecturally, exist as a homology manifold, and we'd still want the approximate fibering.

There is a similar thing one can say for stratified spaces and MSAFs.

In other words, one cannot achieve a fiber bundle or even a block fibration over N, because these notions require a homogeneity of the map from the point of view of closed sets, but one can arrange for this homogeneity to be correct from the point of view of open subsets.

The results of Farrell and Jones mentioned above suffice to verify this conjecture for nonpositively curved manifolds.

Bibliography

[Ad1] J.F. Adams, *On the groups $J(X)$ I,II,III,IV*, Topology **2** (1963), 181–195; **3** (1964), 137–171, 193–222; **5** (1966), 21–71.

[Ad2] ———, *Stable homotopy and generalized homology*, University of Chicago Press, 1974.

[Ad3] ———, *Infinite loop spaces*, Princeton University Press, 1978.

[ACD] A. Adem, R. Cohen, and W. Dwyer, *Generalized Tate homology, homotopy fixed points, and the transfer*, Cont. Math **96** (1989), 1–13.

[ACFP] D. Anderson, F. Connelly, S. Ferry, and E. Pederson, *K-theory continuously controlled at infinity*, J. Pure and Appl. Algebra (1993).

[AH1] D. Anderson and W.C. Hsiang, *The functors K_{-i} and pseudoisotopies of polyhedra*, Ann. of Math. **105** (1977), 201–223.

[AH2] ———, *Extending combinatorial piecewise linear structures on stratified spaces* II, Trans. AMS **260** (1980), 223–253.

[AM1] D. Anderson and H. Munkholm, *Boundedly controlled topology*, Springer LNM, 1988.

[AM2] ———, *Geometric modules and algebraic K-homology theory*, K-Theory **3** (1990), 561–602.

[Ar] M. Armstrong, *Princeton notes on the hauptvermutung*, unpublished notes (1969).

[As] A. Assadi, *Finite group actions on simply-connected manifolds and CW complexes*, Memoirs AMS **257** (1982).

[AV] A.H. Assadi and P. Vogel, *Actions of finite groups on compact manifolds*, Topology **26** (1987), 239–263.

[A1] M. Atiyah, *Bott periodicity and the index of elliptic operators*, Quart. J. of Math. **19** (1968), 113–140.

[A2] ———, *The signature of fiber bundles*, in *Global analysis: papers in honor of K. Kodaira*, University of Tokyo Press and Princeton University Press, 73–84.

[A3] ———, *Thom complexes*, Proc. London Math. Soc. **11** (1961), 291–310.

[A4] ———, *Global theory of elliptic operators*, Proc. Int. Conf. on Functional Analysis (1969), University of Tokyo Press 21–30.

[A5] ———, *Characters and cohomology of finite groups*, Publ. Math. d'IHES (1961), 247–289.

[AB] M. Atiyah and R. Bott, *A Lefshetz fixed point formula for elliptic complexes, II: applications*, Ann. of Math. **88** (1968), 451–491.

[AtH1] M. Atiyah and F. Hirzebruch, *Analytic cycles on complex manifolds*, Topology **1** (1962), 25–45.

[AtH2] ———, *Riemann-Roch theorems for differentiable manifolds*, BAMS **65** (1959), 276–281.

[APS] M. Atiyah, V. Patodi, and I. Singer, *Spectral asymmetry and Riemannian geometry* I, II, III, Math. Proc. Cam. Phil. Soc. **77** (1975), 43–69; **78** (1975), 405–432; **79** (1976), 71–99.

[ASe] M. Atiyah and G. Segal, *Exponential isomorphisms for λ-rings*, Quart. J. of Math. **22** (1971), 371–378.

[AS] M. Atiyah and I. Singer, *The index of elliptic operators* I, III, IV, V, Ann. of Math. **87** (1968), 484–530, 546–604; **93** (1971), 119–138, 139–149.

[At] O. Attie, *1992 Courant Institute Ph.D. thesis*.

[ABW] O. Attie, J. Block, and S. Weinberger, *Characteristic classes, distortion of diffeomorphisms, and amenability*, JAMS **5** (1992), 919–921.

[Au] K.K. Au, *1990 University of California, San Diego, Ph.D. thesis*.

[BGS] W. Ballman, M. Gromov, and V. Schroeder, *Manifolds of nonpositive curvature*, Birkhauser, 1985.

[Ba] H. Bass, *K-theory*, Benjamin, 1968.

[BHS] H. Bass, A. Heller, and R. Swan, *The Whitehead group of a polynomial extension*, Math. Publ. d'IHES **22** (1964), 61–80.

[Bau] P. Baum, *1991 CBMS lecture notes* (in preparation).

[BC] P. Baum and A. Connes, *Chern character for discrete groups*, in *A Fete of Topology*, Academic Press, 1988, 163–232.

[BDO] P. Baum, M. Davis, and C. Ogle, *Novikov conjecture for proper actions of discrete groups*, 1990 preprint.

[BD] P. Baum and R. Douglas, *K-homology and index theory*, Proc. Symp. Pure Math. **38** I, 117–173.

[BG] J. Becker and D. Gottlieb, *The transfer map and fiber bundles*, Topology **14** (1975), 1–13.

[BBD] A. Beilinson, J. Bernstein, and P. Deligne, *Faisceaux pervers, analyse et topologique sur les espaces singuliers*, Asterisque **100** (1982), 1–171.

[Bes] M. Bestvina, *Characterizing k-dimensional universal Menger compacta*, Memoirs AMS **71** (1988).

[Bi] R.H. Bing, *The collected papers*, American Math. Soc., Providence, RI, (1988).

[BiC] J.M. Bismut and J. Cheeger, *Invariants eta et indices des familles pour des variétés a bordes*, Comptes Rendus **305** (1987), 127–130.

[BiL] M. Bismut and J. Lott, *Flat vector bundles, direct images, and higher real analytic torsion*, 1993 preprint.

[Bla] B. Blackadar, *K-theory of operator algebras*, Springer-Verlag.

[BW1] J. Block and S. Weinberger, *Aperiodic tilings, positive scalar curvature and amenability*, JAMS **5** (1992), 907–918.

[BW2] ———, *Uniformly finite and large scale homologies, and group actions* (in preparation).

[BW3] ———, *Index theory for manifolds of bounded geometry I, II* (in preparation).

[BHM] M. Bokstedt, W. C. Hsiang, and I. Madsen, *The cyclotomic trace and algebraic K-theory of spaces*, Inven. Math. **111** (1993), 465–540.

[Bo] A. Borel, *Intersection cohomology*, Birkhauser PM series, 1984.

[Bre] G. Bredon, *Introduction to compact transformation groups*, Academic Press, 1972.

[Br1] W. Browder, *Surgery on simply connected manifolds*, Springer, 1972.

[Br2] ———, *Embedding smooth manifolds*, Proc. ICM Moscow (1966), 712–719.

[Br3] ———, *Isovariant vs. equivariant homotopy equivalences*, Princeton preprint.

[Br4] ———, *Torsion in H-spaces*, Ann. of Math. **74** (1961), 24–51.

[BLe] W. Browder and J. Levine, *Fibering manifolds over a circle*, Comm. Math. Helv. **40** (1965–66), 153–160.

[BLi] W. Browder and R. Livesay, *Fixed point free involutions on homotopy spheres*, Tohoku J. of Math. **25** (1973), 69–88.

[BLL] W. Browder, R. Livesay, and J. Levine, *Putting a boundary on open manifolds*, Amer. J. of Math. **87** (1965), 1017–1028.

[BPW] W. Browder, T. Petrie, and C.T.C. Wall, *The classification of free actions of cyclic groups of odd order on homotopy spheres*, BAMS **77** (1971), 455–459.

[BQ] W. Browder and F. Quinn, *A surgery theory for G-manifolds and stratified sets*, in *Manifolds*, Univ. of Tokyo Press, 1975, 27–36.

[Brne] E. Brown, *Cohomology theories*, Ann. of Math. **75** (1962), 467–484.

[Brnk] K. Brown, *Cohomology of groups*, Springer-Verlag, 1982.

[BG] K.S. Brown and S.M. Gersten, *Algebraic K-theory as generalized sheaf cohomology*, LNM **341** (1972), 266–292.

[BDF] L. Brown, R. Douglas, and P. Fillmore, *Extensions of C^*-algebras and K-homology*, Ann. of Math. **105** (1977), 265–324.

[Brnm] M. Brown, *A proof of the generalised Schoenflies theorem*, BAMS **66** (1960), 74–76.

[BFMW] J. Bryant, S. Ferry, W. Mio, and S. Weinberger, *The topology of homology manifolds*, 1993 preprint.

[Bu] D. Burghelea, *The cyclic homology of group rings*, Comm. Math. Helv. **60** (1983), 354–365.

[BLR] D. Burghelea, R. Lashof, and M. Rothenberg, *Automorphisms of manifolds*, LNM **473** (1975).

[Can] J. Cannon, *Shrinking cell like decompositions of manifolds: codimension three*, Ann. of Math. **110** (1979), 83–112.

[CEHPT] J. Cannon, D. Epstein, D. Holt, S. Levy, M. Paterson, and W. Thurston, *Word Processing in Groups*, Jones and Barlett, Boston and London, 1992.

[Ca1] S. Cappell, *A splitting theorem for manifolds*, Inven. Math. **33** (1976), 69–170.

[Ca2] ———, *On connected sums of manifolds*, Topology **13** (1974), 395–400.

[Ca3] ———, *Unitary nilpotent groups and Hermitian K-theory*, BAMS **80** (1974), 1117–1122.

[Ca4] ———, *Mayer-Vietoris sequences in Hermitian K-theory*, LNM **343** (1972), 478–512.

[Ca5] ———, *On the homotopy invariance of higher signatures*, Inven. Math. **33** (1976), 171–179.

[CS1] S. Cappell and J. Shaneson, *The codimension two placement problem and homology equivalent manifolds*, Ann. of Math. **99** (1974), 27–348.

[CS2] ———, *Pseudofree group actions I*, LNM **763** (1979), 395–447.

[CS3] ———, *Nonlinear similarity*, Ann. of Math. **113** (1981), 311–351.

[CS4] ———, *Piecewise linear embeddings and their singularities*, Ann. of Math. **103** (1976), 163–228.

[CS5] ———, *Singular spaces, characteristic classes, and intersection homology*, Ann. of Math. **134** (1991), 325–374.

[CS6] ———, *Stratifiable maps and topological invariants*, JAMS **4** (1991), 521–551.

[CS7] ———, *The topological rationality of linear representations*, Publ. Math. d'IHES **56** (1982), 309–336.

[CS8] ———, *An introduction to embeddings, immersions, and singularities in codimension two*, Proc. Symp. Pure Math. **32** part 2 (1978), 129–150.

[CS9] ———, *Genera of algebraic varieties and counting of lattice points*, BAMS **30** (1994), 62–69.

[CSSWW1] S. Cappell, J. Shaneson, M. Steinberger, S. Weinberger, and J. West, *The topological classification of linear representations of $\mathbb{Z}_2r$*, BAMS **22** (1990), 51–57.

[CSSWW2] ———, *Topological conjugacy of linear maps in low dimensions*, 1990 preprint.

[CSSW] S. Cappell, J. Shaneson, M. Steinberger, and J. West, *Nonlinear similarity in dimension six*, Amer. J. of Math **111** (1989), 717–752.

[CSW] S. Cappell, J. Shaneson, and S. Weinberger, *Classes topologiques caractéristiques pour les actions de groupes sur les espaces singuliers*, Comptes Rendus **313** (1991), 293–295.

[CW1] S. Cappell and S. Weinberger, *1987 CBMS informal lecture notes*, (distributed at the meeting/revised version, in preparation).
[CW2] ———, *Classification de certaines espaces stratifiés*, Comptes Rendus **313** (1991), 399–401.
[CW3] ———, *A geometric interpretation of Siebenmann's periodicity phenomenon*, Proc. 1985 Georgia topology conference, 47–52.
[CW4] ———, *Homology propagation of group actions*, Comm. Pure and Applied Math. **40** (1987), 723–744.
[CW5] ———, *Atiyah-Singer classes and PL transformation groups*, J. Diff. Geo. **33** (1991), 731–742.
[CW6] ———, *Which H-spaces are manifolds?* I, Topology **27** (1988), 377–386.
[CW7] ———, *Parallelisability of finite H-spaces*, Comm. Math. Helv. **60** (1985), 628–629.
[CW8] ———, *Fixed point set and normal representations of homotopy equivalent group actions*, 1992 preprint.
[CWY] S. Cappell, S. Weinberger, and M. Yan, *Decompositions and functoriality of isovariant structure sets* (tentative title, in preparation).
[Car] G. Carlsson, *Applications of bounded K-theory*, 1990 Princeton preprint.
[CP] G. Carlsson and E. Pederson (in preparation).
[Cs] A. Casson, *Fibrations over spheres*, Topology **6** (1967), 489–499.
[Ch1] T. Chapman, *Lectures on Hilbert cube manifolds*, CBMS Notes **28** (1976).
[Ch2] ———, *Approximation results in topological manifolds*, Memoirs AMS **251** (1981).
[Ch3] ———, *Controlled simple homotopy theory and applications*, LNM **1009** (1983).
[Ch4] ———, *Homotopy conditions which detect simple homotopy equivalences*, Pac. J. of Math. **80** (1979), 13–46.
[Ch5] ———, *Topological invariance of Whitehead torsion*, Amer. J. of Math. **96** (1974), 488–497.
[ChF] T. Chapman and S. Ferry, *Approximating homotopy equivalences by homeomorphisms*, Amer. J. of Math. **101** (1979), 583–607.
[Che1] J. Cheeger, *On the Hodge theory of Riemannian pseudomanifolds*, Proc. Symp. Pure Math. **36** (1980), 91–145.
[Che2] ———, *Spectral geometry of singular Riemannian spaces*, J. Diff. Geo. **18** (1983), 575–657.
[Che3] ———, *Analytic torsion and the heat equation*, Ann. of Math. **109** (1979), 259–322.
[Che4] ———, *A finiteness theorem for Riemannian manifolds*, Amer. J. of Math. **92** (1970), 61–74.
[Che5] ———, *Critical points of distance functions and applications to Riemannian geometry*, LNM **1504** (1992), 1–38.
[CFG] J. Cheeger, S. Fukaya, and M. Gromov, *Nilpotent structures and invariant metrics on collapsed manifolds*, JAMS **5** (1992), 327–372.
[CGM] J. Cheeger, M. Goresky, and R. MacPherson, *L^2-cohomology and intersection homology for singular varieties*, in Ann. Math. Studies **102** (1982), 303–340.
[ChGr1] J. Cheeger and M. Gromov, *Collapsing Riemannian Manifolds I and II*, J. Diff. Geo. **23** (1986), 309–346; **32** (1990), 269–298.
[ChGr2] ———, *On the characteristic numbers of complete manifolds of bounded curvature and finite volume*, in *Differential geometry and complex analysis*, Springer, 1985, 115–154.
[CO] T. Cochran and K. Orr, *Not all links are concordant to boundary links*, Ann. of Math. **138** (1993), 519–554.
[Co] M. Cohen, *A course in simple homotopy theory*, Springer, 1973.
[CoHo] E. Connell and J. Hollingsworth, *Geometric groups and Whitehead torsion*, Trans. AMS **140** (1969), 161–180.

[Cn] A. Connes, *Noncommutative differential geometry*, Publ. Math. d'IHES **62** (1985), 41–144.

[CoGM] A. Connes, M. Gromov, and H. Moscovici, *Conjecture de Novikov et fibres presque plats*, Comptes Rendus (1990), 273–282.

[CM] A. Connes and H. Moscovici, *Cyclic cohomology, the Novikov conjecture, and hyperbolic groups*, Topology **29** (1990), 345–388.

[CK1] F. Connolly and T. Kosniewski, *Rigidity of crystallographic actions* I, II, Inven. Math. **99** (1990), 25–48.

[CK2] ———, *Examples of lack of rigidity in crystallographic groups*, LNM **1474** (1991), 139–145.

[CL] F. Connolly and W. Luck, *The involution on the equivariant Whitehead group*, K-theory **3** (1989), 123–140.

[ConS] F. Connolly and M. da Silva, *The groups $N^i K_0(\mathbb{Z}\pi)$ are finitely generated $\mathbb{Z}N^i$ modules if π is a finite group*, 1990 Notre Dame preprint.

[CF] P. Connorr and E. Floyd, *The relation of cobordism to K-theories*, LNM **28** (1966).

[CD1] D. Coram and P. Duvall, *Approximate fibrations*, Rocky Mountain J. Math. **7** (1977), 275–288.

[CD2] ———, *Approximate fibrations and a movability condition for maps*, Pac. J. of Math. **72** (1977), 41–56.

[Cor] K. Corlette, *Flat G-bundles with canonical metrics*, J. Diff. Geo. **28** (1988), 361–382.

[CoW] S. Costenoble and S. Waner, *G-transversality*, 1990 preprint.

[Cu] S. Curran, *Intersection homology and free group actions on Witt spaces*, Michigan Math. J. **39** (1992), 111–127.

[Dvr] R. Daverman, *Decompositions of manifolds*, Academic Press, 1986.

[Da1] J. Davis, *The surgery semicharacteristic*, Proc. London Math. Soc. **47** (1983), 411–428.

[Da2] ———, *Evaluation of odd dimensional surgery obstructions with finite fundamental group*, Topology **27** (1988), 179–204.

[DM] J. Davis and J. Milgram, *A survey of the spherical space form problem*, Mathematical Surveys, Harwood, 1985.

[DR] J. Davis and A. Ranicki, *Semi-invariants in surgery*, K-Theory **1** (1987), 83–110.

[DS] G. Dula and R. Schultz, *Isovariant homotopy theory and diagram cohomology*, Memoirs AMS (to appear).

[DW] J. Davis and S. Weinberger, *Swan subgroups in L-theory and applications* (in preparation).

[lD] le Dimet, *Cobordisme d'enlacements des disques*, Mem. French MS **32** (1988).

[Do] H. Dovermann, *Almost isovariant normal maps*, Amer. J. of Math. **111** (1989), 851–904.

[DoR1] H. Dovermann and M. Rothenberg, *The generalized Whitehead torsion of a G-fiber homotopy equivalence*, LNM **1375** (1987), 60–88.

[DoR2] ———, *Equivariant surgery and classification of finite group actions on manifolds*, Mem. AMS **71** (1988), 379.

[DoS] H. Dovermann and R. Schultz, *Periodicity in equivariant surgery*, LNM **1443** (1990).

[DF] A.N. Dranishnikov and S. Ferry, *Limits of manifolds in Gromov-Hausdorff space*, 1993 preprint.

[DFW] A.N. Dranishnikov, S. Ferry, and S. Weinberger, *A flexible uniformly contractible manifold*, 1994 preprint.

[Dr] A. Dress, *Induction and structure theorems for orthogonal representations of finite groups.*, Ann. of Math. **102** (1975), 291–325.

[Ed] R. Edwards, *The topology of manifolds and cell-like maps*, Proc. ICM, Helsinki (1978), 111–127.

[Fa1] F.T. Farrell, *The obstruction to fibering a manifold over the circle*, Indiana U. Math. J. **21** (1971), 315–346.

[Fa2] ———, *The obstruction to fibering a manifold over the circle*, Proc. ICM, Nice, **2** (1970), 69–72.

[FH1] F.T. Farrell and W. C. Hsiang, *Topological characterization of flat and almost flat Riemannian manifolds $M^n (n \neq 3, 4)$*, Amer. J. of Math. **105** (1983), 641–672.

[FH2] ———, *The stable topological hyperbolic space form problem for finite volume manifolds*, Inven. Math. **69** (1982), 155–170.

[FH3] ———, *Rational L-groups of Bieberbach manifolds*, Comm. Math. Helv. **52** (1977), 89–109.

[FJ1] F.T. Farrell and L. Jones, *K-theory and dynamics*, I II, Ann. of Math. **124** (1986), 531–569.

[FJ2] ———, *A topological analogue of Mostow's rigidity theorem*, JAMS **2** (1989), 257–370.

[FJ3] ———, *Computations of stable pseudoisotopy spaces for aspherical manifolds*, LNM **1474** (1991) 59–74.

[FJ4] ———, *Classical aspherical manifolds*, CBMS lecture notes #75, 1990.

[FJ5] ———, *Isomorphism conjectures in algebraic K-theory*, JAMS **6** (1993), 249–297.

[Fe] H. Federer, *A study of function spaces by spectral sequences*, Trans. AMS **82** (1956), 340–361.

[Fe1] S. Ferry, *Homotoping ϵ-maps to homeomorphisms*, Amer. J. of Math. **101** (1979), 567–582.

[Fe2] ———, *A simple-homotopy approach to the finiteness obstruction*, LNM **870** (1981), 73–81.

[Fe3] ———, *Mapping manifolds to polyhedra*, 1992 preprint.

[FHP] S. Ferry, I. Hambleton, and E. Pederson, *A survey of bounded surgery theory and applications*, in MSRI Publ. Math., (to appear).

[FP1] S. Ferry and E. Pederson, *Epsilon surgery theory*, Binghamton preprint 1989.

[FP2] ———, *Bounded K-theory and splitting theorems* (tentative title, in preparation).

[FRW] S. Ferry, J. Rosenberg, and S. Weinberger, *Equivariant topological rigidity phenomena*, Comptes Rendus **306** (1988), 777–782.

[FeW1] S. Ferry and S. Weinberger, *Curvature, tangentiality, and controlled topology*, Inven. Math. **105** (1991), 401–414.

[FeW2] ———, *A coarse approach to the Novikov conjecture*, 1992 preprint.

[Fl] A. Floer, *Morse theory for Lagrangian intersections*, J. Diff. Geo. **28** (1988), 513–547.

[Fr1] M. Freedman, *The disk theorem in 4-dimensional topology*, Proc. ICM, Warsaw (1983), 647–663.

[Fr2] ———, *The topology of four-dimensional manifolds*, J. Diff. Geo. **17** (1982), 357–453.

[FrQ] M. Freedman and F. Quinn, *Topology of 4-manifolds*, Princeton University Press, 1990.

[Fu] S. Fukaya, *Collapsing a Riemannian manifold onto one of lower dimension*, J. Diff. Geo. **25** (1987), 139–156.

[Ful1] W. Fulton, *On the topology of algebraic varieties*, Proc. Symp. Pure Math. **46** (1987), 15–48.

[Ful2] ———, *Intersection theory*, Springer, 1988.

[FM] W. Fulton and R. MacPherson, *A categorical framework for the study of singular spaces*, Mem. AMS **243** (1981).

[GaS] D. Galewski and R. Stern, *Classification of simplicial triangulations of topological manifolds*, Ann. Math. **111** (1980), 1–34.

[Ge1] S. Gersten, *A product formula for Wall's obstruction*, Amer. J. of Math. **88** (1966), 337–346.

[Ge2] ———, *Bounded cohomology and group extensions*, 1992 preprint.

[Gi] P. Gilmer, *Topological proof of the G-signature theorem for G finite*, Pac. J. of Math. **97** (1981), 105–114.

[GiL] P. Gilmer and C. Livingston, Topology **31** (1992), 475–492.

[Gn] Donggeng Gong, 1992 Stony Brook Ph.D. thesis.

[G1] T. Goodwillie, *On the Farrell-Jones assembly map* (lectures at the 1991 Hawaii topology conference and 1992 Cornell topology festival).

[G2] ———, *Multiple disjunction and the Morlet disjunction lemma*, Mem. AMS **431** (1990).

[GM1] M. Goresky and R. MacPherson, *Intersection homology* I II, Topology **19** (1980), 135–162; Inven. Math. **72** (1983), 77–130.

[GM2] ———, *Stratified Morse theory*, Springer, 1988.

[GM3] ———, *The topology of complex algebraic maps*, LNM **961** (La Rabida algebraic geometry conference) (1982), 119–129.

[GP] M. Goresky and W. Pardon, *Wu number of singular spaces*, Topology **28** (1989), 325–367.

[GS] M. Goresky and P. Siegel, *Linking pairings on singular spaces*, Comm. Math. Helv. **58** (1983), 96–110.

[GrM] J.P.C. Greenlees and J.P. May, *Generalized Tate, Borel, and coBorel cohomology*, 1992 preprint.

[Gr1] M. Gromov, *Hyperbolic groups*, in *Essays on groups*, MSRI series, Springer-Verlag, 1987.

[Gr2] ———, *Large Riemannian manifolds*, LNM **1201** (1985), 108–122.

[Gr3] ———, *Almost flat manifolds*, J. Diff. Geo. **13** (1978), 231–241.

[Gr4] ———, with J. Lafontaine and P. Pansu, *Structures métriques pour les variétes riemanniennes*, Fernang Nathan, Cedic, 1981.

[GL1] M. Gromov and H.B. Lawson, *Positive scalar curvature in the presence of a fundamental group*, Ann. of Math. **111** (1980), 209–230.

[GL2] ———, *The classification of simply connected manifolds of positive scalar curvature*, Ann. of Math. **111** (1980), 423–424.

[GL3] ———, *Positive scalar curvature and the Dirac operator on complete Riemannian manifolds*, Publ. Math. d'IHES **58** (1983), 295–408.

[GPW] K. Grove, P. Peterson, and J.Y. Wu, *Geometric finiteness theorems and controlled topology*, Inven. Math. **99** (1990), 205–214.

[Ha] A. Haefliger, *Knotted $4k-1$ spheres in $6k$ space*, Ann. of Math. **75** (1962), 452–466.

[HP] A. Haefliger and V. Poenaru, *La classification des immersions combinatoires*, Publ. Math. d'IHES **23** (1964), 75–91.

[Hm] I. Hambleton, with T.T. Dieck, *Notes on surgery calculations*, in *Surgery theory and the geometry of representations*, DMV seminar series, Birkhauser, 1988.

[HMTW] I. Hambleton, R.J. Milgram, L. Taylor, and B. Williams, *Surgery with finite fundamental group*, Proc. London Math. Soc. **56** (1988), 349–379.

[HP] I. Hambleton and E. Pedersen, *Bounded surgery and dihedral group actions on spheres*, JAMS **4** (1991), 105–126.

[HmTW] I. Hambleton, L. Taylor, and B. Williams, *Induction theory in surgery*, Notre Dame preprint.

[HsM] Hasselholt and I. Madsen (in preparation).

[HV] J.C. Hausmann and P. Vogel, *Geometry on Poincaré spaces*, Princeton University Press, 1993.

[Higman] G. Higman, *Units in group rings*, Proc. LMS **46** (1940), 231–248.

[Hig1] N. Higson, *On the relative homology of Baum and Douglas*, 1988 preprint.

[Hig2] ———, *The K-theory of bounded propagation speed operators*, preprint.

[Hig3] ———, *A primer on KK theory*, Proc. Symp. Pure Math. **51**, part 1 (1990), 239–283.

[Hig4] ———, *Signature operator on Lipschitz manifolds and unbounded Kasparov bimodules*, LNM **1132**, 254–288.

[Hil] M. Hilsum, *Signature operator on Lipschitz manifolds and unbounded Kasparov bimodules*, LNM **1132** (1989), 254–288.

[HS] M. Hilsum and G. Skandalis, *Invariance par homotopie de la signature a coefficients dans un fibre presque plat*, Crelle J. **423** (1992), 73–99.

[HM] M. Hirsch and B. Mazur, *Smoothings of piecewise linear manifolds*, Annals of Math. Studies **80** (1974), Princeton University Press.

[Hi] F. Hirzebruch, *Topological methods in algebraic geometry*, Springer-Verlag, 1978 translation of 1962 German edition.

[HsP] W.C. Hsiang and W. Pardon, *When are topologically equivalent linear representations linearly equivalent?* Inven. Math. **68** (1982), 275–316.

[HsSh] W.C. Hsiang and J. Shaneson, *Fake tori*, in *Topology of manifolds*, Markham Press, 1969, 18–51.

[Hud] J.F.P Hudson, *Introduction to PL topology*, Benjamin Books, 1969.

[Hu1] B. Hughes, *Controlled homotopy topological structures*, Pac. J. of Math. **133** (1988), 69–97.

[Hu2] ———, *Approximate fibrations on topological manifolds*, Mich. Math. J. **32** (1985), 167–183.

[Hu3] ———, *Neighborhoods in stratified spaces* (tentative title, in preparation).

[HTWW] B. Hughes, L. Taylor, S. Weinberger, and B. Williams, *The classification of neighborhoods in a stratified space* (in preparation).

[HTW1] B. Hughes, L. Taylor, and B. Williams, *Controlled surgery theory* (in preparation).

[HTW2] ———, *Controlled topology over a manifold of nonpositive curvature*, Vanderbilt University preprint.

[HTW3] ———, *Bundle theories for topological manifolds*, Trans. AMS **319** (1990), 1–65.

[IK] K. Igusa and J. Klein, *Higher Reidemeister torsion*, 1992 preprint.

[Iv] B. Iverson, *Cohomology of sheaves*, Springer, 1986.

[J1] L. Jones, *Patch spaces: a geometric representation of Poincaré duality spaces*, Ann. of Math. **97** (1973), 306–343.

[J2] ———, *Combinatorial symmetries of the m-disk*, BAMS **79** (1973), 167–169.

[KaM1] J. Kaminker and J. Miller, *Homotopy invariance of the analytic index*, J. of Operator Theory **14** (1985), 113–127.

[KaM2] ———, *A comment on the Novikov conjecture*, Proc. AMS **83** (1981), 656–658.

[Kas1] G. Kasparov, *Topological invariants of elliptic operators I: K-homology*, Math. USSR, Izvestia, **39** (1975), 751–792.

[Kas2] ———, *Equivariant KK theory and the Novikov conjecture*, Inven. Math. **91** (1988), 147–201.

[KaS] G. Kasparov and G. Skandalis, *Groups acting on buildings, operator K-theory, and Novikov's conjecture*, K Theory **4** (1991), 303–337.

[Ka] G. Katz, *Some realization theorems in equivarient L-theory and calculations of G-signature*, Amer. J. of Math. **105** (1983), 939–973.

[Ke] M. Kervaire, *Lectures on the theorems of Browder and Novikov and Siebenmann's thesis*, Tata Institute of Fundamental Res., 1969.

[KMdR] M. Kervaire, S. Maumary, and G. DeRham, *Torsion et type simple d'homotopy*, LNM **48** (1967).

[KM] M. Kervaire and J. Milnor, *Groups of homotopy spheres*, Ann. of Math. **77** (1963), 504–537.

[Ki] R. Kirby, *Stable homeomorphisms and the annulus conjecture*, Ann. of Math. **89** (1969), 575–582.

[KS] R. Kirby and L. Siebenmann, *Foundational essays on topological manifolds, smoothings, and triangulations*, Princeton University Press, 1977.

[KwSz] K. Kwun and R. Szczarba, *Product and sum theorems for Whitehead torsion*, Ann. of Math. **82** (1965), 183–190.

[LR1] R. Lashof and M. Rothenberg, *The immersion approach to triangulation*, in *Topology of manifolds* (Hocking, ed.), Prindle, Weber, and Schmidt, 1968.

[LR2] ———, *G-smoothing theory*, Proc. Symp. Pure Math. **32** (1978), 211–266.

[LwM] H.B. Lawson and M.L. Michelson, *Spinor geometry*, Princeton University Press, 1989.

[Le] R. Lee, *Semicharacteristic classes*, Topology **12** (1973), 183–199.

[Ls] J. Alexander Lees, *The surgery obstruction groups of C.T.C. Wall*, Advances in Math. **11** (1973), 113–156.

[Lv1] J. Levine, *Unknotting spheres in codimension two*, Topology **4** (1965), 9–16.

[Lv2] ———, *A classification of differentiable knots*, Ann. of Math. **82** (1965), 15–50.

[Lvt1] N. Levitt, *Poincaré duality cobordism*, Ann. of Math. **96** (1972), 211–244.

[Lvt2] ———, *Immersions of homotopy equivalent manifolds*, Canadian Conference Series **2**, part 2 (1982), 287–421.

[Lvt3] ———, *A necessary and sufficient condition for fibering a manifold*, Topology **14** (1975), 229–236.

[LvR] N. Levitt and A. Ranicki, *Intrinsic transversality structures*, Pac. J. of Math. **129** (1987), 85–144.

[LMS] L.G. Lewis, J.P. May, and M. Steinberger, *Equivariant stable homotopy theory*, LNM, **1213** (1986).

[LdM] S. Lopez de Medrano, *Involutions on manifolds*, Springer-Verlag, 1971.

[Lo] J. Lott *Higher eta invariants*, K Theory **6** (1992), 191–233.

[Luck1] W. Luck, *Transformation groups and algebraic K-theory*, LNM **1408** (1990).

[Luck2] ———, *Reidemeister and analytic torsion for manifolds with boundary and symmetry*, J. Diff. Geo. **37** (1993), 269–322.

[LM] W. Luck and I. Madsen, *Equivariant L-theory* I, II, Math. Z. **203** (1990), 503–526; **204** (1990), 253–268.

[LuR] W. Luck and A. Ranicki, *Surgery obstructions for fiber bundles*, J. Pure Applied Algebra **81** (1992), 139–189.

[Lus] G. Lusztig, *Novikov's higher signature and families of elliptic operators*, J. Diff. Geo. **7** (1971), 229–256.

[M] R. MacPherson, *Global questions in the topology of singular spaces*, Proc. ICM, Warsaw (1983), 213–235.

[MM] I. Madsen and J. Milgram, *The classifying spaces for surgery and cobordism of manifolds*, Princeton University Press, 1979.

[MR] I. Madsen and M. Rothenberg, *On the classification of G-spheres* I II III, Acta Math. **160** (1988), 65–104; Math. Scan **64** (1989), 161–218; and several related Aarhus preprints. See, in particular, Cont. Math. **19** (1983), 193–226, for a summary.

[MTW] I. Madsen, C. Thomas, and C.T.C. Wall, *The topological spherical spaceform problem II*, Topology **15** (1976), 375–382.

[Mar] G. Margulis, *Discrete subgroups of semisimple Lie groups*, Springer, 1991.

[Mau] S. Maumary, *Proper surgery, groups and Wall-Novikov groups*, LNM **343** (1972), 526–539.

[May] J.P. May, *Simplicial objects in algebraic topology*, Van Nostrand Math. Studies **11** 1967.

[MQRT] J.P. May, with contributions by F. Quinn, N. Ray, and J. Tornehave, *E_∞ ring spaces and E_∞ ring spectra*, LNM **577** (1977).

[Maz] B. Mazur, *The method of infinite repetition in pure topology I*, Ann. of Math. **80** (1964), 201–226.

[Mi1] J. Milnor, *A procedure for killing the homotopy groups of a manifold*, Proc. Symp. Pure Math. **3** (1961), 39–55.

[Mi2] ______, *Lectures on the h-cobordism theorem*, Princeton University Press, 1965.

[Mi3] ______, *Whitehead torsion*, BAMS **72** (1966), 358–426.

[Mi4] ______, *Groups which act on S^n without fixed points*, Amer. J. of Math. **79** (1957), 623–630.

[Mi5] ______, *Two complexes which are homeomorphic but combinatorially distinct*, Ann. of Math. **74** (1961), 575–590.

[Mi6] ______, *Introduction to algebraic K-theory*, Princeton University Press, 1971.

[Mi7] ______, *Topology from the differentiable viewpoint*, University of Virginia Press, 1965.

[Mi8] ______, *Morse theory*, Princeton University Press, 1963.

[Mi9] ______, *Infinite cyclic coverings*, in *Topology of manifolds* (Hocking, ed.), Prindle, Weber, and Schmidt 1968, 115–133.

[Mi10] ______, *Some free actions of cyclic groups on the sphere*, in Differential Analysis, Bombay Colloq., 1964, 37–42.

[MS] J. Milnor and J. Stasheff, *Characteristic classes*, Princeton University Press, 1974.

[Ms1] A. Mischenko, *Hermitian K-theory, the theory of characteristic classes, and methods of functional analysis*, Russian Math. Surveys **31** (1976), 71–136.

[Ms2] ______, *Homotopy invariants of nonsimply connected manifolds I: rational invariants*, Izv. Math. Nauk SSSR Ser. Mat. **34** (1970), 501–514.

[Mis] G. Mislin, *Finitely dominated nilpotent spaces*, Ann. of Math. **103** (1976), 547–556.

[Mo] J. Morgan, *A product formula for surgery obstructions*, Mem. AMS **201** 1978.

[MoS] J. Morgan and D. Sullivan, *The transversality characteristic class and linking cycles in surgery theory*, Ann. of Math. **99** (1974), 384–463.

[Mos] D. Mostow, *Strong rigidity of locally symmetric spaces*, Princeton University Press, 1973.

[Mu] W. Muller, *Analytic torsion and R-torsion of Riemannian manifolds*, Advances in Math. **28** (1978), 233–305.

[Na] M. Nagata, *The structure of equivariant F/PL for odd order abelian group actions*, 1987 Chicago Ph.D. thesis.

[Neu] W. Neumann, *Signature related invariants of manifolds I: monodromy and γ-invariants*, Topology **18** (1979), 147–172.

[Ni] A. Nicas, *Induction theorems for groups of manifold homotopy structure sets*, Mem. AMS **267** (1982).

[No] S. Novikov, Izv. Akad. Nauk SSSR Ser. Mat. **34** (1970), 253–288, 478–500.

[Og1] C. Ogle, *Assembly maps, K theory, and hyperbolic groups*, K Theory **6** (1992), 235–265.

[Og2] ______, *Bounded homotopy theory and the Novikov conjecture*, 1992, Ohio State preprint.

[O1] R. Oliver, *Whitehead groups of finite groups*, Cambridge University Press, 1988.

[O2] ______, *Fixed point sets of group actions on finite acyclic complexes*, Comm. Math. Helv. **50** (1975), 145–177.

[Pa] W. Pardon, *Intersection homology, Poincaré spaces and the characteristic variety theorem*, Comm. Math. Helv. **65** (1990), 198–233.

[P] E. Pederson, *On the K_{-i} functors*, J. Algebra **90** (1984), 461–475.

[PR] E. Pederson and A. Ranicki, *Projective surgery theory*, Topology **19** (1980), 239–254.

[PW1] E. Pederson and C. Weibel, *A nonconnective delooping of algebraic K-theory*, LNM **1126** (1985), 166–181.

[PW2] ______, *A model for K theoretic homology*, LNM **1370** (1989), 346–361.

[Pe1] T. Petrie, *Smooth S^1 actions on homotopy complex projective spaces, and related topics*, BAMS **78** (1972), 105–153.

[Pe2] ———, *Pseudoequivalences of G-manifolds*, Proc. Symp. Pure Math. **32** (1978), 169–210.

[Phi] C. Phillips, *Equivariant K-theory for proper actions*, Longman Scientific and Technical, Wiley, 1989.

[Pi] M. Pimsner, *K-groups of crossed products with groups acting on trees* **86** (1986), 603–634.

[PV] M. Pimsner and D. Voiculesciou, *Exact segment as for K group and Ext groups of certain crossed products of C^* algebras*, J. Operator Theory **4** (1980), 93–118.

[Ql1] D. Quillen, *The Adams conjecture*, Topology **10** (1971), 67–80.

[Ql2] ———, *Higher algebraic K theory I*, LNM **341** (1972), 85–147.

[Q1] F. Quinn, *A geometric formulation of surgery*, 1969 Princeton Ph.D. thesis.

[Q2] ———, *Ends of maps* I, II, III, IV, Ann. of Math. **110** (1979), 275–331; Inven. Math. **68** (1982), 353–424; J. Diff. Geo. **17** (1982), 503–521; Amer. J. of Math. **108** (1986), 1139–1162.

[Q3] ———, *Homotopically stratified spaces*, JAMS **1** (1988), 441–499.

[Q4] ———, *Surgery on Poincaré and normal spaces*, BAMS **78** (1972), 262–267.

[Q5] ———, *Resolution of homology manifolds*, Inven. Math. **72** (1983), 267–284; Corrigendrum **85** (1986), 653.

[Q6] ———, *Algebraic K theory of poly- (finite or cyclic) groups*, BAMS **12** (1985), 221–226.

[Q7] ———, *Applications of topology with control*, Proc. ICM, Berkeley (1986), 598–606.

[Q8] ———, *Surgery on Poincaré and normal spaces*, BAMS **78** (1972), 262–267.

[Q9] ———, *Geometric algebra*, LNM **1126** (1985), 182–198.

[Q10] F. Quinn, *Intrinsic skeleta and intersection homology of weakly stratified spaces*, Proc. 1985 Georgia Topology Conference, 233–249.

[Ra1] A. Ranicki, *The total surgery obstruction*, LNM **763** (1979), 275–316.

[Ra2] ———, *The algebraic theory of surgery* I, II, Proc. LMS **40** (1980), 87–192, 193–283.

[Ra3] ———, *Localisation in quadratic L-theory*, LNM **741** (1979), 102–137.

[Ra4] ———, *Exact sequences in the algebraic theory of surgery*, Mathematical Notes, Princeton University Press, 1981.

[Ra5] ———, *Algebraic L theory assembly*, Edinburgh preprint, 1990.

[Ra6] ———, *The algebraic theory of the finiteness obstruction*, Math. Scan. **57** (1985), 105–126.

[RaS] D. Ray and I. Singer, *R-torsion and the Laplacian on Riemannian manifolds*, Advances in Math. **7** (1971), 145–210.

[Re] I. Reiner, *Class groups and Picard groups of group rings and orders*, CBMS lecture notes **26** (1975).

[Ro] V. Rochlin, *A new result in the theory of 4-dimensional manifolds*, Sov. Math. Doklady **8** (1952), 221–224.

[Roe1] J. Roe, *An index theorem on open manifolds* I, II, J. Diff. Geo. **27** (1988), 87–113, 115–136.

[Roe2] ———, *Exotic cohomology and index theory*, BAMS **23** (1990), 447–454.

[Roe3] ———, *Coarse cohomology and index theory on complete Riemannian manifolds*, Memoirs, AMS **497** (1993).

[Roe4] ———, *Elliptic operators, topology and asymptotic methods*, Pitman Research Notes in Mathematics **179** (1988).

[Ros1] J. Rosenberg, *Applications of analysis on Lipschitz manifolds*, in Miniconferences on harmonic analysis and operator algebras (Canberra, 1987), 269–283.

[Ros2] ______, *C^*-algebras, positive scalar curvature and the Novikov conjecture*, Publ. Math. d'IHES **58** (1983), 409–424.

[Ros3] ______, *The KO assembly map and positive scalar curvature*, LNM **1474** (1991), 170–182.

[RsW1] J. Rosenberg and S. Weinberger, *Higher G-signatures on Lipschitz manifolds* K-Theory **7** (1993), 101–132.

[RsW2] J. Rosenberg and S. Weinberger, with an appendix by J. P. May, *An equivariant Novikov conjecture*, K-Theory **4** (1990), 29–53.

[RsW3] ______, *Higher signatures and scalar curvature for manifolds with boundary* (in preparation).

[RsW4] ______, *The signature operator at 2* (in preparation).

[RsW5] ______, *Higher G-indices*, Ann. Sci. d'Ecole Norm. Sup. **21** (1988), 479–495.

[Rot] M. Rothenberg, *Torsion invariants and finite transformation group*, Proc. Symp. Pure Math. **33** (1978), 267–312.

[RoS] M. Rothenberg and J. Sondow, *Nonlinear smooth representations of compact Lie groups*, Pac. J. of Math. **84** (1979), 427–444.

[RtW] M. Rothenberg and S. Weinberger, *Group actions and equivariant Lipschitz analysis*, BAMS **17** (1987), 109–112 (long version in preparation).

[RS1] C. Rourke and B. Sanderson, *Δ-sets* I,II, Quart. J. Math. **22** (1971), 321–338, 465–485.

[RS2] ______, *Block bundles* I,II,III Ann. of Math. **87** (1968), 1–28, 256–278, 431–483.

[RS3] ______, *On topological neighborhoods*, Comp. Math. **22** (1970), 387–424.

[RS4] ______, *Introduction to PL topology*, Springer-Verlag, 1972.

[RSu] C. Rourke and D. Sullivan, *On the Kervaire obstruction*, Ann. of Math. **94** (1971), 397–413.

[Ru] T.B. Rushing, *Topological embeddings*, Academic Press, 1973.

[Sai] M. Saito, *Hodge modules*, Publ. of the RIMS **26** (Kyoto U.) (1990), 221–333.

[SS] L. Saper and M. Stern, *L^2 cohomology of arithmetic manifolds*, Ann. of Math. **132** (1990), 1–70.

[ScY1] R. Schoen and S.T. Yau, *On the structure of manifolds of positive scalar curvature*, Manuscripta Math. **28** (1979), 159–183.

[ScY2] ______, *Compact group actions and the topology of manifolds of nonpositive curvature*, Topology **18** (1979), 361–380.

[Scz1] R. Schultz, *Topological conjugacy for odd p-groups*, Topology **16** (1977), 263–270.

[Scz2] ______, *Spherelike G-manifolds with exotic equivariant tangent bundles*, in *Studies in algebraic topology*, Academic Press, (1979), 1–39.

[Se1] J.P. Serre, *A course in arithmetic*, Springer-Verlag, 1973.

[Se2] ______, *Linear representations of finite groups*, Springer-Verlag, 1977.

[Sh1] J. Shaneson, *Wall's surgery obstruction groups for $\mathbb{Z} \times G$*, Ann. Math. **90** (1969), 296–334.

[Sh2] ______, *Linear algebra, topology, and number theory*, Proc. Int. Cong. Math. Warsaw **1**, (1983), 685–698.

[Si1] L. Siebenmann, *The obstruction to finding a boundary for an open manifold of dimension greater than five*, 1965 Princeton Ph.D. thesis.

[Si2] ______, *Infinite simple homotopy types*, Indag. Math. **32** (1970), 479–495.

[Si3] ______, *Approximating cellular maps by homeomorphisms*, Topology **11** (1972), 271–294.

[Si4] ______, *Deformations of homeomorphisms on stratified sets*, Comm. Math. Helv. **47** (1972), 123–163.

[Si5] ______, *A total Whitehead obstruction to fibering over the circle*, Comm. Math. Helv. **45** (1970), 1–48.

[Sg] P. Siegel, *Witt spaces: a cycle theory for KO-homology at odd primes*, Amer. J. of Math. (1983), 1067–1105.

[Sm] J. Smith, *Complements of codimension two submanifolds,* II: *homology* below the middle dimension, Ill. J. of Math. **25** (1981), 470–497.

[Sp] E. Spanier, *Algebraic topology*, McGraw-Hill, 1966.

[Spi] M. Spivak, *Spaces satisfying Poincaré duality*, Topology **6** (1969), 77–102.

[Sta1] J. Stallings, *Infinite processes leading to differentiability in the complement of a point*, in *Differential and combinatorial topology*, Princeton University Press, 1965, 245–253.

[Sta2] ———, *On fibering certain three manifolds*, in *Topology of 3-manifolds*, Prentice-Hall, 1962, 95–100.

[Sta3] ———, *The Poincaré conjecture for $n \geq 5$*, in *Topology of 3-manifolds*, Prentice-Hall, 1962, 198–204.

[Sts] J. Stasheff, *A classification theorem for fibre spaces*, Topology **2** (1963), 239–246.

[St] M. Steinberger, *The equivariant topological s-cobordism theorem*, Inven. Math. **91** (1988), 61–104.

[StW] M. Steinberger and J. West, *Controlled finiteness is the obstruction to equivariant handles*, in Proc. 1986 Georgia topology conference, 277–295.

[Ste] M. Stern, *L^2 index theorems on locally symmetric spaces*, Inven. Math. **96** (1989), 231–282.

[Stoltz] S. Stoltz, *Simply connected manifolds of positive scalar curvature*, Ann. Math. **136** (1992), 511–540.

[Sto] D. Stone, *Stratified polyhedra*, LNM **252** (1972).

[Su1] D. Sullivan, *Triangulating homotopy equivalences*, 1966 Princeton Ph.D. thesis.

[Su2] ———, *Geometric topology seminar notes* (unpublished notes, Princeton, 1968).

[Su3] ———, *Genetics of homotopy theory and the Adams conjecture*, Ann. of Math. **100** (1974), 1–80.

[Su4] ———, *Geometric topology, part I. Localisation, periodicity, and Galois* symmetry (unpublished notes, MIT, 1970).

[SuT] D. Sullivan and N. Teleman, *An analytic proof of Novikov's theorem on rational Pontrjagin classes*, Publ. Math. d'IHES **58** (1983), 291–293.

[Sw] R. Swan, *Periodic resolutions for finite groups*, Ann. of Math. **72** (1960), 267–291.

[Ta] L. Taylor, *Surgery on paracompact manifolds*, 1972 Berkeley Ph.D. thesis.

[TW] L. Taylor and B. Williams, *Surgery spaces: structure and formulae*, LNM **741** (1979), 170–195.

[Te] N. Teleman, *The index of the signature operator on Lipschitz manifolds* Publ. Math. d'IHES **58** (1983), 251–290.

[Th] R. Thom, *Quelques propriétés globales des variétés differentiables*, Commentarii **28** (1954), 17–86.

[Tho] R. Thomason, *Algebraic K-theory and etale cohomology*, Ann. Sci. d'Ecole Norm. Sup. **18** (1985), 437–552.

[Ti] U. Tillman, *K-Theory of fine topological algebras, Chern character, and assembly*, K-Theory **6** (1992), 57–86.

[V] K. Varadrajan, *The finiteness obstruction of CTC Wall*, Wiley, 1989.

[Wald1] F. Waldhausen, *Algebraic K theory of spaces*, Stanford conference, Proc. Symp. Pure Math. **17** (1978), 35–65.

[Wald2] ———, *On irreducible 3-manifolds which are sufficiently large*, Ann. of Math. **87** (1968), 56–88.

[Wald3] ———, *Algebraic K-theory of amalgamated free products*, Ann. of Math. **108** (1978), 135–256.

[Wa1] C.T.C. Wall, *Surgery on compact manifolds*, Academic Press, 1971.

[Wa2] ———, *On the classification of Hermitian forms,* VI: *group rings*, Ann. of Math. **103** (1976), 1–80.

[Wa3] ______, *Finiteness conditions for CW complexes*, Ann. of Math. **81** (1965), 56–69.

[Wa4] ______, *Nonadditivity of signature*, Inven. Math. **7** (1969), 269–274.

[Wa5] ______, *Poincaré complexes*, I, Ann. of Math. **86** (1967), 213–245.

[Wa6] ______, *Formulae for surgery obstructions*, Topology **16** (1977), 495–496.

[Wal] J. Walsh, unpublished disproof of the Whyburn conjecture, but see [Bes].

[Wei1] S. Weinberger, *Fibering four- and five-manifolds*, Israel J. of Math. **59** (1987), 1–8.

[Wei2] ______, *Constructions of group actions; a survey of some recent developments*, Cont. Math. **36** (1985), 269–298.

[Wei3] ______, *Semifree PL locally linear actions on the sphere*, Israel J. of Math. **66** (1989), 351–363.

[Wei4] ______, *Higher ρ-invariants*, K-theory, (to appear).

[Wei5] ______, *Aspects of the Novikov conjecture*, Contemp. Math. **105** (1990), 281–297.

[Wei6] ______, *Homotopy invariance of Eta invariants*, Proc. Nat. Acad. Sci. USA **85** (1988), 5362–5363.

[Wei7] ______, *G-signatures and cyclotomic units*, Topology and Its Applications **32** (1989), 183–196.

[Wei8] ______, *On smooth surgery*, Comm. Pure and Appl. Math. **43** (1990), 695–696.

[Wei9] ______, *Homologically trivial group actions* I,II, Amer. J. of Math. **108** (1986), 1005–1021, 1259–1275.

[Wei10] ______, *An analytic proof of the topological invariance of rational Pontrjagin classes*, (preprint).

[Wei11] ______, *Group actions and higher signatures* I, II, Proc. Nat. Acad. Sci. USA **82** (1985), 1297–1298; Comm. Pure and Appl. Math. **40** (1987), 179–187.

[Wei12] ______, *Fake G-tori*, Comm. Pure and Appl. Math., (to appear).

[WY] S. Weinberger and M. Yan, *Products in stratified surgery and periodicity*, preprint.

[Ws] M. Weiss, *Visible L-theory*, Forum Math. **4** (1992), 465–498.

[WW] M. Weiss and B. Williams, *Automorphism of manifolds and algebraic K theory* I, II, III, K-Theory **1** (1988), 575–626; II, J. Pure and Appl. Algebra **62** (1989), 47–107; III, Gottigen preprint.

[We] J. West, *Compact ANRs have finite type*, BAMS **187** (1975), 163–165.

[Wh] J.H.C. Whitehead, *Simple homotopy types*, Amer. J. of Math. **72** (1952), 1–57.

[Wi] E. Witten, *Supersymmetry and Morse theory*, J. Diff. Geo. **17** (1982), 661–692.

[Wo] J. Wolf, *Spaces of constant curvature*, Publish or Perish, 1977.

[Ya] M. Yamasaki, *L-groups of crystallographic groups*, Inven. Math. **88** (1987), 571–602.

[Yn] M. Yan, *The periodicity in equivariant surgery*, 1990 Ph.D. thesis, University of Chicago (to appear in Comm. Pure and Appl. Math.).

[Z] E.C. Zeeman, *Unknotting combinatorial balls*, Ann. of Math. **78** (1963), 501–526.

[Zi] R. Zimmer, *Ergodic theory and semisimple groups*, Birkhauser, 1988.

Index